SOCIETY FOR EXPERIMENTAL BIOLOGY
SEMINAR SERIES · 10

THE CELL CYCLE

THE CELL CYCLE

Edited by

P. C. L. JOHN

Botany Department, The Queen's University of Belfast

CAMBRIDGE UNIVERSITY PRESS
Cambridge
London   New York   New Rochelle
Melbourne   Sydney

Published by the Press Syndicate of the University of Cambridge
The Pitt Building, Trumpington Street, Cambridge CB2 1RP
32 East 57th Street, New York, NY 10022, USA
296 Beaconsfield Parade, Middle Park, Melbourne 3206, Australia

First published 1981

Printed in the United States of America by Vail-Ballou Press, Inc., Binghamton, NY

British Library Cataloguing in Publication Data

The Cell cycle. – (Society for Experimental
Biology seminar series; 10)

1. Cellular control mechanisms
2. Cell cycle
I. John, P. C. L.   II. Series
574.8'62   QH604   80-42176

ISBN 0-521-23912-5 hard covers
ISBN 0-521-28342-6 paperback

CONTENTS

CONTRIBUTORS

Brooks, R. F. Cell Proliferation Laboratory, Imperial Cancer Research Fund, Lincoln Inn's Field, London WC2A 3PX, UK

Burns, R. G. The Blackett Laboratory, Imperial College of Science and Technology, Prince Consort Road, London SW1 2BZ, UK

Carter, B. L. A. Department of Genetics, University of Dublin, Trinity College, Dublin 2, Ireland

Cochran, B. H. Department of Microbiology and Molecular Genetics, Harvard Medical School and Department of Tumor Biology, Sidney Farber Cancer Institute, 44 Binney Street, Boston, MA 02115, USA

Donachie, W. D. Department of Molecular Biology, University of Edinburgh, Mayfield Road, Edinburgh EH9 3JR, UK

Fantes, P. A. Department of Zoology, University of Edinburgh, West Mains Road, Edinburgh EH9 3JT, UK

John, P. C. L. Botany Department, The Queen's University of Belfast, Belfast BT7 1NN, UK

Lambe, C. A. Botany Department, The Queen's University of Belfast, Belfast, BT7 1NN, UK

McGookin, R. Botany Department, The Queen's University of Belfast, Belfast BT7 1NN, UK

Matthews, H. R. Department of Biological Chemistry, School of Medicine, University of California, Davis, California 95616, USA.

Mitchison, J. M. Department of Zoology, University of Edinburgh, West Mains Road, Edinburgh EH9 3JT, UK

Nurse, P. School of Biological Sciences, The University of Sussex, Brighton, Sussex BN1 9QG, UK

Orr, B. Botany Department, The Queen's University of Belfast, Belfast BT7 1NN, UK

Sachsenmaier, W. Institut für Biochemie und Experimentelle Krebsforschung der Universität Innsbruck, A-6090 Innsbruck, Austria

Scher, C. D. Department of Pediatrics, Harvard Medical School and Department of Hematology-Oncology, Sidney Farber Cancer Institute, 44 Biney Street, Boston, MA 02115, USA

Stiles, C. D. Department of Microbiology and Molecular Genetics, Harvard Medical School and Laboratory of Tumor Biology, Sidney Farber Cancer Institute, 44 Biney Street, Boston, MA 02115, USA

Yeoman, M. M. Department of Botany, University of Edinburgh, Mayfield Road, Edinburgh EH9 3JH, UK

PREFACE

The accelerating pace of research into the cell cycle has established a wealth of new information since the publication in 1971 of J. M. Mitchison's classic book *The Biology of the Cell Cycle*. His introduction to the present book, appearing ten years later, reviews the new techniques which have been exploited for cycle study and sets the new understanding that they have provided in the context of previous perspectives of the cell cycle. Each of the important new avenues of cycle research is illustrated in the subsequent chapters which deal with a prokaryote and with a diversity of eukaryotic fungal, algal, mammalian and plant cells.

The great impact that genetical analysis has made is well illustrated in a chapter by W. D. Donachie on the prokaryote *Escherichia coli* and in two chapters dealing with eukaryotes, by P. Nurse & P. A. Fantes on *Schizosaccharomyces pombe* and by B. L. A. Carter on *Saccharomyces cerevisiae*. A stimulating taste of different scientific views can be sampled by comparing the chapter of P. A. Fantes & P. Nurse with that of R. Brooks: these chapters explore the areas of disagreement between those who consider that the cell cycle is regulated by the time of attainment of a threshold size for division and those who consider that any influence of cell size is of minor significance and that, especially in mammalian cells, random transitions are the key controls. The debate concerning control of division is illuminated by several other chapters, including W. Sachsenmaier's review of mitotic control in the plasmodium of *Physarum polycephalum* and the appraisal by C. D. Stiles, B. H. Cochran & C. D. Scher of the complex stimulators of growth and division in mammalian cells, while the role and mechanism of cell division in the life of a plant is reviewed by M. M. Yeoman.

Changes in enzyme level and metabolism which may underlie progress through the cell cycle are reviewed by P. C. L. John, C. A. Lambe, R. McGookin & B. Orr who consider the accumulation of enzymes involved in growth, rather than division, and argue that for most of these enzymes neither enzyme activity nor enzyme protein need accumulate periodically in the cell cycle. Enzymes whose levels do change considerably through the cycle are therefore of great potential significance in cycle control and H. H. Matthews reviews evidence that the appearance of enzymes that alter chromatin proteins may initiate mitosis. The movement of chromosomes in mitosis is a major structural event in the cell cycle and the mechanism is discussed by R. G. Burns.

Each of these accounts is presented from laboratories in which important advances in understanding have been made and the authors have been asked to present only the most essential facts. This conciseness is helpful to the reader, but fellow scientists may sometimes feel that their work deserved more attention and of these I beg forgiveness and an appreciation that we had to be brief.

The Queen's University of Belfast   P. C. L. John
October 1980

ABBREVIATIONS

BHK	baby hamster kidney
CHO	Chinese hamster ovary
CHL	Chinese hamster Lung
dThd	2'-Deoxyribosylthymine
dTMP	2'-Deoxyribosylthymine 5'-phosphate
FLM	frequency of Labelled mitosis
FUDR	5-fluoro-2'-deoxyuridine
Glc-6-P	glucose-6-phosphate
G1	gap after division and before DNA synthesis
G2	gap after DNA synthesis and before mitosis
HMG	high mobility group
M	mitosis
MAP	microtubule associated proteins
MTOC	microtubule organising centre
Oro	orotic acid (precursor of uridine)
PDGF	platelet-derived growth factor
RuBP	ribulose 1,5-bisphosphate
Ru5P	ribulose 5-phosphate
S	phase of DNA synthesis
SDS	sodium dodecyl sulphate
Urd	uridine
YEPD	medium containing 1% yeast extract, 2% bactopeptone, 2% glucose

J.M.MITCHISON

Changing perspectives in the cell cycle

A good starting point for a quick trip through modern cell cycle research is the 'late Precambrian' of 1952, just before the 'Double Helix'. Hughes (1952) had published a book on what we might now call the cell cycle but which was significantly called *The Mitotic Cycle*. There was good reason for this since most of the book was concerned with cells in mitosis, a subject with a background of more than 50 years of research. In contrast, very little was known about how cells and their components grew between one mitosis and the next. But interest was awakening in this, and Hughes said 'The evidence for the synthesis of DNA between mitotic periods is increasing' – a statement that may be strange to us now but has to be taken against a background of earlier beliefs that RNA could be converted into DNA which then condensed onto the mitotic chromosomes. It was also of course uncertain whether or not DNA was a substance of real importance to the cell.

Starting from this time, we can distinguish two broad areas of research. One of them has been concerned with models for the control of division, a theme which we will return to later. The second has been concerned with growth during the cycle and the patterns of macromolecular synthesis. Here DNA holds pride of place. It became clear in the early fifties, both from microspectrophotometry and from autoradiography, that DNA synthesis took place during a restricted period of interphase in a limited range of higher eukaryotic cells. The names for the period of synthesis (S period) and the gaps before and after (G1 and G2) were coined by Howard & Pelc (1953) and are now firmly embedded in the literature. Work on the DNA cycle accelerated in the late fifties and became for more than a decade the dominant aspect of cell cycle research not only because of an increasing awareness of the importance of DNA as the genetic material but also because of the advent of tritiated thymidine. This provided a specific label for DNA which could be used efficiently in autoradiographs. In particular the FLM (Frequency of Labelled Mitoses) method, first developed in detail by Quastler & Sherman (1959), made it possible to measure the lengths of G1, S and G2 in tissues where only some of the cells were passing through the cycle.

It was established by the mid-sixties that nearly all eukaryotic cells, both

1

higher and lower, show periodic rather than continuous synthesis of DNA. G1, however, is absent in quite a number of lower eukaryotes. In addition there is a tendency for S + G2 to be the least variable part of the cycle both between individual cells in a culture and between different tissues. Put the other way round, G1 is the most variable part of the cycle.

Superficially, the position in prokaryotes seems different. There was a false start by Lark & Maaløe (1956) showing periodic synthesis of DNA, which is only worth mentioning because it is a striking example of the artefacts that can occur in synchronous cultures. By now there is ample evidence of continuous synthesis in fast-growing cells of *Escherichia coli* and a widely accepted model of chromosome replication (Cooper & Helmstetter, 1968). But this model emphasises the point that although bulk synthesis of DNA is continuous, the initiation of rounds of chromosome replication is periodic. Bulk synthesis is continuous only because the rounds overlap at fast growth rates. They do not do so at slow growth rates and synthesis then becomes discontinuous. The major difference between eukaryotes and prokaryotes is the number of replication forks per chromosome. There are many more in eukaryotes and their number also varies inversely with the length of the S period. It was thought in the late sixties that another difference between eukaryotes and prokaryotes was that replication was unidirectional in *E. coli* and bidirectional in mammalian cells. The evidence in *E. coli* seemed very strong since it came from four independent sets of experiments. But in one of the most remarkable reversals in the history of molecular biology, further experiments in the early seventies showed that *E. coli* replication is bidirectional, as in eukaryotes, and that all the earlier work had to be reinterpreted.

As well as defining the patterns of DNA synthesis through the cycle, workers in the sixties also produced good evidence that there is an initiator of DNA synthesis present in the cytoplasm of S-phase cells. This had been suspected from the existence of synchronous initiation in multinucleate cells but harder evidence came from a series of experiments involving nuclear transplantation and cell fusion, well reviewed by Johnson & Rao (1971). But the nature of the initiator was not revealed.

If the sixties were the golden age of DNA in the cell cycle (as they were for DNA in a broader context), the seventies have been a silver age. The volume of work diminished once the main patterns of DNA synthesis had been established. Not that recent work lacks interest and importance, as we can see from two examples. Liskay (1977) found a strain of Chinese hamster cells that has no G1 or G2 and therefore shows continuous DNA synthesis throughout interphase – breaking what had previously been a universal rule of periodic DNA synthesis in all eukaryotic cells except perhaps some early embryos. Using fusion with mitotic cells to produce 'premature chromosome condensation', Hittelman & Rao

(1978) have produced evidence which suggests, though does not definitely prove, that chromosomes extend and decondense throughout G1.

But despite recent work, we are left with questions that were unanswered ten years ago and are still unanswered. What is the cytoplasmic initiator? Why is protein synthesis needed for initiation of DNA synthesis? Why is the initiation of heterochromatin delayed after euchromatin? Even more important is: why is there a G1 and a G2 in many cells but not in all? It is attractive to think of a continuous process, the 'chromosome cycle', in which there is a decondensation of the chromosomes during G1 until they reach a fully extended state at the start of S. When S is completed, the chromosomes then condense again during G2 until they are apparent to the light microscope in prophase. Yet this deterministic sequence does not accord with the absence of G1 and G2 in Liskay's Chinese hamster cells or with the absence of a G1 in many lower eukaryotes unless the decondensation process is unusually rapid. In *Physarum,* for example, DNA synthesis has achieved its full rate within five minutes of the end of telophase. Even more important, the great variability in G1 in mammalian cells is not what we would expect if this period was simply one that allowed a continuous morphological change to be completed.

Parallel to the river of DNA there ran a narrower stream of work on the patterns of increase in dry mass, volume, total protein and RNA. Apart from the conspicuous stop in RNA synthesis at mitosis, these patterns are nearly all continuous and do not show the marked periodicity of DNA synthesis. This made their analysis more difficult since methods such as FLM curves could not be used. The early work was done on single cells, using cytochemistry, interference microscopy and autoradiography, though there was a pioneer paper by Prescott (1955) in which single living *Amoebae* were weighed on a Cartesian diver balance. The culmination of this approach was the series of papers by Zetterberg & Killander (e.g. 1965) on mouse L cells.

Although single cell methods can give very precise measurements of growth patterns, the components that can be measured are strictly limited. Synchronous cell cultures are necessary, if the methods of modern biochemistry are to be used. Such cultures were available from the early fifties but they could only be made with a few cell types. During the sixties, a whole series of methods were developed for making synchronous cultures of a wide range of cells. The earlier ones used 'induction synchrony' in which the cells of a normal asynchronous culture are induced to divide synchronously. This can be done by environmental changes (e.g. of temperature or light) or by blocking a stage of the cycle (DNA synthesis or mitosis) and then releasing the block. The yield is large since all the cells of the initial culture are used but there have always been worries about possible artefacts caused by the forced synchronisation – worries fully justified in some cases such as the experiments of Lark & Maaløe mentioned above. An important point is that cell cycle blocks do not stop the growth of the cells which therefore become unusually large during the course of the treatment.

The second main method of producing synchronous cultures is by selection. Cells at a particular stage of the cycle are separated off from a normal asynchronous culture and then grown up as a synchronous culture. This was first done by Terasima & Tolmach (1961) using 'wash-off' or 'selective detachment' of mitotic mammalian cells growing as monolayers. Two other methods were developed in the mid-sixties: membrane elution worked well with *E. coli,* and size separation on sucrose density gradients provided a powerful tool for a wide variety of cells. In contrast to induction methods, the yield from non-induction methods is small though it can be very much increased by a modification of gradient separation in which the whole of the initial culture is 'age fractionated' on the gradient and successive samples of increasing cell size represent cells of increasing age in the cycle. Selection synchrony should in principle produce much less perturbation of synthetic patterns than induction synchrony. Nevertheless, the handling that cells undergo in some of the selection methods does seem to produce perturbations and the search for better methods continues (Creanor & Mitchison, 1979). Before leaving synchronous cultures, we should remember the excellent natural mitotic synchrony in the multinucleate slime mould *Physarum polycephalum.* Following the pioneer work of Harold Rusch in the early sixties, this organism has become a prime material for cell cycle studies.

Synthronous cultures have been used extensively for the last fifteen years for following biochemical changes through the cell cycle. These studies have confirmed and extended our knowledge of the patterns of DNA synthesis which were first determined by the earlier single cell methods. They also showed that many of the bulk components (e.g. total protein and RNA) increase continuously through interphase. This is not unexpected or very dramatic, but there is one important point of detail. The curve of increase may be smooth and approximately exponential or it may have linear segments with a point of rate doubling between the segments. If such a linear pattern can be established (and it is difficult to do so), the rate doubling point is an interesting 'marker' in the cycle and may be correlated with other cell cycle events such as the doubling in the gene dosage during the S period.

The pattern of continuous increase in total protein could encompass periodic patterns of synthesis of individual proteins. Indeed it is now widely accepted that histones are synthesised periodically during the S period. But it might be that many of the cell proteins are synthesised in steps at differing parts of the cycle and the sum of these steps gives the continuous increase in total protein. This idea gained considerable credibility from two important papers in the mid-sixties which showed step-wise rises in enzyme activity in synchronous cultures of budding yeast (Gorman, Tauro, La Berge & Halvorson, 1964) and bacteria (Masters, Kuempel & Pardee, 1964). There followed a burst of papers in the late sixties on enzyme activity through the cycle and some hypotheses to explain what was assumed to be periodic and sequential synthesis of enzyme proteins. One of the most exciting was 'linear reading' or 'sequential transcription' in which it was proposed that the genome was transcribed sequentially once per cycle and the enzyme steps in the cycle appeared in the same order as their structural genes.

Work on enzymes during the cell cycle continued through the seventies and close on a hundred papers have now been published. It is now clear that periodic rises in activity of what I have called 'step enzymes' are by no means the rule. Out of 19 enzymes examined in synchronous cultures of the fission yeast *Schizosacharomyces pombe,* nearly all show continuous rises in activity (Mitchison, 1977*a*). In *Physarum,* where synchrony is natural, eight out of 14 enzymes that have been assayed also show continuous rises (refs. in Mitchison, 1977*a*). The presence of so many 'continuous' enzymes makes 'linear reading' unlikely. A second point is that workers in the field have become increasingly aware that the techniques of making synchronous cultures can produce perturbations and that it is vital to run controls. Although selection synchrony causes fewer perturbations than induction synchrony, they can still be striking and long-lasting (Mitchison, 1977*a*). But the really interesting dilemma at present is the situation in budding yeast. More enzymes have been assayed through the cycle than in any other organism. A review by Halvorson, Carter & Tauro (1971) lists 30 enzymes all of which behave as step enzymes. In some of the earlier work there were no controls for the possible perturbing effects of selection synchrony. But six of the enzymes were also assayed after age fractionation on a zonal rotor (refs. in Mitchison, 1977*a*) where the perturbations should be minimal. In addition, one enzyme was assayed in single cells and also showed a step pattern (Yashpe & Halvorson, 1976). So the predominant picture is one of step patterns with some worries about perturbations. We now have to contrast this with the results of Elliott & McLaughlin (1978) who have made a two-dimensional electrophoretic analysis of protein synthesis through the cycle of budding yeast. Of the more abundant proteins on the gels, 111 showed a continuously increasing rate of synthesis and not a pattern of periodic increase. There are various ways round this dilemma. Either the gels or, more likely, the enzyme assays may be in error because of perturbations and other technical problems. The gel spots may also exclude the enzyme proteins that have been assayed. But the most likely explanation at present is, as Elliott & McLaughlin suggest, that the changes in enzyme activity do not follow the changes in the amount of enzyme protein. If so, we are faced with a deep problem. I have pointed out earlier (Mitchison, 1973) that it needs a fairly sophisticated control to ensure that the specific activity per unit of enzyme protein falls while the total activity remains constant in the first part of a step pattern when the amount of enzyme protein is rising with continuous synthesis. The specific activity must then rise sharply during the step, and thereafter fall for the rest of the cycle. This is not a simple system, especially if the steps for different enzymes are at different points in the cycle. Sequential activation, if confirmed, may be as interesting as sequential transcription.

Turning back into history, one of the most dramatic cell cycle discoveries of the mid-fifties was that cultures could be made synchronous by repetitive changes of their environment. *Chlorella* can be synchronised by light – dark cycles (Tamiya *et al.*, 1953) and *Tetrahymena* by temperature changes (Scherbaum & Zeuthen, 1954). With so many methods now available for making synchronous cultures, it is sometimes hard to realise how striking were these pioneer results. In both cases, they raised the question of what is the temporal control of division and how is it modified to generate synchrony. They are the start of the second broad area of research which I mentioned earlier – models for division control.

Division control models in *Chlorella* and other algae have always tended to be somewhat separate from those in other systems. In many cases, this is because the cell cycle is about 24 h and the models involve circadian oscillators which are not applicable to cells, particularly micro-organisms, with much shorter cycles. Nevertheless, there are interesting bridges to be built between the circadian rhythm field and that of the cell cycle (Edmunds, 1978).

The question of division control was actively followed up in *Tetrahymena* by Zeuthen and his colleagues and they formulated the concept of 'division proteins'. Although the concept was not expressed in quantitative terms, it was important in emphasising that the trigger for mitosis could be the completion of a structure rather than the attainment of a critical concentration of an effector. In this way, the number of molecules in a cell can be 'counted'.

'Division proteins' was the main division control model for eukaryotes in the sixties. But prokaryotes, as often happens, were in advance of eukaryotes. The model of Cooper & Helmstetter (1968) provided a satisfactory explanation of chromosome replication during the cell cycle of *E. coli*. It was and is an important advance, particularly since it contains elements that are surprising to those who work with higher cells, for example that the time for a complete round of DNA replication is greater than the cycle time in fast growing cells and that cell division takes place while the chromosomes are in the process of replication. It also became clear in the late sixties that an elegant explanation of the variation in the division size of *E. coli* at different growth rates could be provided by a model in which initiation of DNA synthesis takes place at a constant mass (per chromosome origin), irrespective of growth rate. But whether this mass measuring mechanism operates by the accumulation of an initiator (Donachie, 1968) or the dilution of an inhibitor (Pritchard, Barth & Collins, 1969) remains unresolved.

These were the main models of the sixties and it is obvious that this was a poorly developed field in the eukaryotes. In contrast, models have been one of the main growth areas of the seventies, as is evident from the fact that half the chapters in this book are concerned in whole or in part with control models. I shall not therefore say much about them here, except to make a few general points. Many of the eukaryotic models and the techniques used to test them have an ancestry in the earlier work on *E. coli*. Size controls, in the sense of mechanisms which monitor cell size and produce a signal when a critical size is reached, are becoming increasingly important in yeasts and *Physarum*. But they have their origin ten years ago in *E. coli*. Nutritional shifts, also started in *E. coli*, are now proving powerful tools in the yeasts. 'Transition probability' has no bacterial ancestry since it was originally developed to explain the variability of G1 in mammalian cells. Although it is a model which is lacking in mechanism, it has one particular value: it concentrates on variability, which most other models ignore or slide over. Contrary to some impressions, bacterial cell cycle times are as variable as those of mammalian cells (refs. in Mitchison (1977b). Another question about the models is their universality. It would be a happy circumstance if all cells had the same controls, and it is not unreasonable to believe that the immediate signals for initiating DNA synthesis and mitosis will be the same, at any rate in eukaryotic cells. But the main control or 'trigger' upstream from the immediate signals may well differ from one cell type to another. It may even vary in one cell type according to circumstance. According to the sophisticated models developed for fission yeast, the main size control operates at a different time and on a different process in *wee* mutants hhan in normal cells (Nurse & Fantes, this volume). Finally all the models, including the bacterial ones, are still at the level of cell biology rather than molecular biology. The language is about initiators, inhibitors, receptors and sites, and in no case have the molecules been identified – though not for want of trying. The fact that there has been little progress in finding the molecules in *E. coli* where there are powerful genetic and biochemical tools and where the models have their longest history suggests that it may prove an even more difficult search in eukaryotes.

This introduction would be incomplete wihout a mention of one of the most striking advances of the seventies – the development of genetical methods for analysing the eukaryotic cell cycle. This was pioneered by Hartwell using budding yeast (the first paper being Hartwell, Culotti & Reid, 1970) but it was extended later to fission yeast, *Chlamydomonas, Aspergillus, Tetrahymena* and, with greater difficulty, to mammalian cells. A good recent review of the whole field is by Simchen (1978). The most popular technique has been the isolation of temperature-sensitive *cdc* (cell division cycle) mutants. These progress normally through the cycle at the permissive (usually lower) temperature but become blocked at some point in the cycle when transferred to the restrictive temperature. In use these mutants are not unlike chemical inhibitors of the cycle, such as hydroxyurea or colchicine, and they have some disadvantages compared to inhibitors; for instance the mode of action of inhibitors is often better known and the temperature shift may have prolonged side effects in physiological experiments. But their advantages outweigh their disadvantages, at any rate in organisms where mutants are easy to isolate and analyse. In fission yeast, for example, we have found only four chemical inhibitors that block the cycle, and the block is transient. By contrast, we have mutants in about 25 genes that block the cycle in DNA synthesis, nuclear division or cell division (Nurse & Fantes, this volume). In addition, there are mutants that are smaller (*wee*) and larger than wild type and are powerful tools for investigating the role of cell size in cycle controls. This genetic armoury is far better stocked than the inhibitor one, and the situation is similar in budding yeast. Given this armoury, it is reasonable to expect that genetic dissection will take us a lot deeper into the control mechanisms of the cell cycle, particularly if it can be combined with biochemical identification of the gene products.

References

Only a few references have been given in this article, especially in the case of the earlier work. Much fuller accounts and bibliographies can be found in the two books on the cell cycle by Prescott (1976) and Mitchison (1971).

Cooper, S. & Helmstetter, C. E. (1968). Chromosome replication and the division cycle of *Escherichia coli* B/r. *Journal of Molecular Biology*, **31**, 519–40.
Creanor, J. & Mitchison, J. M. (1979). Reduction of perturbations in leucine incorporation in synchronous cultures of *Schizosaccharomyces pombe* made by elutriation. *Journal of General Microbiology*, **112**, 385–8.
Donachie, W. D. (1968) Relationship between cell size and the time of initiation of DNA replication. *Nature, London*, **219**, 1077–9.
Edmunds, L. N. (1978). Clocked cycle clocks: implications toward chronopharmacology and aging. In *Aging and Biological Rhythms*, ed. H. V. Samis & S. Capobianco, pp. 125–84. New York: Plenum.
Elliott, S. G. & McLaughlin, C. S. (1978). Rate of macromolecular synthesis through the cell cycle of the yeast *Saccharomyces cerevisiae*. *Proceedings of the National Academy of Sciences, USA*, **75**, 4384–8.
Gorman, J., Tauro, P., La Berge, M. & Halvorson, H. O. (1964). Timing of enzyme synthesis during synchronous division in yeast. *Biochemical and Biophysical Research Communications*, **15**, 43–9.

Halvorson, H. O., Carter, B. L. A. & Tauro, P. (1971). Synthesis of enzymes during the cell cycle. *Advances in Microbial Physiology,* **6,** 47–106.

Hartwell, L. H., Culotti, J. & Reid, B. (1970). Genetic control of the cell-division cycle in yeast. I. Detection of mutants. *Proceedings of the National Academy of Sciences, USA,* **66,** 352–9.

Hittelman, W. N. & Rao, P. N. (1978). Mapping G1 phase by the structural morphology of the prematurely condensed chromosomes. *Journal of Cellular Physiology,* **95,** 333–42.

Howard, A. & Pelc, S. R. (1953). Synthesis of desoxyribonucleic acid in normal and iradiated cells and its relation to chromosome breakage. *Heredity, London (Supplement),* **6,** 261–73.

Hughes, A. (1952). *The Mitotic Cycle.* London: Butterworths.

Johnson, R. T. & Rao, P. N. (1971). Nucleo-cytoplasmic interactions in the achievement of nuclear synchrony in DNA synthesis and mitosis in multinucleate cells. *Biological Reviews,* **46,** 97–155.

Lark, K. G. & Maaløe, O. (1956). Nucleic acid synthesis and the division cycle of *Salmonella typhimurium. Biochimica et Biophysica Acta,* **21,** 448–58.

Liskay, R. M. (1977). Absence of a measurable G2 phase in two Chinese hamster cell lines. *Proceedings of the National Academy of Sciences, USA,* **74,** 1622–5.

Masters, M., Kuempel, P. L. & Pardee, A. B. (1964). Enzyme synthesis in synchronous cultures of bacteria. *Biochemical and Biophysical Research Communications,* **15,** 38–42.

Mitchison, J. M. (1971). *The Biology of the Cell Cycle.* London: Cambridge University Press.

Mitchison, J. M. (1973). Differentiation in the cell cycle. In *The Cell Cycle in Development and Differentiation,* ed. M. Balls & F. S. Billett, pp. 1–11. London: Cambridge University Press.

Mitchison, J. M. (1977a). Enzyme synthesis during the cell cycle. In *Cell Differentiation in Microorganisms, Plants and Animals,* ed. L. Nover & K. Mothes, pp. 377–401. Jena: VEB Gustav Fischer Verlag.

Mitchison, J. M. (1977b). The timing of cell cycle events. In *Mitosis, Facts and Questions,* ed. M. Little, N. Paweletz, C. Petzelt, H. Postingl, D. Schroeter & H-P. Zimmerman, pp. 1–19. Berlin: Springer Verlag.

Prescott, D. M. (1955). Relations between cell growth and cell division. 1. Reduced weight, cell volume, protein content, and nuclear volume of *Amoeba proteus* from division to division. *Experimental Cell Research,* **9,** 328–37.

Prescott, D. M. (1976). *Reproduction of Eukaryotic Cells.* New York: Academic Press.

Pritchard, R. H., Barth, P. T. & Collins, J. (1969). Control of DNA synthesis in bacteria. *Symposium of the Society for General Microbiology,* **19,** 263–97.

Quastler, H. & Sherman, F. G. (1959). Cell population kinetics in the intestinal epithelium of the mouse. *Experimental Cell Research,* **17,** 420–38.

Scherbaum, O. & Zeuthen, E. (1954). Induction of synchronous cell division in mass cultures of *Tetrahymena pyriformis. Experimental Cell Research,* **6,** 221–7.

Simchen, G. (1978). Cell cycle mutants. *Annual Review of Genetics,* **12,** 161–91.

Tamiya, H., Iwamura, T., Shibata, K., Hase, E. & Nihei, T. (1953). Correlation between photosynthesis and light-independent metabolism of *Chlorella*. *Biochimica et Biophysica Acta*, **12**, 23–40.

Terasima, T. & Tolmach, L. J. (1961). Changes in X-ray sensitivity of HeLa cells during the division cycle. *Nature, London*, **190**, 1210–11.

Yashpe, J. & Halvorson, H. O. (1976). β-D-galactosidase activity in single yeast cells during cell cycle of *Saccharomyces lactis*. *Science*, **191**, 1283–4.

Zetterberg, A. & Killander, D. (1965). Quantitative cytochemical studies on interphase growth. II. Derivation of synthesis curves from the distribution of DNA, RNA and mass values of individual mouse fibroblasts *in vitro*. *Experimental Cell Research*, **39**, 22–32.

P.A.FANTES & P.NURSE

Division timing: controls, models and mechanisms

Introduction

In this chapter we shall be concerned with the way in which the timing of events in the cell division cycle of eukaryotic cells is controlled. These events include division itself, and the cycle stages that lead up to it: the G1, S and G2 phases, and mitosis. In particular, the relationship between growth and division processes will be examined. The relevance of cellular growth to the division cycle arises because cell mass increases through the cycle. The timing of a cycle process therefore bears a strong relationship to the cell mass at which it occurs. The term 'mass' as used here is intended to include such diverse parameters as protein, RNA and small molecule content, volume, and so on. The growth of individual cells will be considered to be a simple continuous increase in mass, except for those components directly involved in division cycle events, such as the nucleus and its chromosomes. (It should be noted that the term 'growth' has two meanings: in the broad sense it refers to the increase in biomass and cell number in a cell population, but we shall use it mainly to refer to the increase in mass of individual cells.)

The specific propositions we shall consider are: (i) that the growth and division of cells are co-ordinated, (ii) that cell mass is often important in controlling the timing of division cycle events, (iii) that the co-ordination between growth and division occurs either in G1 or G2, depending on the cell type, and (iv) that despite the number of mechanisms proposed which could account for the co-ordination, in no experimental system is there enough information to say whether any particular mechanism actually operates. A final section deals with the contentious topic of the role of variation between cells in the control of the cell cycle.

Because of the limited space available, this chapter will be concerned mainly with providing a theoretical framework for some of the concepts and models currently favoured. Specific experimental systems are described elsewhere in the book and will be referred to occasionally.

Are growth and division co-ordinated?

When cells are grown under constant conditions, the number of cells in the population increases exponentially, doubling in a time which is nearly equal to the average cycle time of individual cells. The total mass of the population also increases exponentially, with the same doubling time as cell number. The average cell mass (the ratio of total mass to number), remains constant from one generation to the next. Inequality of the division and growth rates would lead to progressive changes in mean mass or size (Fig. 1). The experimentally observed constancy of cell size (Fantes, 1977; Tsuboi, Kurotsu & Terasima, 1976) therefore suggests that the growth and division of cells are co-ordinated.

When we consider individual cells in a steady state population (where growth and division run in parallel), the situation is seen to be more complex. Cells are not all the same size at corresponding cycle stages, nor are their cycles all of the same length. Both parameters can take a range of values, distributed around a mean. The amount of variation is often expressed as the coefficient of variation, defined as the standard deviation divided by the mean. Coefficients of variation of 10–20% are typical for cycle duration, cell size at division being rather less variable (see Mitchison, 1977, for review). The degree of variation in both cycle time and division size remains constant through steady-state growth (Fantes,

Fig. 1. Effect on mean cell mass of growth and division rate. Division rates greater than (- - -) equal to (– – –) or less than (– · –) growth rate.

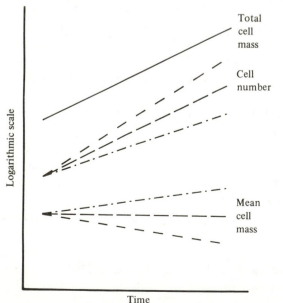

Time

1977; Yen *et al.*, 1975*a*). This constancy is an important factor which must be explained by any model for the control of the cell cycle.

We now consider what effect the variation in cycle time might have on the variation in cell division size. A hypothetical group of new-born cells of the same size embark on their cycles simultaneously. Some cells will divide earlier, some later. Early dividers will be smaller at division than late dividers, provided all cells grow in a similar way. Thus within one cycle, variation in division size will have been generated. During the next cycle, more variation will be generated, and it would appear that the longer the population is maintained, the greater the variation in size (Fig. 2). This does not happen, so one of our assumptions must be wrong.

Fig. 2. Generation of size variation from cycle time variation. Abscissa: units of the mean cycle time. Ordinate: 1 unit is the mean birth mass, and 2 units the mean division mass. *Top*. Cells of the same initial mass divide at various times, thus generating variation in cell division mass. Typical division times and masses shown at a–e (●). Cell c has the population mean cycle time and division mass (○). *Bottom*. Cells a–e have divided to produce daughters of half the division mass. The progress of the subsequent cycles is shown. Each cell has the same cycle time distribution as its mother. This increases the total division size variation. (←——→) indicates the range of time and size over which division can occur. The size distribution of dividing cells is exaggerated in the bottom diagram. The variance of the lower distribution is twice that of the upper: hence the breadths should be in the ration $\sqrt{2} : 1$.

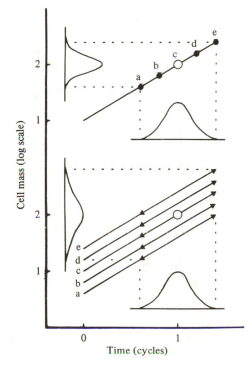

One way out of the paradox is to assume that completion of some cycle event is dependent on the attainment of a critical size or mass (Fig. 3*a*). This will ensure constancy of the average division size, but to account for the existence of variation, it must also be assumed that the size monitoring mechanism is to some degree inaccurate (see p. 24).

Alternatively, there might be a requirement for a particular mass increase between the occurrence of an event in one cycle and its occurrence in the next. The increase in total mass *per se* might not be important, but there could be a requirement for synthesis of a critical amount of some control molecule, produced in proportion to overall growth (Zeuthen & Williams, 1969). In this scheme, the mean cell size will be maintained, since cells of abnormal division size can only pass on half the difference to each daughter (Fig. 3*b*). This produces a regression towards the normal size every cycle, tending to reduce the variation in size. Other factors, such as random variation in cycle time and unequal division of cells will increase the variation, and a balance will be reached where the overall variability in the population will be constant.

A third way out of the paradox is to assume that growth processes are irrelevant to the timing of division, and that some mechanism controls the mean cycle time, inaccurately, to account for cycle time variation. This class of models where the cycle is controlled by a timing mechanism alone includes 'oscillator' models, such as the limit cycle model proposed by Kauffman & Wille (1975). In addition it is necessary to assume that the cell increases its mass in a particular way: a linear pattern is acceptable but an exponential increase is not (Fig. 3*b*, *c*). If growth is linear, the model is very similar to the 'critical growth increment' scheme, since the amount of growth equals the product of the cycle time and the linear growth rate. Constancy of the mean cell size and of its variability will thus be maintained.

We have now proposed three models which can account for the constancy of mean cell size and of its variability, in exponentially proliferating populations. The models will from now be referred to as 'sizer', 'incremental' and 'clock' models, respectively.

Importance of cell mass

So far we have considered only populations of cells proliferating under constant conditions. In this section we shall examine whether each of the models can account for observations on cell populations growing at different rates, and on cells that have been artificially made unusually large or small.

Most types of cells are able to maintain not only one, but a number of different steady states of growth at different proliferation rates, depending on the culture conditions. Micro-organisms can be made to grow at different rates by altering the composition of the medium, which limits the availability of one or more

Fig. 3. Control models and cell size variation. Abscissa: units of the mean cycle time. Ordinate: the mean birth mass is 1 unit, and the mean division mass 2 units. In each of the diagrams the fates of three cells are considered. The centre one (bold) represents a cell of average birth size; the other two are initially 20% smaller or larger than this value. Dashed lines represent the parameter that is controlled in each model.

(a) *Cell division mass*. Cell mass is monitored once every cycle, and cells larger or smaller than average at birth all divide at the normal size. Thus large cells have short cycle times, and small cells, long cycles.

(b) *Increment in mass between divisions*. Cells increase in mass by a fixed amount between divisions. A cell of 1.2 mass units at birth will be $1.2 + 1.0 = 2.2$ units at division. Its daughters will therefore be 1.1 units at birth. The deviation from the mean size is therefore reduced each cycle. The 'clock' model also works this way if growth is strictly linear (see text).

(c) *Cycle time*. Cells grow exponentially for a fixed time between birth and division. A cell of 1.2 mass units at birth will be $1.2 \times 2 = 2.4$ units at division, and its daughters will have a birth size of 1.2. There is therefore no mechanism for reducing variation in size.

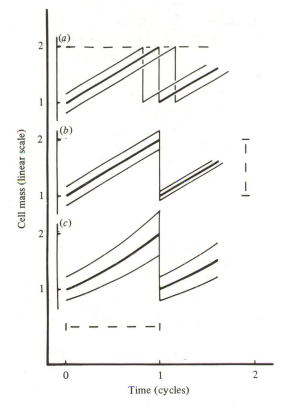

nutrients. The same is true of mammalian cells, where restricting the glucose supply (Tovey & Brouty-Boye, 1976) or providing alternative carbon sources (Johnson & Schwartz, 1976) have been used. A more common way of manipulating the growth of mammalian cells, however, is to alter the concentration of serum, since the cells require this for growth in culture (Brooks, 1976). It is not clear whether reduction of the serum level is similar in effect to nutritional limitation, since serum may contain factors that specifically affect division cycle processes, rather than the rate of metabolism. Assuming for the moment that growth rate is determined by the availability of nutrients, we now consider how the three models can account for the existence of different steady states. The 'sizer' model has no difficulty, since division (or whichever cycle event is controlled) is consequent on the cell reaching a critical size. Reduction of the rate of mass accumulation means that it will take longer to double in mass between birth and division. The time between divisions will automatically be increased to match the mass doubling time. In a similar way, in the 'incremental' model, the time taken for the critical growth increment will be extended, and so will the time between divisions. The 'clock' model, however, faces difficulties: to balance cycle time with mass doubling time if mass accumulates at faster or slower rates then the clock must also run faster or slower. Such a balance could be the result of two mechanisms which respond in very similar ways to culture conditions, one regulating rate of growth, the other the speed of the clock. This seems unlikely, but is theoretically possible. Alternatively, one sensing mechanism could serve to control both processes. This, however, would be tantamount to admitting the existence of a co-ordinating mechanism between growth and division, contrary to the principle of the 'clock' model. Thus the two models in which division is coupled to growth can easily account for the existence of different steady states of growth, while the 'clock' model can do so only if further, rather unlikely, assumptions are made.

Another aspect of the behaviour of cells which the models must account for is the effect of cell size on division kinetics. There is a substantial body of evidence that cell size affects the length of division cycles in such a way as to reduce the size deviation from the mean, that is, homeostatically. This effectively means that the larger the cell at birth, the smaller the growth increment and the shorter the time before division. Experiments to show this are in principle straightforward: all that is needed is to observe the growth and division of individual cells, either in steady state growth or after experimental perturbation of the normal cycle. In practice there are rather few reports of such experiments, largely because of the technical difficulty of measuring the size of individual cells with sufficient accuracy. Several studies with the fission yeast *Schizosaccharomyces pombe,* which because of its linear growth mode is suited for single cell measurements (see Nurse & Fantes, this volume), have shown that cells that are large at birth grow by less than the normal amount before dividing. Furthermore, the cycles of such cells are shorter than average, while their rate of growth is scarcely different from normal-sized cells (Fantes, 1977; Miyata, Miyata & Ito, 1978). Similar experiments with the multi-nucleate plasmodium of *Physarum* have shown that experimental alteration of the nuclear concentration (number of nuclei per unit mass, equivalent to the inverse of cell size in a uninucleate cell) is followed by restoration of the normal value after the experimental treatment is stopped. Reducing the nuclear concentration in any of several ways results in shortened subsequent cycles (Brewer & Rusch, 1968; Sachsenmaier, Dönges, Rupff & Czihak, 1970; Sudbery & Grant, 1975, 1976; discussed by Sachsenmaier, this volume). There is less growth than normal during such cycles, which tends to restore the steady state nuclear concentration (Sudbery & Grant, 1976). There are rather few detailed reports from other systems, but some investigations have yielded results which show a shortening of the cycle in large cells: blocking the division of mammalian cells which causes an increase in volume (Rosenberg & Gregg, 1969; Ross, 1976) and protein content (Fournier & Pardee, 1975), is followed by shortened cycles after release of the block (Galavazi & Bootsma, 1966); and large *Tetrahymena* or *Amoeba* cells have cycles which are shorter than average (Lövlie, 1963; Prescott, 1956). Experimentally-generated small cells of budding yeast have a cycle which is longer than normal (Johnston, Pringle & Hartwell, 1977), and within this group there is an inverse relationship between initial size and cycle time. Similar results have been obtained with mammalian cells when serum starved cells were stimulated to proliferate by re-addition of serum (Shields *et al.*, 1978).

The above observations are all consistent with the 'sizer' model, since the attainment of a critical size is predicted to require less growth in cells of large initial size, and more growth in small cells. Some studies have shown this directly (Sudbery & Grant, 1976; Fantes, 1977; Miyata *et al.*, 1978). Provided that the growth rate is not strongly dependent on cell size, a corollary is that large cells should have short cycles, and small cells long cycles, a prediction supported by all the above reports.

In contrast, the 'incremental' model predicts that initial cell size should not affect the amount of growth between birth and division, while according to the 'clock' model the cycle time should be independent of initial size. The evidence therefore strongly favours the 'sizer' model.

There are a few cases where cell size was found to have no effect on division cycle progress (Fox & Pardee, 1970; Fournier & Pardee, 1975). These experiments were done with perturbed cell populations, and it is possible that the experimental treatment generated side effects which masked any effect of cell size.

There is evidence for this assertion in one case (Fournier & Pardee, 1975), where a slight modification to the experimental procedure brought to light an effect most easily ascribable to one of cell size (Cress & Gerner, 1977).

One final and rather striking experiment showing the importance of cell size was done first by Hartmann (1928) and subsequently by Prescott (1956). An amoeba progressing towards division was cut in two: the part without the nucleus died, but the nucleate part continued to grow. Division was delayed, suggesting that the cell could not divide because it could not attain the critical size as soon as an untreated cell. Furthermore, the treatment of amputation followed by re-growth could be repeated many times without division occurring. When the treatment was stopped, and the amoeba allowed to grow, the cell divided.

To conclude, there is strong evidence that cell size is an important factor in cell cycle control. In some instances it is possible to say that cells monitor their absolute size, and can only complete some cell cycle process upon which division depends when a critical size has been attained.

When do the controls act?

Having shown that cell size is an important factor in the control of the division cycle, we now consider which cycle event or events are subject to control. As will be discussed, there appear to be controls which act over entry into mitosis, and others which control the initiation of DNA replication. Before continuing with this aspect of the problem, however, we shall digress briefly to consider the concept of the 'timer'. A timer, as we shall use the term here, is a cycle control element which ensures that a certain time elapses between an early event and a later one. This is different from a 'clock', which runs from a particular event in one cycle up to the same event in the next cycle. A timer that is especially important in some cells connects the initiation of DNA replication to mitosis, including the S and G2 phases. The timer is proposed to be of constant duration at constant temperature, and relatively independent of total cycle time. There may however be some variation around the mean timer period for individual cells within a population.

There are in principle two ways in which a 'sizer' could control cell size at division. There might be a 'sizer' directly controlling entry into mitosis, division occurring shortly after this event. Alternatively, some earlier event, such as the initiation of DNA replication, could be subject to size control, and mitosis and division would follow after a period defined by a timer. This second possibility is similar to one model originated for the control of division in bacteria (Donachie, 1968; Pritchard, Barth & Collins, 1969; illustrated by Donachie, this volume). The need to propose such a hybrid 'sizer-timer' model arises because division in some cell types is regulated at the time of entry into S phase. A cell operating on this model will, after birth, grow for a time through G1 until it has attained the critical size needed to initiate DNA replication. Once this has happened, the rest of the cycle is determined by operation of the timer. Thus the decision whether to divide is taken during G1, though its ultimate effect is not observed till later. Operation of this particular hybrid model makes several predictions. (a) As the proliferation rate falls, the total cycle time is extended, while the period included in the timer, (S + G2), remains constant. Thus G1 is extended in absolute duration and as a fraction of the total cycle. Ultimately, in non-proliferating populations, all cells should arrest in G1. These predictions are supported by observations on mammalian cells (Bürk, 1970; Temin, 1971) and budding yeast (Jagadish & Carter, 1977). (b) Mean cell size should fall at low proliferation rates. This is because cell size at the initiation of S phase is constant for slow- or fast-growing cells, but fast-growing cells will increase more in mass than slow-growing cells during the constant timer period (Fig. 4). Fast-growing cells will therefore be larger at division than slow-growing cells. This prediction is supported by observations on budding yeast (Jagadish, Lorincz & Carter, 1977), and to a lesser extent by the few mammalian cell studies carried out (Yen et al., 1975b; Tovey & Brouty-Boye, 1976; Shields et al., 1978). (c) Growth rate shifts will not affect the division rate until a period equal to the length of the timer has elapsed (Pritchard, 1974; Carter, this volume). Such results have been obtained for budding yeast (Carter, Lorincz & Johnston, 1978), but the mammalian cell evidence is not as conclusive (Sisken & Kinosita, 1961).

It is not certain that mammalian cells have a critical size requirement for entry into S phase, though one study agreed with this conclusion (Killander & Zetter-

Fig. 4. Dependence of cell division size on growth rate for a hybrid 'sizer–timer' model. Cells initiate a timer when they attain a critical mass, and divide a fixed time later. Fast-growing cells increase in mass more before division than slow-growing cells. Division size therefore increases with growth rate. The three doubling times show are in the ratio 2 : 3 : 4 (● : ▲ : ◆), and the corresponding division sizes are 1.74 : 1.45 : 1.32. Open symbols represent cell birth; closed symbols represent division. M_c, critical cell mass.

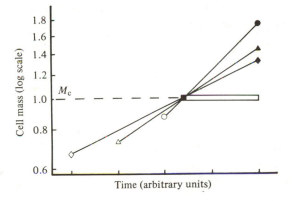

berg, 1965*a*) and others are consistent with an effect of size on G1 duration (Killander & Zetterberg, 1965*b*; Shields *et al.*, 1978). There is no doubt about the existence of a control point in G1, where non-proliferating cells arrest (Ley & Tobey, 1970; Bürk, 1970; Temin, 1971). A number of treatments which stop proliferation all cause cells to arrest in the same part of G1, termed the 'restriction point' by Pardee (1974), though whether a size control acts here is at present unknown. This arrest is discussed in this volume by Stiles, Cochran & Scher. A similar unique stage early in the G1 phase of the budding yeast cycle, termed 'start' (Hartwell, 1974), appears to be associated with a size control (Johnston *et al.*, 1977).

Not all types of cell have a primary control point in G1. In several cell types, particularly lower eukaryotes, the G1 phase is short or non-existent (Nygaard, Guttes & Rusch, 1960; Ron & Prescott, 1969; Bostock, 1970). In these cells a primary control which acts in G2 might be expected, and there is good evidence for this in several systems. The best examples of G2 'sizers' which determine entry into mitosis come from fission yeast and *Physarum*. In fission yeast the control has been identified genetically, by the isolation of mutants altered in size, and physiologically, since size depends on growth conditions (Nurse, 1975; Fantes & Nurse, 1977; Thuriaux, Nurse & Carter, 1978; Nurse & Fantes, this volume). The alteration of size under different conditions is mediated by a mechanism completely distinct from that operating in budding yeast. In fission yeast, it appears that the critical size required for mitosis is directly affected by the growth conditions (Fantes & Nurse, 1977). Similar results have been obtained in *Tetrahymena* (Zalkinder, 1979). It is of interest that the G1 size control in budding yeast is also directly affected by growth conditions in a similar way (Johnston, Ehrhardt, Lorincz & Carter, 1979).

To conclude this section, it seems that some cycle event is subject to size control in lower eukaryotes, and probably also in mammalian cells. Some cell types have a G1 'sizer' which controls entry into S phase, and division is determined by a timer which runs from this stage up to mitosis and division. Other cell types have a G2 'sizer' which directly controls entry into mitosis. Both G1 and G2 critical sizes may be affected by growth conditions.

How do the controls work?

The concepts of 'sizer' and 'timer' derive from the behaviour of whole cells. It is the ultimate aim of research into division controls to define them in molecular terms, though this aim is a long way from achievement. We shall first briefly consider possible molecular mechanisms which could account for the properties of sizers and timers, and then turn to ways of distinguishing between them.

Timers are perhaps the easier of the two controls to understand. The simplest explanation of a timer is a biochemical reaction that takes a definite time to complete. More realistically, a fixed sequence of events needed for the cell's passage through S and G2 phases might occupy a fairly constant time, at constant temperature, and act as a timer. There is good evidence that the process of DNA replication is a temporally organised process: certain parts of the genome always replicate early in S phase, and certain parts always late (Mueller & Kajiwara, 1966; Muldoon, Evans, Nygaard & Evans, 1971). It may be that the duration of S phase is fixed by the need to maintain this orderly sequence. Less is known about the events which occur in G2, though it is clear that there are changes in chromatin structure, for instance condensation of the chromosomes (Johnson & Rao, 1970). These and other processes may take a definite time to complete, thus providing the second part of the (S + G2) timer. It is not, however, necessary to assume that the timer is composed of structural changes which occur in S and G2: there could be a purpose-built timer mechanism which operates independently, especially as G2 can be shortened in certain circumstances in mammalian cells (Galavazi & Bootsma, 1966). More information is needed before any conclusions can be reached about timer mechanisms.

A number of molecular mechanisms have been proposed for the control of timing of mitosis. It is not our purpose here to describe these exhaustively or in any great detail; the original publications should be consulted if more information is required. Some mechanisms do not contain any element of size control, such as the oscillator model of Kauffman & Wille (1975); other mechanisms are not specifically concerned with cell size control, but still function as size controllers provided appropriate assumptions are made (reviewed in Fantes et al., 1975; Sudbery & Grant, 1975).

Molecular mechanisms for sizers fall into two broad categories. (a) Mechanisms in which there is a sensing element (receptor) in the nucleus which responds when the concentration of some effector molecule either rises above a threshold value, or falls below one (Donachie, 1968; Pritchard et al., 1969; Rosenberg, Cavalieri & Ungers, 1969; Ycas, Sugita & Bensam, 1965; Sompayrac & Maaløe, 1973). Such models control both the size and time at which an event occurs. One case where this is particularly easy to see is the inhibitor dilution model of Pritchard et al. (1969). A constant pulse of inhibitor is produced once every cycle at a discrete time; the concentration of inhibitor then falls by twofold as the volume of the cell increases. Reduction of the inhibitor concentration to this critical value triggers the cycle event. Cell size at triggering is directly controlled, while the timing of the event will also be controlled, depending on the growth rate. The effector molecules in these models should be in the cytoplasm, since a common feature of multinucleate cells is that all the nuclei mitose synchronously (Johnson & Rao, 1971). An effector which is in the cytoplasm can transmit the same signal to many nuclei simultaneously, while a nuclear-restricted effector cannot. (*b*) Mechanisms in which control molecules accumulate continuously through the cell cycle, and are counted by the cell (Zeuthen & Rasmussen, 1971; Sachsenmaier, Remy & Plattner-Schobel, 1972; Sudbery & Grant, 1975). The absolute number of control molecules is counted by their stoichiometric binding to a fixed number of sites (Sachsenmaier *et al.,* 1972), or by the assembly of a fixed number of molecules into a structure needed for mitosis (Zeuthen & Rasmussen, 1971). Operation of the stoichiometric binding model is discussed and illustrated by Sachsenmaier in this volume. These mechanisms for controlling the time of division will also control cell size at division if the rate of production of control molecules is related to the rate of mass accumulation.

This is a brief summary of several different mechanisms proposed for the control of mitosis: fuller descriptions and mathematical formulations are considered elsewhere (Fantes *et al.,* 1975; Sudbery & Grant, 1975). Perhaps the most useful conclusion to emerge from analysis of such mechanisms is that despite great differences in their biological bases, the behaviour of hypothetical cells endowed with the different mechanisms is often very similar (Fantes *et al.,* 1975). Most models that control size in any way predict that size deviations from the normal are corrected within one cycle, provided the internal contents of the cell have not been grossly affected by the process which caused the initial deviation. This is an important conclusion, since it means that the behaviour of whole cells under particular conditions cannot in general be used to derive conclusions about the underlying molecular mechanisms. Such conclusions may be 'correct' in the sense that they are consistent with the observations, but it is often possible to explain the results in entirely different ways. This last argument applies with even greater force to experiments in which very non-specific treatments, such as stopping protein synthesis, or sublethal heat shocks, are used to perturb the normal cycle. Such experiments certainly show effects on the cell cycle, but there is in general no evidence that a cell cycle process is the primary target of the treatment. Even if such evidence were available, unambiguous interpretation at the molecular level would require demonstration that a controlling or rate-limiting process were the target.

In view of the difficulty of interpreting physiological experiments in molecular terms, what is needed is a more direct approach to identifying the molecules involved in the control. So far, in only two systems has much progress been made in this direction. One is fission yeast, in which no control molecules have been identified but the genes which code for some of them have been (Thuriaux *et al.,* 1978; Nurse & Fantes, this volume). It has been possible, by a largely genetical approach, to show that one gene product is some kind of inhibitor of mitosis, and that another may be an activator (Nurse, 1977; Nurse & Fantes, this volume). The other case where a potential control molecule has been identified is that of H1 histone kinase in *Physarum* (Inglis *et al.*, 1976; Matthews, this volume). This enzyme increases in activity in the cell shortly before mitosis, and falls away sharply afterwards (Bradbury, Inglis & Matthews, 1974). Furthermore, a small advancement of mitosis is observed when a similar enzyme preparation is fed to premitotic cells (Inglis *et al.*, 1976). One possibility is that the enzyme rapidly phosphorylates histone H1, which then brings about chromosome condensation and mitosis (Louie & Dixon, 1973). What determines the sharp increase in enzyme activity at the appropriate cycle stage is not known, and another level of control may be involved.

To summarise, much of the work done to unravel the molecular basis of cell cycle control is not unambiguously interpretable because of the non-specific experimental tools often used. Much more investigation is needed, which should preferably adopt an approach that combines cell physiology, biochemistry, genetics and molecular biology.

Variability in the cell cycle

So far we have treated the cell cycle as a deterministic sequence of events. Cells undergo mitosis or enter S phase when they attain a particular size, or when a particular time has elapsed since some previous event. This way of regarding the cell cycle works well provided we are only concerned with the 'average' cell. When individual cells are considered, however, it is clear that not every cell in a population is exactly the same size or the same age at division: in all cell types there is some variation. The amount of variation is surprisingly independent of the cell type, the coefficient of variation being of the order of 10–20% (Mitchison, 1977). So much is generally agreed, but much more contentious is the question of the source and significance of the variability. There is a broad division of opinion between groups of researchers about this. We shall first outline the two contrasting points of view and examine their consistency with observation. Finally we shall consider whether any reconciliation between the apparently opposite viewpoints is possible.

Variation and deterministic models

According to the deterministic view of the cell cycle, every cell in a population should ideally behave in an identical way. Cell size and cell age at division, for instance, should be invariant. The problem, then, is to explain why this is not so. The answer generally given is that the deterministic controls are 'sloppy', and this gives rise to variation around the population mean values. This does not mean that there is no control mechanism, merely that the mechanism is imprecise. A good analogy is that of height in humans. There is a well defined population average value for the height of adult males, and there is also variation around this mean. No one would argue that the existence of variation implies an absence of control. In the same way, the existence of different critical cell sizes according to the growth conditions and genetic constitution of fission yeast (Nurse & Fantes, this volume) does not negate the existence of a control. The control is alterable in the sense that the mean can be set at a range of values, and there will still be variation around the mean. To pursue the analogy with human height, adult females are *on average* shorter than males, despite the fact that both male and female heights are distributed around a mean, so that some men are shorter than some women. The point is that sex controls height in the two subpopulations: in the same way, growth conditions control the mean cell size of a group of fission yeast cells. The variation between individual cells is often unimportant, since the behaviour of a large group can be adequately predicted according to deterministic control models.

How then can 'sloppiness' be generated? Deterministic control models, as we have seen, rely on an interaction between effector molecules and receptor sites, or on the stoichiometric binding of control molecules to sites. The interaction between an effector molecule and its receptor is ultimately a stochastic process, and is subject to statistical fluctuation. In addition, local variations in the concentration of effector molecules will increase the variability of the interaction. This means that even in identical cells with the same number of effector molecules, there will be some spread in the time at which the receptor system will respond to the effector concentration reaching a level around the critical value. The spread will be large if the number of effector molecules per cell is small, because statistical fluctuations will then be relatively greater. A small number of control molecules might be expected, by analogy with effector molecules for bacterial operons (Maloney & Rotman, 1973). The above argument does not hold in quite the same way for mechanisms where the *number,* rather than the *concentration,* of control molecules is the triggering factor, but statistical fluctuations will still play some part in generating inaccuracy.

Other factors contributing to the variability are the asymmetric division of cells, and unequal distribution of control molecules to the daughters. The former effect is partly responsible for the latter, but statistical fluctuations in the spatial distribution of control molecules at the time of division may increase the inequality of partitioning between the daughter cells. Daughter cells may therefore be born with different numbers of control molecules, or different sizes, or both. In general, the critical number or concentration of control molecules will not be attained simultaneously in the two sister cells. Thus two different factors will generate variation in a cell population: the differences between 'identical' cells just described, and variability in the stimulus–response behaviour of the trigger mechanism. On average, cell size or division time will nevertheless be controlled, because statistical effects and random division asymmetries will average out over a large number of cells (Fig. 5). It is also possible to account for the amount of variation and the shape of the distribution for cycle time or division size, though extra assumptions are needed (Koch, 1966; Kubitschek, 1971).

Transition probability models

There is an alternative way of considering the variability of the cell cycle, in which the variation itself is a fundamental part of the cycle. This group of transition probability models is described in detail in the chapter by Brooks. In the original version of the idea (Smith & Martin, 1973) the cell cycle is divided into two parts. A cell at birth, in G1, is in the A state, which is of indeterminate length. It remains there until a transition occurs, and then enters the B phase. The transition is proposed to occur at random, so that the probability of the transition occurring in unit time is constant, irrespective of the cell's age, size or past history. Changes in the proliferation rate are proposed to be mediated by alteration of the transition probability K_A. Once in the B phase, the cell progresses through S and G2 and on to mitosis and division. The model predicts a particular distribution of cell cycle times provided the B phase is of constant length (T_B). A semilogarithmic plot of the proportion of undivided cells against cell age (the α curve) should consist of a horizontal line for cell ages between zero and T_B, and a line of constant downward slope thereafter. In practice the changeover at T_B is not so abrupt, and a substantial amount of curvature is found.

Fig. 5. Accurate and inaccurate cell mass control. Abscissa: cell mass, in arbitrary units. The critical mass is indicated by the solid triangle. *Top*. Probability of a cell having divided at a given mass. A completely accurate control would appear as a step (– – –). Inaccuracy in the mass monitor will produce a curve which rises from 0 to 1 over a range of sizes (——). *Bottom*. The distributions of cell size at division are shown, for accurate or inaccurate size monitoring.

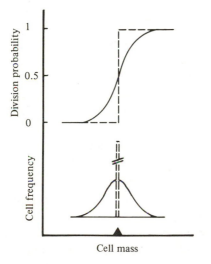

This was ascribed to variation in the determinate B phase, in a similar way to the 'sloppiness' of deterministic controls. Another postulate was added to the model to account for the observed correlation between sister cell cycle times. Such a correlation should not exist if all cells behaved independently. The additional proposal was that sister cells had identical B phases, while unrelated cells in general did not. This assumption seemed arbitrary at the time, but the linear β curves observed were consistent with the idea. (The β curve shows the cumulative frequency distribution of the differences between sister cell cycle times, on a semilogarithmic plot; see Brooks, this volume). To provide a satisfactory explanation of the identity of B phases in sister cells, and for other reasons, a new, more complex, transition probability model has recently been proposed, outlined in the chapter by Brooks. The main new feature is that the old B phase is replaced by a new Q state and a timer L. The Q state, like the A state, is indeterminate in length, and the cell's exit from Q is followed by its embarking on the timer L. The new model is therefore mathematically described by three parameters K_A, K_Q and T_L.

The above outline of the transition probability models is necessarily rather brief: for a fuller account the reader should refer to the chapter by Brooks and the references cited therein.

Comparison of deterministic and transition probability models

The two types of model have different origins, in two respects. First, deterministic models were proposed mainly by workers on microorganisms, while the idea of transition probability emerged from work on mammalian cells. Second, deterministic models arose from investigations into the physiological and molecular mechanisms underlying the control of cell cycle events. Such an approach by its very nature stresses the similarities between cells and tends to ignore the differences. Transition probability, on the other hand, was proposed specifically to explain the differences between the cycles of individual cells.

Rather few attempts have been made to bridge the gap between the two 'camps' either theoretically or experimentally (Pardee, Shilo & Koch, 1979). It is our intention now to discuss the good and bad points of both classes of model. We shall ask whether either class can be generalised to systems other than that for which each model was originally devised.

The transition probability model relies heavily on precise knowledge of cycle kinetics, and the exact shapes of distributions have played a key role in establishing and modifying the assumptions made. The problem of using only kinetic data to investigate the control of the complex biological events of the cell cycle is rather well exemplified by two reports from the same laboratory. In one (Minor & Smith, 1974), the authors claimed that the final slope of the α curve was equal to that of the β curve, for two sets of data. Such an interpretation of this and later data was not questioned until recently, but Brooks now claims that in fact the slopes of α and β curves are slightly different, at least in some cases. That the small differences were not previously noticed demonstrates the difficulties of obtaining enough reliable data to permit unambiguous curve fitting. This problem is so severe that the foundation of any model supported largely by the shape of distributions must be considered with caution. This is especially so in the new model since K_A and K_Q are proposed to vary independently with growth conditions, and L with size. The assignment of values to these parameters can only be made indirectly from the observed distributions by making the assumption that the model's axioms are correct. Furthermore, given three independent parameters, a great many curves can be generated. Against this, it must be said that independent estimates of some parameters can be made (see Brooks, this volume), and they are in good agreement with one another. The exact shape of α curves can also be formulated. There is therefore at least internal consistency.

The transition probability model is by no means the only one which can generate curves which are a good fit to experimentally observed cycle time distributions. In particular, experimentally obtained α curves can be adequately simulated by log-normal or reciprocal-normal distributions (Koch, 1966; Kubitschek, 1971; Wheals, 1977), consistent with size control models. Acceptable β curves can also be obtained (Pardee et al., 1979). The argument presented by Brooks that the distribution of G1 does not follow the predictions of either a log- or reciprocal-normal distribution is not strictly correct. His procedure is to derive the G1 distribution by subtracting a constant $(S + G2 + M)$ period from the total cycle distribution. This procedure is only correct if the distribution of $(S + G2 + M)$ is known, and if this cycle period is not correlated with the length of G1. There is little evidence about this, but one report (Killander & Zetterberg, 1965b) suggests a negative correlation between the durations of G1 and G2.

To explain why β curves for sister cells are straight lines on a semilogarithmic plot, Minor & Smith (1974) proposed that the B phases of sisters were for some reason identical, while unrelated cells would have B phases that were in general different. This line of argument has been continued, and Brooks now suggests a reason for the identity of sister cell B phases. Again, the arguments employed have always been indirect, based on shapes of distributions. Van Wijk and colleagues (1977) have more directly examined the cell cycles of hepatoma cells with respect to the durations of G1 and G2. They report that the G2 phases of sisters differ by more, on average, than the G1 phases. In contrast, the much larger differences between unrelated cells are primarily due to G1 differences. In the absence of any other direct information about the duration of cell cycle phases in sisters, it is hard to see how the claim of identical B phases can be upheld.

Deterministic models can readily account for the correlation between sister cycle times. Sister cells have a common mother, and however asymmetric its division, sisters are bound to be on average more similar in birth size than unrelated cells. With the weight of evidence suggesting that birth size has a strong influence on cycle duration, it is evident that sister cycle times are likely to be positively correlated.

Effects of cell size have long been ignored by transition probability theory, but recently Shields and coworkers (1978) reported that small cells showed a longer cycle than large cells. This is interpreted by Brooks as the duration of the timer L being influenced by cell size. The proposition that size might have some effect on the rate of a division-controlling process represents a major revision of the transition probability theory. If cells can respond to their size by altering the length of their cycle, they must be able to monitor their size by some means (however indirect), a long-standing postulate of size-control models. Alteration in the length of L in response to size may well be a satisfactory explanation for mammalian cells, but such a hypothesis cannot account for many observations in lower eukaryotes. The problem lies in the response time of the system: an alteration in the length of L would be cumulative with time, and would only be manifest after a lag of the order of one cell cycle. Certain experiments, especially with fission yeast, show that an almost instantaneous change in the division rate occurs after certain shifts (Nurse & Fantes, this volume). Similar results have been obtained with *Tetrahymena* (Zalkinder, 1979) and related observations made in budding yeast (Johnston *et al.*, 1979).

The transition probability model takes as implicit the constancy of the post-G1 part of the cell cycle, which is at least approximately true for some mammalian cell types. However, there is good evidence for some cell types that the period between the start of S phase and division is not constant, as discussed earlier. For some cells there is strong support for a primary cycle control acting over mitosis (see p. 20). It is therefore doubtful whether the transition probability model, in its present form, can account for the behaviour of such cell types.

As regards molecular mechanisms, a number have been proposed for cell size monitoring, though none so far is adequate to explain all data available for all cell types (Fantes *et al.*, 1975). Molecular mechanisms have been proposed to account for a random transition which occurred with constant probability per unit time (Smith & Martin, 1973), though we do not consider them particularly plausible. (The new model would of course require two such transitions and two mechanisms.) However, the proposal by Brooks (this volume) that behaviour of mitotic centres, possibly identified with centrioles, might be the biological counterpart of the random transitions, is an exciting one. It should in principle lead to rigorous testing of the transition probability model by examination of centriole duplication, so that the kinetic basis of the model could be broadened. One possible problem is hinted at by Brooks: if centrioles themselves do not act as mi-totic centres, there is no reason for centrioles to behave in the way predicted by the new transition probability model. Such a finding would not negate the model, since the real mitotic centres would be as elusive as ever. This would however restore the theory to its former state of being supported by kinetic data alone.

A possible reconciliation

We now return to a fundamental question: can a model be constructed which will account for the size-controlled behaviour of lower eukaryotes (and some mammalian cells) and the random transitions which may best explain the cycle kinetics of mammalian cells? We are sure that such models exist: here is one example. Suppose that the size-control mechanism in a cell involves the completion of a structure by assembly from subunits, and that initiation of mi-tosis requires a completed structure. If the subunits are synthesised at a rate proportional to cell mass, then the structure will be completed at a particular cell size. The initiation of mitosis will on the other hand involve an interaction be-tween the structure and some other cellular component, perhaps in the nucleus. This interaction will by definition involve only a small number of molecules (e.g. a single structure) and binding sites. The probability of such an interaction will be constant with time, and initiation of mitosis might well follow first order kinetics. If in one cell type the accumulation of subunits and assembly of the structure is rapid, then the event limiting the rate of cell cycle progress will be the random initiation event, giving kinetics consistent with the transition proba-bility model. If on the other hand the build-up of subunits is slow, and assembly of the structure the rate-limiting factor in the cycle, the division control will appear deterministic, consistent with size-control models. The two apparently different types of behaviour observed in different cell types may simply be a question of the balance of rates of two continuing processes. This view has been further discussed by Nurse (1980). In conclusion, therefore, we suggest that there need not be any major conflict between the two viewpoints as they may be describing different facets of the same underlying mechanism.

References

Bostock, C. J. (1970). DNA synthesis in the fission yeast *Schizosaccharo-myces pombe*. *Experimental Cell Research*, **60**, 16–26.

Bradbury, E. M., Inglis, R. J. & Matthews, H. R. (1974). Control of cell di-vision by very lysine rich histone (F1) phosphorylation. *Nature, London*, **247**, 257–61.

Brewer, E. N. & Rusch, H. P. (1968). Effects of elevated temperature shocks on mitosis and on the initiation of DNA replication in *Physarum polyce-phalum*. *Experimental Cell Research*, **49**, 79–86.

Brooks, R. F. (1976). Regulation of the fibroblast cell cycle by serum. *Nature, London,* **260,** 248–50.

Bürk, R. R. (1970). One-step growth cycle for BHK 21/13 hamster fibroblasts. *Experimental Cell Research,* **63,** 309–16.

Carter, B. L. A., Lorincz, A. & Johnston, G. C. (1978). Protein synthesis, cell division and the cell cycle in *Saccharomyces cerevisiae* following a shift to richer medium. *Journal of General Microbiology,* **106,** 222–5.

Cress, A. E. & Gerner, E. W. (1977) Hydroxyurea treatment affects the G1 phase in next generation CHO cells. *Experimental Cell Research,* **110,** 347–53.

Donachie, W. D. (1968). Relationship between cell size and the time of initiation of DNA replication. *Nature, London,* **219,** 1077–9.

Fantes, P. A. (1977). Control of cell size and cycle time in *Schizosaccharomyces pombe. Journal of Cell Science,* **24,** 51–67.

Fantes, P. A., Grant, W. D., Pritchard, R. H., Sudbery, P. E. & Wheals, A. E. (1975). The regulation of cell size and the control of mitosis. *Journal of Theoretical Biology,* **50,** 213–44.

Fantes, P. A. & Nurse, P. (1977). Control of cell size at division in fission yeast by a growth-modulated size control over nuclear division. *Experimental Cell Research,* **107,** 377–86.

Fournier, R. E. & Pardee, A. B. (1975). Cell cycle studies of mononucleate and cytochalasin-B-induced binucleate fibroblasts. *Proceedings of the National Academy of Sciences, USA,* **72,** 869–73.

Fox, T. O. & Pardee, A. B. (1970). Animal cells: noncorrelation of length of G1 phase with size after mitosis. *Science,* **167,** 80–2.

Galavazi, G. & Bootsma, D. (1966). Synchronization of mammalian cells *in vitro* by inhibition of the DNA synthesis. II. Population dynamics. *Experimental Cell Research,* **41,** 438–51.

Hartmann, M. (1928). Über experimentelle Unsterblichkeit von Protozoen-Individuen. Ersatz der Fortpflanzung von *Amoeba Proteus* durch fortgesetzte Regenerationen. *Zoologisches Jahrbuch,* **45,** 973–87.

Hartwell, L. H. (1974). *Saccharomyces cerevisiae* cell cycle. *Bacteriological Reviews,* **38,** 164–98.

Inglis, R. J., Langan, T. A., Matthews, H. R., Hardie, D. G. & Bradbury, E. M. (1976). Advance of mitosis by histone phosphokinase. *Experimental Cell Research,* **97,** 418–25.

Jagadish, M. N. & Carter, B. L. A. (1977). Genetic control of cell division in yeast cultured at different growth rates. *Nature, London,* **269,** 145–7.

Jagadish, M. N., Lorincz, A. & Carter, B. L. A. (1977). Cell size and cell division in yeast cultured at different growth rates. *FEMS Letters,* **2,** 235–7.

Johnson, G. S. & Schwartz, J. P. (1976). Effects of sugars on the physiology of culture fibroblasts. *Experimental Cell Research,* **97,** 281–90.

Johnson, R. T. & Rao, P. N. (1970). Mammalian cell fusion: induction of premature chromosome condensation in interphase nuclei. *Nature, London,* **226,** 717–22.

Johnson, R. T. & Rao, P. N. (1971). Nucleo-cytoplasmic interactions in the achievement of nuclear synchrony in DNA synthesis and mitosis in multinucleate cells. *Biological Reviews,* **46,** 97–155.

Johnston, G. C., Ehrhardt, C. W., Lorincz, A. & Carter, B. L. A. (1979). Regulation of cell size in the yeast *Saccharomyces cerevisiae. Journal of Bacteriology,* **137,** 1–5.

Johnston, G. C., Pringle, J. R. & Hartwell, L. H. (1977). Co-ordination of growth with cell division in the yeast *Saccharomyces cerevisiae*. *Experimental Cell Research*, **105**, 79–98.

Kauffman, S. & Wille, J. J. (1975). The mitotic oscillator in *Physarum polycephalum*. *Journal of Theoretical Biology*, **55**, 47–93.

Killander, D. & Zetterberg, A. (1965*a*). Quantitative cytochemical studies on interphase growth. I. Determination of DNA, RNA and mass content of age determined mouse fibroblasts *in vitro* and of intercellular variation in generation time. *Experimental Cell Research*, **38**, 272–84.

Killander, D. & Zetterberg, A. (1965*b*). A quantitative cytochemical investigation of the relationship between cell mass and initiation of DNA synthesis in mouse fibroblasts *in vitro*. *Experimental Cell Research*, **40**, 12–20.

Koch, A. L. (1966). Distribution of cell size in growing cultures of bacteria and the applicability of the Collins-Richmond principle. *Journal of General Microbiology*, **45**, 409–17.

Kubitschek, H. E. (1971). The distribution of cell generation times. *Cell and Tissue Kinetics*, **4**, 113–22.

Ley, K. D. & Tobey, R. A. (1970). Regulation of initiation of DNA synthesis in Chinese hamster cells. II. Induction of DNA synthesis and cell division by isoleucine and glutamine in G1-arrested cells in suspension culture. *Journal of Cell Biology*, **47**, 453–9.

Louie, A. J. & Dixon, G. H. (1973). Kinetics of phosphorylation and dephosphorylation of testis histones and their possible role in determining chromosomal structure. *Nature New Biology*, **243**, 164–8.

Lövlie, A. (1963). Growth in mass and respiration rate during the cell cycle of *Tetrahymena pyriformis*. *Comptes Rendus des Travaux du Laboratoire Carlsberg*, **33**, 377–413.

Maloney, P. C. & Rotman, B. (1973). Distribution of suboptimally induced β-D-galactosidase in *Escherichia coli*. *Journal of General Microbiology*, **73**, 77–84.

Minor, P. D. & Smith, J. A. (1974). Explanation of degree of correlation of sibling generation times in animal cells. *Nature, London*, **248**, 241–3.

Mitchison, J. M. (1977). The timing of cell cycle events. In *Mitosis, Facts and Questions*, ed. M. Little, N. Paweletz, C. Petzelt, H. Ponstingel, D. Schroeter & H-P. Zimmermann, pp. 1–18. Berlin: Springer-Verlag.

Miyata, H., Miyata, M. & Ito, M. (1978). The cell cycle in the fission yeast, *Schizosaccharomyces pombe*. I. Relationship between cell size and cycle time. *Cell Structure and Function*, **3**, 39–46.

Mueller, G. C. & Kajiwara, K. (1966). Early- and late-replicating deoxyribonucleic acid complexes in HeLa nuclei. *Biochimica et Biophysica Acta*, **114**, 108–15.

Muldoon, J. J., Evans, T. E., Nygaard, O. F. & Evans, H. H. (1971). Control of DNA replication by protein synthesis at defined times during the S period in *Physarum polycephalum*. *Biochimica et Biophysica Acta*, **247**, 310–21.

Nurse, P. (1975). Genetic control of cell size at cell division in yeast. *Nature, London*, **256**, 547–51.

Nurse, P. (1977). Cell cycle control in yeasts. *Transactions of the Biochemical Society*, **5**, 1191–3.

Nurse, P. (1980). Cell cycle control – both deterministic and probabilistic. *Nature, London*, **286**, 9–10.

Nygaard, O. F., Guttes, S. & Rusch, H. P. (1960). Nucleic acid metabolism in a slime mold with synchronous mitosis. *Biochimica et Biophysica Acta*, **38**, 298–306.

Pardee, A. B. (1974). A restriction point for control of normal animal cell proliferation. *Proceedings of the National Academy of Sciences, USA*, **71**, 1286–90.

Pardee, A. B., Shilo, B. & Koch, A. L. (1979). Variability of the cell cycle. In *Hormones and Cell Culture*, ed. G. Sato & R. Ross. New York: Cold Spring Harbour Laboratory.

Prescott, D. M. (1956). Relation between cell growth and cell division. (*a*) II. The effect of cell size on cell growth rate and generation time in Amoeba proteus. (*b*) III. Changes in nuclear volume and growth rate and prevention of cell division in Amoeba proteus resulting from cytoplasmic amputations. *Experimental Cell Research, 11*, 86–98.

Pritchard, R. H. (1974). On the growth and form of a bacterial cell. *Philosophical Transactions of the Royal Society of London, Series B*, **267**, 303–36.

Pritchard, R. H., Barth, P. T. & Collins, J. (1969). Control of DNA synthesis in bacteria. *Symposium of the Society for General Microbiology*, **19**, 263–97.

Ron, A. & Prescott, D. M. (1969). The timing of DNA synthesis in *Amoeba proteus*. *Experimental Cell Research*, **56**, 430–4.

Rosenberg, B. H., Cavalieri, L. F. & Ungers, G. (1969). The negative control mechanism for *Escherichia coli* DNA replication. *Proceedings of the National Academy of Sciences, USA*, **63**, 1410–17.

Rosenberg, H. M. & Gregg, E. C. (1969). Kinetics of cell volume changes of murine lymphoma cells subjected to different agents *in vitro*. *Biophysical Journal*, **9**, 592–606.

Ross, D. W. (1976). Cell volume growth after cell cycle block with chemotherapeutic agents. *Cell and Tissue Kinetics*, **9**, 379–87.

Sachsenmaier, W., Dönges, K. H., Rupff, H. & Czihak, G. (1970). Advanced initiation of synchronous mitosis in *Physarum polycephalum* following ultraviolet-irradiation. *Zeitschrift für Naturforschung*, **25b**, 866–71.

Sachsenmaier, W., Remy, U. & Plattner-Schobel, R. (1972). Initiation of synchronous mitosis in *Physarum polycephalum*. *Experimental Cell Research*, **73**, 41–8.

Shields, R., Brooks, R. F., Riddle, P. N., Capellaro, D. F. & Delia, D. (1978). Cell size, cell cycle and transition probability in mouse fibroblasts. *Cell*, **15**, 469–74.

Sisken, J. E. & Kinosita, R. (1961). Timing of DNA synthesis in the mitotic cycle *in vitro*. *Journal of Biophysical and Biochemical Cytology*, **9**, 509–18.

Smith, J. A. & Martin, L. (1973). Do cells cycle? *Proceedings of the National Academy of Sciences, USA*, **70**, 1263–7.

Sompayrac, L. & Maaløe, O. (1973). Autorepressor model for control of DNA replication. *Nature New Biology*, **241**, 133–5.

Sudbery, P. E. & Grant, W. D. (1975). The control of mitosis in *Physarum polycephalum*. The effect of lowering the DNA: mass ratio by UV irradiation. *Experimental Cell Research*, **95**, 405–15.

Sudbery, P. E. & Grant, W. D. (1976). The control of mitosis in *Physarum polycephalum:* the effect of delaying mitosis and evidence for the operation of the control mechanism in the absence of growth. *Journal of Cell Science*, **22**, 59–65.

Temin, H. M. (1971). Stimulation by serum of multiplication of stationary chicken cells. *Journal of Cellular Physiology*, **78**, 161–70.

Thuriaux, P., Nurse, P. & Carter, B. (1978). Mutants altered in the control co-ordinating cell division with cell growth in the fission yeast *Schizosaccharomyces pombe*. *Molecular and General Genetics*, **161**, 215–220.

Tovey, M. & Brouty-Boye, D. (1976). Characteristics of the chemostat culture of murine leukaemia L 1210 cells. *Experimental Cell Research*, **101**, 346–54.

Tsuboi, A., Kurotsu, T. & Terasima, T. (1976). Changes in protein content per cell during growth of mouse L cells. *Experimental Cell Research*, **103**, 257–61.

Wheals, A. E. (1977). Transition probability and cell-cycle initiation in yeast. *Nature, London*, **267**, 647.

van Wijk, R., van de Poll, K. W., Amesz, W. J. C. & Geilenkirchen, W. L. M. (1977). Studies on the variations in generation times of rat hepatoma cells in culture. *Experimental Cell Research*, **109**, 371–9.

Ycas, M., Sugita, M. & Bensam, A. (1965). A model of cell size regulation. *Journal of Theoretical Biology*, **9**, 444–70.

Yen, A., Fried, J., Kitahara, T., Strife, A. & Clarkson, B. D. (1975*a*). The kinetic significance of cell size. I. Variation of cell cycle parameters with size measured at mitosis. *Experimental Cell Research*, **95**, 295–302.

Yen, A., Fried, J., Kitahara, T., Strife, A. & Clarkson, B. D. (1975*b*). The kinetic significance of cell size. II. Size distributions of resting and proliferating cells during interphase. *Experimental Cell Research*, **95**, 303–10.

Zalkinder, V. (1979). Correlation between cell nutrition, cell size and division control. Part I. *Biosystems*, **11**, 295–307.

Zeuthen, E. & Rasmussen, L. (1971). Synchronised cell division in protozoa. In *Research in Protozoology*, ed. T. T. Chen, vol. 4, pp. 9–145. Oxford: Pergamon Press.

Zeuthen, E. & Williams, N. E. (1969). Division-limiting morphogenetic processes in *Tetrahymena*. In *Nucleic Acid Metabolism, Cell Differentiation and Cancer Growth*, ed. E. V. Cowdry & S. Seno, pp. 203–16. Oxford: Pergamon Press.

R.F.BROOKS

Variability in the cell cycle and the control of proliferation

Introduction

The age at which individual cells divide is very variable, even when the population as a whole is growing exponentially in steady-state (Kelly & Rahn, 1932; Prescott, 1959; Siskin & Kinosita, 1961; Cook, J. R. & Cook, B., 1962; Dawson, Madoc-Jones & Field, 1965; Siskin & Morasca, 1965). This has sometimes been attributed to a lack of uniformity within the culture or to heterogeneity of the cells. However, the variability has persisted despite many improvements in culture technique. Indeed, it is a property of proliferation *in vivo* (Quastler & Sherman, 1959) and so is not an artefact of cell culture. Variability also occurs within the first few generations of a clone (the progeny of a single cell). It is unlikely, therefore, to be a reflection of genetic heterogeneity (Shields, 1977). Rather, unpredictability appears to be an intrinsic feature of every cell cycle.

An important consequence of variable generation times is the well-known difficulty of keeping synchronised cell populations synchronous. Since cells refuse to do the same thing at the same time for anything more than part of a cycle after synchronisation, many experimental approaches to the study of cell proliferation are rendered inherently difficult – for instance, the attempt to discover biochemical markers of the cycle. For this reason alone it is desirable to have a satisfactory statistical description of the distribution of generation times. A more important reason has emerged with the demonstration that alterations in the proliferation rate of mammalian cells are brought about almost entirely by changing the *degree* of variability of generation times, the minimum age at division being relatively little affected (Shields & Smith, 1977). Variability thus appears to be connected with the regulation of proliferation and it seems not unlikely that it arises as a direct consequence of the way in which the cell cycle is controlled. As will be seen, there is every reason to think that an analysis of this variation may tell us a good deal about the nature of the controls.

35

Before proceeding, it is necessary to admit that several new or unfamiliar terms will be introduced during the course of this chapter. To help alleviate any confusion, a list of definitions is provided in Table 1. It is also necessary to admit that this discussion draws heavily on data obtained for mammalian cells. To some extent this reflects the bias and experience of the author, but it is also partly dictated by the relative paucity of suitable kinetic data for the cell cycles of other eukaryotes. Nevertheless, comparisons with other cell types are made whenever possible.

Explanations of variability
Log-normal and rate-normal distributions

It is commonly believed that variability of age at division is merely a reflection of the underlying complexities of the cell cycle, and to that extent unremarkable. It is argued that small random 'errors' at each of many steps accumulate as cells progress through the cycle. The summation of many independent random variables would tend towards a normal distribution (from the central limit theorem of statistics), though in practice the distribution of generation times is positively skewed. If, however, each step influenced the *rate* of a succeeding step (for example) then random variation at each one would become amplified over the course of the cycle and the distribution skewed. As discussed by Koch (1966, 1969) this is one way (though not the only way) in which a log-normal distribution is generated, i.e. where the logarithm of some variable is normally distributed. Empirically, it is found that the logarithms of cell ages at division do sometimes approximate to a normal distribution quite well (Nachtwey & Cameron, 1968).

An alternative explanation of the skewed distribution of generation times has been put forward by Kubitschek (1962, 1971). He suggested that individual cells traversed the cycle at different, normally distributed rates. If the cell cycle is regarded as a 'race' run over a fixed 'distance' then this predicts that the reciprocal of generation time should be normally distributed since rate = distance/ time. Such rate-normal (or reciprocal-normal) distributions, as they are called, appear to fit the data for a variety of cell types rather better than the log-normal distribution (Kubitschek, 1971).

Although both log-normal and rate-normal distributions sometimes provide reasonable qualitative descriptions of the distribution of generation times, the assumption is made that the variability is spread over the whole cell cycle. For mammalian cells (and possibly others) this is not the case: numerous studies have shown that the sum of the duration of G1, G2 and M has negligible variance. This is indicated by the superimposable curves for entry into S phase and mitosis (Siskin & Morasca, 1965; Tobey, 1973; Shields *et al.*, 1978). The variability of cycle times is therefore confined almost entirely to the G1 phase: virtually no additional variability arises between the initiation of DNA synthesis and cell division. When the duration of S + G2 + M is subtracted from the overall generation time, the resulting distribution (of G1 times) is neither log-normal nor reciprocal-normal.

The small variance in the duration of S + G2 + M (whatever its underlying basis) demonstrates that complex biological processes are not intrinsically variable. The extreme variability in the length of G1 is therefore all the more significant. This becomes even more so in view of repeated demonstrations that proliferation rate is regulated in G1 (Baserga 1965; Prescott, 1968). Thus, not only

Table 1. *Glossary of terms*

α curve	Fraction of undivided cells plotted against cell age. The ordinate frequently has a logarithmic scale.
β curve	Fraction of sister cell pairs with differences between generation times greater than or equal to time t plotted against t. The ordinate frequently has a logarithmic scale.
A state	Indeterminate state in G1 in which cells await the random transition which triggers the start of B phase. In the revised transition probability model, exit from A state starts processes leading to DNA synthesis and mitosis and is accompanied by entry into another indeterminate state Q.
B phase	The period between successive A states containing the S, G2 and M phases of the classical cell cycle together with part of G1. Originally considered to be entirely deterministic, but now thought to be composed of a deterministic process L together with the indeterminate Q state.
K_A	The rate constant for exit from A state.
K_Q	The rate constant for exit from Q state.
L	A process initiated on exit from Q state which cells must complete in order to reach A state. The duration of L accounts for most of the lag between the addition of a growth stimulus and any increase in the rate of initiating DNA synthesis. It is also more or less equivalent to the minimum cycle time. The process L can be initiated at any stage of the conventional cell cycle but not from A state.
Q state	Another indeterminate state which cells enter on leaving A state. Exit from Q is random and triggers the initiation of the process L. Although cells enter Q at much the same time as the start of S phase, Q state can extend over any part of the cycle (i.e. S, G2, M or G1) except A state.
T_A	Time spent by a cell in A state.
T_B	Duration of B phase. The same as $T_Q + T_L$.
T_Q	Time spent by a cell in Q state.
T_L	Duration of the process L.

does the duration of G1 vary from cell to cell in a given population growing at steady-state, but when proliferation rate changes, this is accomplished by changing the mean duration of G1. When net proliferation ceases, it is in G1 that the cells accumulate.

The transition probability model

In attempting to account for the variability of G1 and at the same time explain how the cycle might be regulated, Smith & Martin (1973, 1974) were led to the idea that G1 contained a single critical event which took place at random with a probability dependent on the conditions. According to this view – which is similar to a model put forward earlier by Burns & Tannock (1970) – the existence of this event effectively divides the cell cycle into two fundamentally different parts: 'A state', located somewhere in G1, in which cells pause indefinitely while waiting for the critical event, or transition, to occur; and 'B phase' which includes the rest of the cycle (i.e. $S + G2 + M +$ the remainder of G1). Unlike A state, B phase was regarded as deterministic in that once initiated it was thought to progress inevitably towards completion. Because the transition from A to B was considered to be random, the model predicts an exponential distribution of times spent in A state. If B phase were invariant then this would be reflected in an exponential distribution of cycle times. No cells would divide at ages less than the duration of B phase (T_B), but thereafter the fraction of undivided cells (α) should decline exponentially. A plot of the logarithm of the fraction of cells remaining undivided against age (the so-called α plot) should therefore be linear, though any variation in T_B would tend to introduce curvature into the initial portion of the plot. If proliferation rate were regulated mainly by alterations of the transition probability, as the theory proposed, then this would lead to a change in the slope of the α plot, the minimum interdivision time being largely unaffected.

An example of what happens in practice is shown in Fig. 1. The data are for 3T3 cells (mouse 'fibroblasts') growing at different rates in media supplemented with different serum concentrations. Although there is some initial curvature, the log α plots are approximately linear over most of their length. Furthermore, as the serum concentration is reduced, the predominant effect is a marked change of slope. In contrast, the minimum interdivision time changes very little. Thus, 3T3 cells, and also other mammalian cells (Robinson, Smith, Totty & Riddle, 1976; Shields & Smith, 1977), regulate their proliferation rate mainly by changing the variation in cell cycle times rather than by altering the minimum possible generation time. This, in particular, is difficult to account for except in terms of some sort of random step in the cell cycle. Nevertheless, the initial curvature of the log α plots means that there is more variability than can be ascribed to a single constant transition probability. If there is a random transition, then this means that the duration of B phase must also be variable, or that the transition probability changes with cell age (Svetina, 1977).

The β curve. A more rigorous demonstration of an age-independent transition probability in the cell cycle has come from what, in principle, might have provided convincing evidence against the theory. This is the observation that the generation times of sister cells are usually more alike than random pairs, i.e. are correlated (Cook, J. R. & Cook, B. 1962; Dawson *et al.*, 1965). If each cell cycle were initiated at random then there is no reason why this should be so. Nevertheless, Minor & Smith (1974) found that, despite the correlation, the division times of siblings differed sufficiently to be compatible with the existence

Fig. 1. α curves for Balb/c 3T3 cells growing exponentially at different rates in medium supplemented with either 1%, 3% or 10% calf serum. The percentage of cells remaining undivided (α) is plotted on a logarithmic scale. The data, previously unpublished, were obtained by R. Shields and are shown here with his permission.

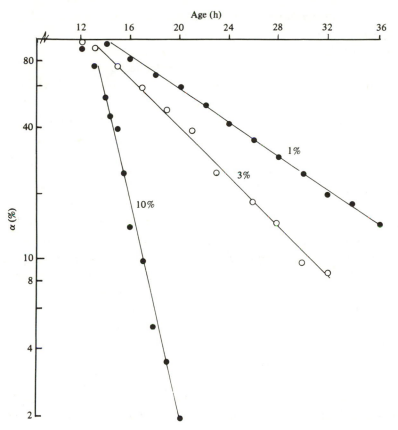

of a random transition in the cycle. This suggested that the duration of B phase, variable in the population as a whole, was more alike in siblings than unrelated pairs of cells. What was surprising was that the distribution of differences between sibling generation times – the so-called β curve – proved to be perfectly exponential and almost (though as will be seen, not quite) parallel to the exponential 'tail' of the α plot (Minor & Smith, 1974; Smith & Martin, 1974; Shields & Smith, 1977; Shields, 1977). Exponential β curves could arise only if there is a random transition in the cell cycle and if the remainder of the cycle (i.e. B phase) is of virtually identical duration in sister cells (Shields, 1978). Thus, the differences between sibling generation times are not merely compatible with the transition probability model. Rather, the β curves demonstrate that these differences can be attributed almost entirely to a single random transition. Furthermore, exponential β curves have been obtained for bacteria (Shields, 1978), *Euglena* (Smith & Martin, 1974) and mammalian cells (Shields & Smith, 1977) suggesting that random transitions may be a rather general feature of cell cycle organisation.

Some observations not explained by the transition probability model

Variability of B phase

Although the transition probability model accounts for the differences between the generation times of sibling cells, it leaves unanswered the question of why T_B should be variable in the population at large, but identical in siblings. It has been found that the duration of B phase has some relation to cell size, big cells having a slightly shorter T_B, on average, than small cells (Shields *et al.*, 1978). This suggested that variability in T_B might be due to differences in size at the time of division, the similarity of T_B in sister cells arising because mass is partitioned more or less equally between them. However, the only way a cell could become large, and give rise to large daughters with short B phases, would be to have a longer than average generation time. This means that the generation times of mother and daughter cells should be negatively correlated. Slight negative correlations have been found for the fission yeast, *Saccharomyces pombe* (Fantes, 1977; Miyata, Miyata & Ito, 1978), but for mammalian cells, the correlation is either insignificant (Shields & Smith, 1977) or weakly positive (Dawson *et al.*, 1965). It seems, therefore, that some other property of the parental cell besides size at division must be responsible for setting the duration of B phase in the daughter cells – in mammals, at least.

The problem of why T_B should be virtually identical in siblings but variable in general, though interesting in itself, has acquired added importance in the light of more recent experiments (Brooks, unpublished). These have suggested that variability in T_B plays a much greater part in the regulation of proliferation than was originally supposed. If proliferation rate were controlled solely by changing the probability of the A to B transition, then the variance in T_B should become a much smaller part of the total variance in cycle times at low growth rates because of the increased dispersion of times spent in A state. Since the correlation between sibling generation times arises from the near identity of their B phase durations, this correlation should also become less as proliferation rate decreases. However, in two populations of 3T3 cells growing exponentially with doubling times of 14 h or 24 h in media supplemented with different serum concentrations, the correlation coefficients for sibling generation times were not significantly different (0.51 and 0.58 respectively). It follows that on changing the doubling time from 14 h to 24 h, the variability in T_B must have increased to the same extent as the variability of times spent in A state. Evidently, the idea that proliferation rate is governed only by the probability of the A to B transition can no longer be sustained.

Mitogenic stimulation

Another feature of cell proliferation not accounted for by the transition probability model (or any previous theory) is the lag which invariably precedes an increase in proliferation rate after mitogenic stimulation. A typical example of this is provided by the restoration of serum to serum-starved cultures of fibroblast-like cells. When these are deprived of serum (a source of mitogenic growth factors: Brooks, 1976a; Gospodarowicz & Moran, 1976; Stiles, Cochran & Scher, this volume) proliferation rate becomes negligible and the cells accumulate in G1. On adding back serum the cells begin to re-enter S phase, but only after a long lag comparable to the doubling time of the cells growing at their maximum rate (8–18 h depending on cell type). Within a short time of the first cells entering S phase, the rate of exit from G1 becomes approximately first-order (Brooks, 1975, 1976b). For 3T3 cells this is achieved within as little as 4 h of the end of the 12–14 h lag, and most of the 4 h can be accounted for by size heterogeneity within the population, big cells have a slightly shorter lag than small cells (Shields et al., 1978). The change in the probability of initiating S phase is thus even more abrupt for individual cells than appears in the population as a whole. After exponential kinetics have been attained, the value of the rate constant depends on how much serum has been added back, though the duration of the lag does not.

Superficially, the approximately first-order kinetics and the serum dependence of the rate constant might seem to support the concept of a transition probability. The cells may be supposed to accumulate in A state when the serum is removed because the transition probability falls to a low value. This increases again when the serum is restored and the cells enter S phase with first-order kinetics. The difficulty is to explain the abruptness with which the rate constant increases at the end of a long lag, the duration of which is not affected by the dose of mitogen. The simplest explanation is that the transition probability increases almost immediately after the serum is added and that a lengthy, deterministic, pre-replicative period exists between the transition and the start of S phase. Such an idea would seem to be supported by the observation that the lag is a characteristic of almost any increase to proliferation rate and is not confined to the stimulation of purely quiescent cultures. For instance, if quiescent cells are first stimulated with a low level of serum and the concentration raised further several hours later (either before or after the end of the first lag), there follows another lag similar in duration to the first before the rate of entry into S phase increases again (Brooks, 1976b). But other facts argue that the transition must be rather closer to the start of S phase. The lag, being comparable to the doubling time of rapidly growing cells, is clearly very much longer than the minimum duration of G1, and hence longer than the minimum interval normally found between the transition and the start of DNA synthesis. Furthermore, if the serum concentration is reduced again in cultures of stimulated 3T3 cells, the rate of entry into S phase declines about 4–5 h later – a third of the 12–14 h lag shown by these cells on serum step-up (Brooks, 1976b). This means that cells are not committed to enter DNA synthesis until a maximum of 4–5 h before they actually do so, leaving most of the lag unaccounted for.

It seems difficult to avoid the necessity of placing a random transition towards the end of the lag. Yet this interpretation is also not without serious problems. To explain the lag it becomes necessary to assume that serum acts not directly on the mechanism responsible for the transition probability but on some other process separated from it by many intervening steps. The transition probability, of course, is not a discrete event but the *chance* of something happening. As such, it must reflect a process that is continuous in time. Consequently, any intervening steps must also be continuous processes rather than events (the analogy of a metabolic pathway comes to mind – a chain of linked, ongoing, enzymatic reactions). To account for the duration of the lag would require a very large number of intermediate processes, yet it can easily be shown that the more there are, the less abrupt will be the eventual change in the transition probability. This does not seem to be compatible with the rather sudden change actually encountered.

If the lag is not a chain of biochemical processes, could it be the time required to accumulate some substance to a critical threshold? The hypothetical substance would not determine the transition probability itself but could be necessary in a permissive sense in order that transitions might occur at all. Unfortunately, this idea does not account for why the lag duration is independent of serum concentration. Since this affects the rate of numerous biochemical processes, including the accumulation of RNA and protein, it is difficult to see how the time taken to reach the threshold could always be the same. It is also ruled out by the 'double step-up' experiment in which a second serum step-up is followed by another lag identical in duration to the first (Brooks, 1976b). If the first lag ends when some substance reaches a critical threshold, then the second lag cannot be due to a lack of the same substance.

An extension of the transition probability model

Many of the problems posed by the independence of the lag duration on the intensity of the mitogenic stimulus are also shared by the minimum cycle time (i.e. minimum T_B) which is surprisingly insensitive to proliferation rate. Both, it seems, are relatively independent of the metabolic activity of the population. There is also a remarkable tendency for them to have similar durations, the lag generally being about 10–20% longer. For example, quiescent BHK cells begin to enter DNA synthesis 7–8 h after serum stimulation (Brooks, 1975) and have a minimum T_B of 6–7 h in growing cultures (Minor & Smith, 1974), whereas 3T3 cells have a lag of 12–14 h (Brooks, 1976b) and a minimum T_B of 10–12 h (Shields et al. 1978). The similarity extends to the budding yeast, Saccharomyces cerevisiae. Here, the earliest step in the cycle is the emergence of a bud, an event closely followed by the start of DNA synthesis. When starved cells (unbudded and in G1) are refed, the lag before the rate of bud initiation changes is approximately 90 min compared to a minimum cycle time of about 85 min in rapidly growing cultures (Johnston, Pringle & Hartwell, 1977). It is possible that the correlation between the lag duration and T_B is mere coincidence. Nevertheless, it seemed worth exploring the possibility that underlying both was the same fundamental cell cycle 'clock'. Pursuing this idea, J. Smith, D. Bennett and myself arrived at a simple model for the lag which accounts for all its puzzling aspects (Brooks, Bennett & Smith, 1980). At the same time, the model explains why B phase should be variable in general, but identical in sister cells.

The model is shown in Fig. 2. The main postulates are that the cell cycle contains not one but *two* random transitions and that quiescent cells are located not in A state but in another indeterminate state called Q. In order to pass from Q to A state, cells must complete a process L, duration T_L, which is initiated at random (rate constant K_Q) and which occupies most of the lag. Like the duration of S + G2, T_L is regarded as being relatively insensitive to the environmental conditions, thus accounting for the constancy of the lag. After completing L, the cells enter A state (the same as Smith & Martin's A state) which they leave at random with rate constant K_A. A short time later (perhaps immediately) the cells enter S phase.

The similarity between the duration of the lag and the minimum cycle time is explicable if both L and Q are integral parts of the normal cell cycle and not confined to quiescent cells. Suppose that the consequences of leaving A state include not only the initiation of events leading to DNA synthesis and mitosis but also re-entry into Q state. If the probability of leaving Q is reasonably high then most cells will re-initiate L before mitosis. As a result, sister cells will be at the same stage of L at birth and will reach the next A state at the same time. Any difference in the time at which siblings enter S phase (or mitosis, since the duration of S + G2 has negligible variance) will then be due solely to the random exit from A state, which accounts for the exponential β curves. Different sister pairs will reach A state at different times however, depending on when L was initiated in the mother cell. Consequently, the division times of siblings will be correlated. The duration of what was formerly called B phase – the interval between successive A states – is now seen to be variable because it includes both the deterministic process L as well as the time spent in the indeterminate state Q.

Although cells enter Q at much the same time as they enter S phase, variation

Fig. 2. A model for the mammalian cell cycle. A and Q are indeterminate states which cells leave at random, rate constants K_A and K_Q respectively. L is a lengthy, deterministic process which cells must complete in order to pass from Q state to A state. The diagram begins with a quiescent cell (arrested in Q state) just after the values of K_Q and K_A have been raised by mitogenic stimulation. It continues with the first complete cycle (mitosis to mitosis) for both daughter cells arising from the first cell division. After leaving Q state (at random) cells embark upon the process L. On completion of L, cells enter A state which again they leave at random. Exit from A state has two consequences: the first is the initiation of events leading to DNA synthesis and mitosis; the second is re-entry into Q. Having re-entered Q state, cells are immediately free to initiate L once more (at random). Provided the value of K_Q is reasonably high, then in most cells this will occur before mitosis. However, completion of L will take place in the daughter cells which accordingly reach the next A state at the same time. Differences in the time at which sister cells initiate DNA synthesis (or mitosis) are therefore due only to random exit from A state. However, different pairs of sister cells reach A state at different times, depending on when L was initiated in the mother cell. Hence the generation times of sister cells are correlated.

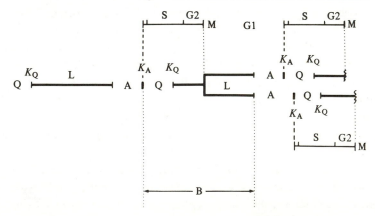

due to the time spent in Q will always be expressed in the next G1. This is because the time of exit from Q governs how soon after mitosis the daughter cells reach A state, and it is from this that cells initiate events leading to DNA synthesis. The minimum cycle time is shown by a cell that spends zero time in A state and whose mother spent zero time in Q state. Hence the duration of L, which accounts for most of the lag, is also equivalent to the minimum generation time. For mammalian cells the minimum generation time (T_L) is slightly longer than the duration of S + G2 + M, the difference being the minimum length of G1. This means that even when the mother cell spends zero time in Q, the process L will be completed after division in the daughter cells. Nevertheless, it is worth pointing out that if T_L were shorter than the interval between the initiation of DNA synthesis and division then overlapping cell cycles like those of bacteria would result. Perhaps a model postulated in order to explain mammalian cell cycle kinetics is also capable of being generalised to include prokaryotes.

Both K_Q and K_A are considered to depend on the environmental conditions. Since one of the transitions is located towards the end of the lag, this accounts for why the rate of entry into S phase declines relatively quickly when growth factors are removed, as in a serum step-down. However, K_Q must be more sensitive and reach the lower value because the cells eventually accumulate predominantly in Q. This means that after growth factor deprivation, the rate of entry into S should not decrease immediately to the very low rate found in quiescent populations but should remain somewhat higher, as indeed it does in practice (Brooks, 1976b). Eventually, all the cells 'trapped' in A state by the step-down would 'escape' (rate constant K_A), initiate S phase and re-enter Q where they accumulate, though since K_A has also been reduced this would take some time to complete. This could well be the reason why cultures are generally deprived of growth factors for several days before being used, in order to allow them to become 'fully quiescent'.

On adding back growth factors to cells trapped in Q state, K_Q is considered to increase quite rapidly, thereby explaining the abrupt change in the rate of entry into S phase at the end of the lag. The value of K_A may also increase quite quickly, but it would be without effect until the cells reach A state. Indeed, if K_Q is more sensitive to growth factor supply, as postulated, then at low concentrations the rate of entry into S phase will be governed more by K_Q than K_A, and the kinetics pseudo-first-order. This also means that when growth factor concentration is raised for a second time – the double serum step-up experiment – there will be another lag before the major change in the rate of entry into S phase occurs.

One further point deserves comment. It has been suggested that both the lag and the minimum cycle time have the same underlying basis, yet the lag is consistently 10–20% longer. This could mean that cells do not enter S phase immediately on leaving A state but an hour or two later after an obligatory pre-replicative period. Alternatively, it is not unreasonable that a short but finite time elapses between the administration of a mitogen and the increase in the value of K_Q, or that T_L is slightly longer in cells that have been starved of growth factors for some time.

Quantitative predictions of the two-transition model

Qualitatively, the two-transition model accounts for many seemingly unrelated features of cell proliferation. As has been discussed at length elsewhere (Brooks et al., 1980) it also gives good quantitative agreement with the distribution of cycle times found in steady-state cultures. The model predicts that the differences between sibling generation times (β) should be exponentially distributed (as is the case) whereas the distribution of generation times (α curve) should be a convolution of two exponentials. Although α and β curves were originally considered to be parallel, this means that they should, in fact, diverge slightly. Closer inspection of published data (Shields & Smith, 1977; Shields et al., 1978) as well as more recent unpublished experiments, shows this to be so (see also Fig. 3).

The correlation between sibling generation times arises because one of the random transitions took place in the mother cell and is thus shared. Now the correlation coefficient for two variables x and y can be regarded as the fraction of the total variance common to both x and y (Snedecor, 1946). Therefore, if T_L and $T_{(S+G2+M)}$ are invariant we may write:

$$r_{ss} = \frac{\text{variance } T_Q}{\text{total variance}} = \frac{\text{var } T_Q}{\text{var } T_Q + \text{var } T_A}$$

where r_{ss} is the correlation coefficient for sibling generation times; T_Q, the time spent in Q by the mother cell – the variable common to both sisters; and T_A, the time spent in A state – the variable different in siblings. The two-transition model predicts that both T_A and T_Q are exponentially distributed. But the variance of an exponential distribution is $1/K^2$. Hence, we may write:

$$r_{ss} = \frac{1/K_Q^2}{\text{total variance}} = \frac{1/K_Q^2}{1/K_Q^2 + 1/K_A^2}.$$

Since r_{ss} and the total variance can be readily calculated, estimates of K_Q and K_A are easily obtained. The latter (K_A) gives the slope of the β curve directly, since $\beta_t = e^{-K_A t}$. The equation for the α curve is:

$$\alpha_t = \frac{1}{K_A - K_Q} \left[K_A e^{-K_Q(t-T_L)} - K_Q e^{-K_A(t-T_L)} \right]$$

$$\text{for } K_A \neq K_Q, t \geq T_L.$$

Thus, it is a simple matter to calculate the expected form of the α curve and to compare it with what is actually obtained. In doing so, the value assigned to T_L is arbitrary (it does not affect the shape of the curve), but it must be close to the minimum generation time observed. As can be seen in Fig. 3, the agreement between the curve predicted by the theory and the experimental data is extraordinarily good. For a more extensive analysis, together with the derivation of the equations, see Brooks *et al.*, (1980).

Fig. 3. α and β curves for Swiss 3T3 cells growing exponentially in medium containing 12.5% calf serum. The α curve refers to the percentage of cells remaining undivided at ages greater than or equal to the times given on the abscissa. The β curve refers to the percentage of sibling pairs whose generation times show differences of more than or equal to the times given on the abscissa. The step diagram represents the experimental data with N being the number of cells or sibling pairs. The heavy lines drawn through the data were calculated using the sibling correlation and the total variance, as described in the text. The variance of generation times was $3.59\,h^2$; the sister/sister correlation coefficient (r_{ss}) was 0.63. The data are taken from Brooks *et al.* (1980) and replotted here with a logarithmic ordinate.

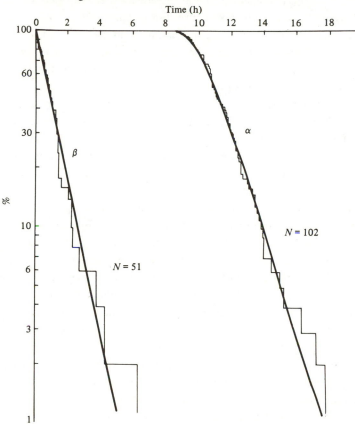

The α curve drawn through the data in Fig. 3 was predicted assuming the combined variance of T_L and $T_{(S+G2+M)}$ to be zero. Although the fit is impressive on a logarithmic ordinate, in a linear plot it is clear that either T_L or $T_{(S+G2+M)}$ must vary slightly (Brooks *et al.*, 1980). This residual variation can probably be ascribed to size heterogeneity within the population (Shields *et al.*, 1978). If so, then of the two, it is most likely to be T_L that varies since $T_{(S+G2+M)}$ is independent of cell size (Shields *et al.*, 1978). The limits of the variation are roughly $\pm$ 20 min of a mean T_L of approximately 9 h, compared to a range of cycle times from 8.5 to 18 h. It would seem that two random transitions account for by far the greater part of the variation in mammalian cell generation times.

A possible biological counterpart of the revised transition probability model

Apart from the economy with which the revised transition probability model accounts for so many aspects of cell proliferation, one of its major attractions is that a possible biological mechanism more or less suggests itself. Many of the properties of the hypothetical process L are most readily understood in terms of the assembly of a complex structure. For instance, the rate of assembly of a structure, once started, could easily depend more on the nature of the architecture than on the supply of building materials, thereby accounting for both the length of the process and its relative independence of the metabolic state of the cell. Furthermore, it is easy to imagine why the initiation of assembly might be random if, like microtubule formation, it depended on a nucleation step. It is also not difficult to imagine that some other event could be tied to the completion of assembly. Because L is frequently initiated in one cell cycle and completed in the next, it is evident that the hypothetical structure must duplicate during the cycle and identical copies be segregated to the daughter cells. This, in turn, recalls to a remarkable degree the behaviour of what Mazia termed the 'mitotic centres' – the structures responsible for determining the spindle poles.

Working with fertilised sea urchin eggs, Mazia, Harris & Bibring (1960) concluded that at the time of mitosis, each spindle pole was a duplex structure consisting of a mature 'centre' together with an immature 'daughter' joined to it. Maturation, in the sense of gaining the capacity to form an independent spindle pole, involved the separation of the daughter from the parent and was normally accomplished shortly after cell division. As a consequence, when mitosis was delayed experimentally relative to maturation, the cell formed a tetrapolar spindle and divided directly into four cells instead of two. Soon after separation, the old and the new centres both initiated duplication of further centres, the new daughters reaching maturity in turn after the subsequent cleavage. The period between the initiation of duplication and the completion of maturation is thus lengthy (equivalent to most of the cycle time) and spans cell division, each daughter cell receiving a mature centre together with a daughter centre of identical immaturity. This behaviour is illustrated schematically in Fig. 4. Though inferred originally from the effects of transiently preventing mitosis in sea urchin eggs, the scheme seems likely to be more widely applicable, since multipolar spindles are also formed in both mammalian and plant cells after delaying mitosis (Stubblefield, 1968; Coss & Pickett-Heaps, 1974). The parallels with the hypothetical process L are obvious and striking. Furthermore, the initiation and separation steps stand out as potential candidates for the two random transitions – a possibility reinforced by the fact that DNA synthesis in sea urchin eggs is initiated at roughly the same time as the centres separate (Bucher & Mazia, 1960).

In the above I have deliberately adhered to Mazia's phrase, the mitotic centre, in preference to using a morphological term such as the centriole. Though centrioles are undoubtedly located at the spindle poles of many organisms, there is much to suggest that they are 'passengers' rather than endowed with the capacity to organise spindle microtubules in their own right (Pickett-Heaps, 1969, 1971). Many species, including all higher plants, lack centrioles at all stages of the life histories, yet are perfectly capable of forming bipolar spindles. Other organisms regularly form centrioles *de novo* specifically when required as basal bodies during gametogenesis. Even when centrioles are located at the spindle poles, microtubules do not impinge directly on to them but rather converge towards the halo of amorphous material that surrounds the centriole (Pickett-Heaps, 1969). Nevertheless, the biogenesis of centrioles follows a pattern extraordinarily close to that expected of the structures operationally defined by Mazia as mitotic centres. Each daughter centriole arises in association with the base of a mature centriole and grows out perpendicularly from it. In mammalian cells this begins

Fig. 4. The duplication of the mitotic centres during cleavage of fertilised sea urchin eggs as inferred by Mazia *et al*. (1960). Parallels with the two-transition model of the cell cycle (Fig. 2) are indicated in parentheses.

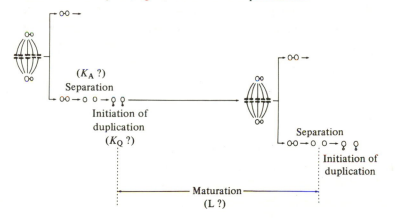

at about the same time as DNA synthesis is initiated but does not appear to be complete by mitosis (Stubblefield, 1968; Robbins, Jentzsch & Micali, 1968; Rattner & Phillips, 1973). Some time during the subsequent interphase the perpendicular orientation is lost, the distance between mother and daughter centrioles increases slightly and the next duplication begins. Even if centrioles are not strictly the mitotic centres themselves, it is possible that in metazoans they have acquired the task of organising the amorphous pericentriolar material which is more directly responsible for nucleating spindle microtubules. This could help explain why centrioles are so meticulously replicated and segregated during each cell cycle even when they are not required as basal bodies. At the very least, centrioles may provide a convenient marker for the activity of the less tangible, but possibly more important pericentriolar material.

In many organisms that lack centrioles, the spindle forms in association with a discrete body composed of rather amorphous, electron-dense material (Pickett-Heaps, 1969; 1971). This may well be a more organised homologue of the diffuse pericentriolar material found in mammalian cells. In the budding yeast (which has an intranuclear mitosis) it takes the form of a dense plaque on the nuclear envelope. Using the range of cell cycle mutants isolated by Hartwell, duplication of this plaque has been found to be the first identifiable step in the cell cycle, shortly preceding the initiation of DNA synthesis (Byers & Goetsch 1974, 1975). The so-called 'start' mutations, which prevent the initiation of the cycle, affect the duplication of this plaque, and it is possible that they may involve some structural component of it. If so, then these mutations may provide direct evidence for the importance of spindle pole bodies in the control of the cell cycle. It remains to be established that the two-transition model applies to yeasts, of course, and if so, to what degree the spindle plaques are homologous to the mitotic centres of higher cells.

If the process L does correspond to the duplication of the mitotic centres then the formalism of the model itself becomes intelligible, at least teleologically. Although L may overlap with any or none of S, G2 or M, it is also the case that each cell division may be viewed as the end result of a linear sequence that begins with L and continues with S, G2 and M. Put this way, it appears that DNA synthesis is allowed to begin only after 'something' has been completed. It makes good sense if this 'something' is the assembly of the principle component of the machinery used to segregate the chromatids, the eventual products of the subsequent S phase.

Probablistic models of cycle regulation and the control of cell size
Criticisms of the transition probability model

One of the most frequent criticisms of probabilistic models of cycle control has been that they make no overt allowance for the control of cell size

(Fantes & Nurse, this volume). It is argued that if initiation of each cell cycle were random then the variance of cell size at division would increase progressively, which does not happen. Indeed, both the mean and variance of size at division remain constant over many generations (in steady state) and this, it is claimed, implies that cells must have a mechanism for knowing how big they are and for ensuring that cell division follows after every doubling in mass (Koch & Schaechter, 1962; Fantes *et al.*, 1975; Fantes, 1977; Fantes & Nurse, this volume). In apparent support of this it has been found that the coefficient of variation of size at division is about half that for age at division suggesting, it would seem, that dividing cells pay more attention to their size than to their age (Schaechter, Williamson, Hood & Koch, 1962; see also data of Fantes, 1977 and Miyata, Miyata & Ito, 1978). Furthermore, it has been shown directly that big cells at birth do have shorter subsequent generation times, on average, than small cells (Prescott, 1956; Fantes, 1977; Miyata *et al.*, 1978; Johnston *et al.*, 1977; Shields *et al.*, 1978).

If cell division did follow deterministically as a consequence of growth to a critical size, then it is necessary to explain why cells do not all divide at the *same* size. Proponents of size control theories usually answer this by saying that the size control is not absolutely precise but slightly 'sloppy', as might be expected of a biological system. By this, it is meant that the probability of the size-controlled cycle event increases not as an exact step-function with increasing size but rather more gradually over a range of cell sizes, the range being a measure of the 'sloppiness'. The 'critical size' is then taken to be the mid-point of the range. Regardless of the precise relationship between age and size, since cells grow continuously through the cycle, the probability of division must increase progressively with age, being very low for young cells and rising fairly steeply around the average cycle time. Because of momentary variation in age for cells of a given size – itself a consequence of the 'sloppy' size control – the actual distribution of age at division would be more spread than that of size and, it is claimed, indistinguishable from the α curve found in practice (Wheals, 1977).

Difficulties of size control theories

Mammalian cells. Although, broadly speaking, deterministic size control theories may be compatible with the shape of the α curve, they present no obvious way of predicting the exact curve analytically. In fact, the types of distribution said to be compatible with critical size hypotheses, namely log-normal or reciprocal-normal distributions (Wheals, 1977), have already been ruled out because they assume the variation to be spread throughout the cycle instead of being confined to G1. Deterministic theories also have difficulty in explaining the exponential distribution of differences between sibling generation times (the β curve), despite the attempt by Pardee, Shilo & Koch (1979). Furthermore, contrary to the usual claims, critical size hypotheses do not account adequately for the behaviour of cell populations with different rates of cellular growth (i.e. mass increase). At low rates, cells should take longer to attain the critical size so that the minimum generation time should increase. However, if the rate of mass increase is the sole factor controlling the timing of division, then the coefficient of variation of size at division, and hence the coefficient of variation of cycle times, should remain unchanged. For mammalian cells, at least, the reverse seems to be the case: when the doubling time of 3T3 cells was lengthened from 14 h to 24 h by reduction of the serum concentration in the medium, the minimum recorded cycle time changed only slightly, from 10.3 h to 11.5 h (both cultures were in steady state). In contrast, the coefficient of variation of age at division increased from 16.8% to 29.1% (Brooks, unpublished). The latter observation, if there is a size control, points to an increase in the degree of 'sloppiness' as the major factor responsible for the change in growth rate. That is, alteration of proliferation rate depends less on changes in the rate of mass increase (the deterministic part of the theory) than on changes in the efficiency with which cell size is 'perceived' (i.e. the probabilistic element). As for the poor dependence of the minimum cycle time on growth rate, then to entertain the notion of a size control at all demands the additional ad-hoc hypothesis that the critical size must change. Indeed, this has been suggested to be the case for the fission yeast *Schizosaccharomyces pombe,* based on the observation that mean size at division *is* a function of growth rate (Fantes & Nurse, 1977). However, if this is so, then not only does the hypothetical size control fail to ensure that cells actually divide at the 'critical' size in steady state, but it also fails to compensate for changes in the rate at which cells increase their mass. As a control said to be necessary to maintain cell size, it clearly leaves much to be desired.

Budding yeast. Although the probablistic element appears to be the most significant factor in the control of mammalian cell cycles, it could be argued that a more deterministic coupling between cell size and division exists in other cell types. Indeed, some of the best evidence suggesting a role for cell size in the timing of cycle events has been obtained with lower eukaryotes, especially yeasts (Fantes & Nurse, this volume). Nevertheless, the behaviour of the budding yeast, *Saccharomyces cerevisiae,* does not fit readily within the constraints of a critical size hypothesis. In this organism, a cell cycle represents the period between the initiation of a bud and its cytological separation from the mother cell. After separation, the time taken by the new cell to bud in turn depends, in part, upon how big it is (Johnston *et al.,* 1977). However, if bud initiation were triggered solely by growth to a critical size, then having budded once, a cell should continue to bud thereafter with the minimum possible cycle time under all growth conditions. Instead, the cycle times of mother cells are variable, the mean depending on (among other things) the nutrient supply (Shilo, Shilo & Simchen, 1976, 1977; and see the discussion by Carter, this volume). Clearly, it is necessary to admit that something other than growth to a critical size is responsible for determining the cycle duration here. As for the variability itself, this is such as to suggest the involvement of a random transition. Thus, like mammalian cells, control of proliferation rate in *S. cerevisiae* appears to contain a pronounced probabilistic element that cannot be explained simply by inaccuracy of a size sensing mechanism. It would seem reasonable to suppose that the influence of size in daughter cells is indirect and superimposed on a more fundamental cell cycle control.

Maintenance of mean cell size in the absence of a size control

In the absence of a direct coupling between cell division and cell size, how is it possible to keep the average cell size constant (under steady-state conditions) or arrive at a smaller coefficient of variation for size than for age? This, in fact, is very easy, as may be seen with a simple example. Suppose we take a population of 50 'small' cells, each 1 mass unit, and a population of 50 'large' cells, each 10 mass units. We then assign a generation time to each by random selection of integers between 1 and 10 using a table of random numbers. For simplicity, we then suppose that the increase in mass during the cell cycle is linear so that a generation time of, say, 7 units will increase mass by 7 units. The total mass is halved at each 'division' and the process continued for 10 generations (for convenience, only one daughter cell is followed after each 'division'). Table 2 gives the mean mass of daughters after each generation together with the coefficient of variation. As can be seen, the average size rapidly converges towards the theoretical population mean (5.5) both for initially large and initially small cells. This occurs despite the lack of any specific control over size because small cells are most likely to more than double their mass during a generation whereas the large cells will most probably achieve less than a doubling. Because of this convergence, the coefficient of variation of mass is less than that for generation times. In this example, growth was assumed to be linear during the cycle, though the arguments apply equally well to more complicated relationships, e.g. growth rate proportional to DNA content which gives a doubling of rate some time during the cycle. All that is required is that the rate of mass increase is not proportional to mass, i.e. not exponential. Evidence that growth is exponential is, to say the least, equivocal (Mitchison, 1971). There are also good reasons why it cannot possibly be in the long term: sooner or later something is bound to become limiting. To explain the relationship between growth rate and division rate requires only that the probability of division be roughly proportional to growth rate. The co-ordination does not need to be exact, so that mean size at division may change with growth rate, as happens in practice (Fantes & Nurse, 1977).

Transition probability and the experimentally determined relationship between cell size and generation time

The above example demonstrates that there are no theoretical reasons why cell division need be controlled by cell size: probabilistic models with no overt concern for a cell's mass are equally compatible with the maintenance of a constant average cell size at division, and with the observation of a smaller coefficient of variation in size compared to age. Nevertheless, as already mentioned, there exists a considerable body of experimental evidence that cell size does influence the timing of division (Fantes & Nurse, this volume). In addition, mutants of fission yeast, *Schizosaccharomyces pombe,* have been isolated which divide at half the size of the wild type, and these have been considered to be defective in the mechanism responsible for monitoring cell size (Nurse, 1975; Nurse & Fantes, this volume). Also in *S. pombe* (as mentioned earlier), the average size of cells at division is proportional to growth rate, and when the latter is increased there is a transient delay in division which lasts until the cells reach the size characteristic of the new growth rate (Fantes & Nurse, 1977).

Table 2. *A simulation of the change in mean mass of 50 'small cells' and 50 'large cells' during 10 consecutive generations in which each division time is a random integer beween 1 and 10 units.*

	Number of generations										
	0	1	2	3	4	5	6	7	8	9	10
'Small cells'											
Mean mass	1	3.44	4.50	4.85	5.10	5.64	5.30	5.58	5.61	5.69	5.54
Coef. var. (%)	0	39.0	28.4	29.2	28.7	27.6	26.4	28.2	29.8	27.8	31.0
Mean gen. time	—	5.88	5.52	5.20	5.34	6.18	4.96	5.84	5.60	5.78	5.30
Coef. var. (%)	—	45.7	48.7	50.2	50.3	47.0	55.3	47.7	53.3	44.3	50.2
'Large cells'											
Mean mass	10	7.73	6.26	5.91	5.83	5.46	5.63	5.78	5.92	5.65	5.95
Coef. var. (%)	0	19.7	21.1	34.1	34.6	39.2	29.5	28.3	27.2	26.8	26.3
Mean gen. time	—	5.46	4.78	5.56	5.74	5.08	5.8	5.92	6.06	5.4	6.24
Coef. var. (%)	—	55.7	53.3	59.7	50.5	58.7	47.9	46.5	48.8	53.9	47.4

When growth rate is decreased, the converse happens: there is a transient acceleration of cells through the cycle, resulting in a smaller size at division. This has been interpreted as evidence of a size control over division which is modulated by growth rate and which delays or accelerates division to ensure that this takes place at the 'correct' size (on average) for the new growth rate.

Yet how compelling is all this as evidence of a control over cell size? To take the size modulation in S. pombe first, the comment has already been made that a size control which fails to maintain size when growth rate changes barely deserves to be called a control. The data can also be explained without any reference to a size control, simply by a transient perturbation in the running of a cell cycle timer. For example, a sudden increase in growth rate could cause a transient set-back in the progress of cells towards division. Because growth continues (indeed, at the higher rate), then when division resumes the cells will be larger than they would have been at division under the previous conditions. The reasons for the set-back are most intriguing, but the observations as they stand cannot be said to provide evidence that cells monitor their size and divide accordingly.

The so-called size control mutants of S. pombe (Nurse, 1975; Nurse & Fantes, this volume) also do not provide unequivocal evidence that cells have a mechanism for monitoring their size. The observations are that the mutant has the same doubling time as the wild type but divides at half the size, which implies that its average growth rate during the cycle is also half that of the wild type in absolute terms. In addition, DNA synthesis takes place later in the cycle, compared to the wild type, leading to a longer G1 and a shorter G2. There are probably many explanations of the mutant phenotype that do not depend on size monitoring, but the following is compatible with the revised transition probability model. The model, it will be recalled (Fig. 2), postulates the existence of what amounts to two cell-cycle timers: one (the process called L) which spans cell division and determines the minimum possible interval between successive initiations of DNA synthesis, and another which may begin only on completion of L and which determines the interval of S + G2 + M. The mutant phenotype would result if the second timer is speeded up (effectively a shortening of G2) coupled to the observed lower rate of growth during the cycle. The shortening of G2 advances mitosis relative to progress through L, but since the latter still determines how soon the next S phase may take place, this results merely in an elongation of G1. The smaller size at division would result because the cycle time is unchanged (since T_L is unchanged) but the average growth rate during the cycle is lower. The lower growth rate is presumably one of the consequences of the mutation itself, though it could be partly explained if growth rate during the cycle of S. pombe were proportional to DNA content, which predicts a doubling in rate as

DNA is replicated. Since the mutant spends a greater proportion of its cycle in G1, and therefore at the lower, pre-S phase growth rate, it would reach division at a slightly smaller size than the wild type for this reason alone.

As for the experimentally determined relationship between cell size and the timing of division, it is certainly incontrovertible that small cells at birth do (sometimes) take longer to reach division (on average) than large cells (Prescott, 1956; Fantes, 1977; Miyata *et al.*, 1978; Johnston *et al.*, 1977; Shields *et al.*, 1978). The point at issue is whether this implies that cells monitor primarily their size and organise their replication activities around this, or whether, instead, there is a process controlling proliferation which does not depend on the cell being a particular size, but whose rate of completion is influenced incidentally by how big the cell is – for instance, because the concentration of some key substance depends not only on the net rate of synthesis but also on the volume of the cell. The second alternative is clearly compatible with a number of possibilities, including the revised transition probability model. In terms of the latter, cell size must have some influence on either the transition probabilities them-selves or on the duration of T_L. For mammalian cells, the variability of generation times is not markedly different for big and small cells making it likely that the effect of size is confined to T_L; (K_A, at least, is size independent since the β curves are identical (Shields *et al.*, 1978). For yeasts, in which size appears to be a more significant determinant of generation time, the variability is also similar (Fantes, 1977; Miyata *et al.*, 1978), again suggesting an effect on T_L. However, it should be borne in mind that, as yet, there is no firm evidence that the revised transition probability model applies to cells other than of mammalian origin.

If T_L does show some dependence on size in mammalian cells, then one might wonder why the calculated α curve in Fig. 3 fits the data so well without taking size heterogeneity (and hence T_L variability) into account. That T_L *is* slightly variable has already been discussed both here and elsewhere (Brooks, *et al.*, 1980). The effect of size, measured experimentally, is also small, a twofold difference in mean size being associated with a 25% difference in the lag duration (i.e. T_L) after serum stimulation of quiescent 3T3 cells (Shields *et al.*, 1978). Since most cells differ in mass by less than twofold (coefficient of variation 10–15%, Killander & Zetterberg, 1965) then the expected variation in T_L due to size heterogeneity would be relatively small. With some non-mammalian cell types (if the model applies), it is possible that the effect of size on the duration of L is very much greater, perhaps because size heterogeneity is greater. In these cases, cell size might appear to be the most important factor determining cycle time – more so than the transition probabilities – even though the underlying control mechanism might be the same as for mammalian cells.

To sum up, there are no experimental observations concerning the importance of cell size that are incompatible with the transition probability model, though this in itself does not mean that it must necessarily apply to all cell types. For mammalian cells, the kinetic evidence in favour of the model is substantial, but it is still possible that a more deterministic control over cell proliferation operates in other organisms. If we are ever to decide between the rival models in a particular case, then it will be necessary to know how the distribution of generation times changes with proliferation rate. Should it be found (as with mammalian cells) that alterations in proliferation rate are brought about mainly by changing the *degree* of variability, then an analysis of the variation is called for, and an explanation of it demanded.

Concluding remarks

In this chapter I have discussed various explanations for the variability of cell age at division. The emphasis has been mainly on mammalian cells for which the kinetic data is most extensive. For these, it seems that almost all of the variability can be accounted for by the existence of two random transitions which are separated by a lengthy deterministic process whose duration is more or less equivalent to the minimum generation time. At high growth rates, the first of these transitions takes place in the mother cell cycle which explains the correlation between sibling generation times. The scheme also accounts for the lag which precedes any increase in proliferation rate after mitogenic stimulation. The most intriguing aspect, however, is the remarkable parallel with the duplication of the mitotic centres. This raises many interesting and fundamental questions, not least of which is the relationship between the mitotic centres and centrioles.

Whether the two-transition model applies to cells other than of mammalian origin remains to be seen. Though the cell cycles of many lower eukaryotes share a number of the characteristics of mammalian cell cycles – for instance, comparable subdivision into G1, S, G2 and M phases, variable generation times, and a lag after refeeding starved cells – the kinetic data currently available are inadequate. Indeed, the asymmetric division of the budding yeast, *S. cerevisiae,* means that meaningful β curves, upon which so much relies, cannot be obtained for this organism: daughter buds at the time of cell separation are generally smaller than the mother cell and, for whatever reason, have longer generation times (Hartwell & Unger, 1977). In view of the large number of cell cycle mutants isolated in *S. cerevisiae* (Hartwell, 1974) this is most regrettable. In other cases, the organisation of the cell cycle differs markedly from the mammalian pattern. Thus, to cite only two examples, *Chlorella* regularly divides directly into either two, four, eight or even sixteen cells (Mitchison, 1971; Pickett-Heaps, 1975) whereas *Amoeba proteus* initiates DNA synthesis in telophase (Mitchison, 1971). There is no reason *a priori* to expect the two-transition model to apply to these organisms without extensive modification in detail. Clearly, further comparative studies would be most profitable in helping to decide how far the model should be generalised.

Many of the ideas discussed here owe their existence to Jim Smith. This chapter could never have been written without the innumerable conversations with him over the past few years. The aggregate consumption of tea, coffee and beer during these occasions must run to many hundreds of gallons.

I am also greatly indebted to Robert Shields for selflessly allowing me to present Fig. 1. In addition, my thanks are due to Peter Riddle, Diana Brown and Glynn Widowson for their past and continuing help in making and analysing the time-lapse films upon which so much of our work depends.

References

Baserga, R. (1965). The relationship of the cell cycle to tumor growth and control of cell division: a review. *Cancer Research*, **25**, 581–95.

Brooks, R. F. (1975). The kinetics of serum-induced initiation of DNA synthesis in BHK 21/C13 cells, and the influence of exogenous adenosine. *Journal of Cellular Physiology*, **86**, 369–78.

Brooks, R. F. (1976a). Growth regulation *in vitro* and the role of serum. In *Structure and Function of Plasma Proteins*, ed. A. C. Allison, vol. 2, pp. 239–89. New York & London: Plenum.

Brooks, R. F. (1976b). Regulation of the fibroblast cell cycle by serum. *Nature, London*, **260**, 248–50.

Brooks, R. F., Bennett, D. C. & Smith, J. A. (1980). Mammalian cell cycles need two random transitions. *Cell*, **19**, 493–504.

Bucher, N. L. R. & Mazia, D. (1960). Deoxyribonucleic acid synthesis in relation to duplication of centres in dividing eggs of the sea urchin, *Strongylocentrotus purpuratus*. *Journal of Biophysical and Biochemical Cytology*, **7**, 651–5.

Burns, F. J. & Tannock, I. F. (1970). On the existence of a G0-phase in the cell cycle. *Cell and Tissue Kinetics*, **3**, 321–34.

Byers, B. & Goetsch, L. (1974). Duplication of spindle plaques and integration of the yeast cell cycle. *Cold Spring Harbor Symposia on Quantitative Biology*, **38**, 123–31.

Byers, B. & Goetsch, L. (1975). Behaviour of spindles and spindle plaques in the cell cycle and conjugation of *Saccharomyces cerevisiae*. *Journal of Bacteriology*, **124**, 511–23.

Cook, J. R. & Cook, B. (1962). Effect of nutrients on the variation of individual generation times. *Experimental Cell Research*, **28**, 524–30.

Coss, R. A. & Pickett-Heaps, J. D. (1974). The effects of isopropyl N-phenyl carbamate on the green alga *Oedogonium cardiacum*. I. Cell division. *Journal of Cell Biology*, **63**, 84–98.

Dawson, K. B., Madoc-Jones, H. & Field, E. O. (1965). Variations in the generation times of a strain of rat sarcoma cells in culture. *Experimental Cell Research*, **38**, 75–84.

Fantes, P. A., Grant, W. D., Pritchard, R. H., Sudberry, P. E. & Wheals, A. E. (1975). The regulation of cell size and the control of mitosis. *Journal of Theoretical Biology*, **50**, 213–44.

Fantes, P. A. (1977). Control of cell size and cycle time in *Schizosaccharomyces pombe*. *Journal of Cell Science*, **24**, 51–67.

Fantes, P. & Nurse, P. (1977). Control of cell size at division in fission yeast by a growth-modulated size control over nuclear division. *Experimental Cell Research*, **107**, 377–86.

Gospodarowicz, D. & Moran, J. S. (1976). Growth factors in mammalian cell culture. *Annual Reviews of Biochemistry*, **45**, 531–58.

Hartwell, L. H. (1974). *Saccharomyces cerevisiae* cell cycle. *Bacteriological Reviews*, **38**, 164–98.

Hartwell, L. H. & Unger, M. W. (1977). Unequal division in *Saccharomyces cerevisiae* and its implications for the control of cell division. *Journal of Cell Biology*, **75**, 422–35.

Johnston, G. C., Pringle, J. R. & Hartwell, L. H. (1977). Co-ordination of growth with cell division in the yeast *Saccharomyces cerevisiae*. *Experimental Cell Research*, **105**, 79–98.

Kelly, C. D. & Rahn, O. (1932). The growth rate of individual bacterial cells. *Journal of Bacteriology*, **23**, 147–53.

Killander, D. & Zetterberg, A. (1965). Quantitative cytochemical studies on interphase growth. I. Determination of DNA, RNA and mass content of age determined mouse fibroblasts *in vitro* and of intercellular variation in generation time. *Experimental Cell Research*, **38**, 272–84.

Koch, A. L. & Schaechter, M. (1962). A model for statistics of the cell division process. *Journal of General Microbiology*, **29**, 435–54.

Koch, A. L. (1966). The logarithm in biology. I. Mechanisms generating the log-normal distribution exactly. *Journal of Theoretical Biology*, **12**, 276–90.

Koch, A. L. (1969). The logarithm in biology. II. Distributions simulating the log-normal. *Journal of Theoretical Biology*, **23**, 251–68.

Kubitschek, H. E. (1962). Normal distribution of cell generation rate. *Experimental Cell Research*, **26**, 439–50.

Kubitschek, H. E. (1971). The distribution of cell generation times. *Cell and Tissue Kinetics*, **4**, 113–22.

Mazia, D., Harris, P. J. & Bibring, T. (1960). The multiplicity of the mitotic centres and the time-course of their duplication and separation. *Journal of Biophysical and Biochemical Cytology*, **7**, 1–20.

Minor, P. D. & Smith, J. A. (1974). Explanation of degree of correlation of sibling generation times in animal cells. *Nature, London*, **248**, 241–3.

Mitchison, J. M. (1971). *The biology of the cell cycle*. Cambridge University Press.

Miyata, H., Miyata, M. & Ito, M. (1978). The cell cycle in the fission yeast, *Schizosaccharomyces pombe*. I. Relationship between cell size and cycle time. *Cell Structure and Function*, **3**, 39–46.

Nachtwey, D. S. & Cameron, I. L. (1968). Cell cycle analysis. *Methods in Cell Physiology*, **3**, 213–59.

Nurse, P. (1975). Genetic control of cell size at cell division in yeast. *Nature, London*, **256**, 547–51.

Pardee, A. B., Shilo, B. & Koch, A. L. (1979). Variability of the cell cycle. In *Hormones and Cell Culture*, ed. G. Sato & R. Ross. New York: Cold Spring Harbor Laboratory.

Pickett-Heaps, J. D. (1969). The evolution of the mitotic apparatus: an attempt at comparative ultrastructural cytology in dividing plant cells. *Cytobios,* **1,** 257–80.

Pickett-Heaps, J. D. (1971). The autonomy of the centriole: fact or fallacy? *Cytobios,* **3,** 205–14.

Pickett-Heaps, J. D. (1975). *Green algae: structure, reproduction and evolution in selected genera.* Sunderland, Massachusetts: Sinauer Associates.

Prescott, D. M. (1956). Relation between cell growth and cell division. II. The effect of cell size on growth rate and generation time in *Amoeba proteus. Experimental Cell Research,* **11,** 86–98.

Prescott, D. M. (1959). Variations in the individual generation times of *Tetrahymena geleii* H.S. *Experimental Cell Research,* **16,** 279–84.

Prescott, D. M. (1968). Regulation of cell reproduction. *Cancer Research,* **28,** 1815–20.

Quastler, H. & Sherman, F. G. (1959). Cell population kinetics in the intestinal epithelium of the mouse. *Experimental Cell Research,* **17,** 420–38.

Rattner, J. B. & Phillips, S. G. (1973). Independence of centriole formation and DNA synthesis. *Journal of Cell Biology,* **57,** 359–72.

Robbins, E., Jentzsch, G. & Micali, A. (1968). The centriole cycle in synchronized HeLa cells. *Journal of Cell Biology,* **36,** 329–39.

Robinson, J. H., Smith, J. A., Totty, N. F. & Riddle, P. N. (1976). Transition probability and the hormonal and density-dependent regulation of proliferation. *Nature, London,* **262,** 298–300.

Schaechter, M., Williamson, J. P., Hood, J. R. & Koch, A. L. (1962). Growth, cell and nuclear divisions in some bacteria. *Journal of General Microbiology,* **29,** 421–34.

Shields, R. (1977). Transition probability and the origin of variation in the cell cycle. *Nature, London,* **267,** 704–7.

Shields, R. & Smith, J. A. (1977). Cells regulate their proliferation through alterations in transition probability. *Journal of Cellular Physiology,* **91,** 345–56.

Shields, R. (1978). Further evidence for a random transition in the cell cycle. *Nature, London,* **273,** 755–8.

Shields, R., Brooks, R. F., Riddle, P. N., Capellaro, D. F. & Delia, D. (1978). Cell size, cell cycle and transition probability in mouse fibroblasts. *Cell,* **15,** 469–74.

Shilo, B., Shilo, V. & Simchen, G. (1976). Cell cycle initiation in yeast follows first-order kinetics. *Nature, London,* **264,** 767–70.

Shilo, B., Shilo, V. & Simchen, G. (1977). Transition probability and cell cycle initiation in yeast. *Nature, London,* **267,** 648–9.

Siskin, J. E. & Kinosita, R. (1961). Variations in the mitotic cycle *in vitro. Experimental Cell Research,* **22,** 521–5.

Siskin, J. E. & Morasca, L. (1965). Intrapopulation kinetics of the mitotic cycle. *Journal of Cell Biology,* **25(2),** 179–89.

Smith, J. A. & Martin, L. (1973). Do cells cycle? *Proceedings of the National Academy of Sciences, USA,* **70,** 1263–7.

Smith, J. A. & Martin, L. (1974). Regulation of cell proliferation. In *Cell Cycle Controls,* ed. G. M. Padilla, I. L. Cameron & A. Zimmerman, pp. 43–60. New York: Academic Press.

Snedecor, G. W. (1946). *Statistical Methods,* 4th edn, chapter 7. Iowa State College Press.

Stubblefield, E. (1968). Centriole replication in a mammalian cell. In *The*

Proliferation and Spread of Neoplastic Cells, pp. 175–89. Baltimore: Williams & Wilkins.

Svetina, S. (1977). An extended transition probability model of the variability of cell generation times. *Cell and Tissue Kinetics*, **10**, 575–81.

Tobey, R. A. (1973). Production and characterization of mammalian cells reversibly arrested in G1 by growth in isoleucine deficient medium. *Methods in Cell Biology*, **6**, 67–112.

Wheals, A. E. (1977). Transition probability and cell cycle initiation in yeast. *Nature, London*, **267**, 647.

W.D.DONACHIE

The cell cycle of *Escherichia coli*

Introduction

Escherichia coli has a very simple cell cycle; perhaps as simple as that of any organism. The cylindrical cell doubles in length and then separates into two cells of equal length by transverse fission. There is no mitosis and the genome consists of a single very large covalently closed circle of DNA which is folded on itself to form a closely packed 'nucleoid' in the cell centre. The nucleoid duplicates and then segregates into two at a fixed time before cell division.

Having said that, we must then admit that the mechanism of this cycle is not at all understood. We do not know how cell shape is maintained, how the cell grows in length, what determines the time and site of division, how septum formation differs from growth of the cell surface during cell elongation, how the DNA is folded, how it is replicated and segregated in the *in vivo* folded state or what events initiate its replication. This is the case despite the fact that the molecular biology and genetics of *E. coli* are more completely known than for any other organism. One reason for this ignorance is probably because research with *E. coli* has until now mostly concerned fundamental molecular processes and not the overall organisation of the cell *per se*. Nevertheless there has been a considerable effort to understand certain aspects of the *E. coli* cell cycle, such as the molecular basis for periodic DNA replication, cell shape and septum formation. However, these efforts have so far failed to provide satisfactory explanations. Thus almost everything about the mechanisms of the cell cycle of *E. coli* remains to be discovered. Even so, I am willing to claim that more is known about the biology of the cell cycle of *E. coli* than is yet known for any other cell type; largely because of the ease with which these cells can be grown and manipulated. The challenge now is to use the marvellous tools of molecular biology to provide a molecular explanation for the well-regulated duplication cycle of these simplest of cells.

An 'ideal cell'

The *E. coli* cell cycle has now been sufficiently well studied and the behaviour of the cells themselves is sufficiently regular and predictable, that it is possible to describe an 'ideal cell', growing and dividing according to a number of precise rules, which will behave, on paper, very like a real, living cell. What follows is such a set of rules of growth for an ideal cell. The predicted behaviour of the model cell will then be compared with some observations on *E. coli* itself.

Rule 1: geometry. The ideal cell is a cylinder with hemispherical poles.

Rule 2: minimum dimensions. The ideal cell has a minimum volume, V_u, and a minimum length, L_u. These define the dimensions of a new-born cell in a population for which the growth rate approaches zero.

Rule 3: growth in volume. Growth in cell volume (and mass) is exponential (i.e. $dV/dt = V/T$; where T is the doubling time in minutes).

Rule 4: growth in length. The rate of cell elongation is proportional to cell volume and to the growth rate, such that $dL/dt = (K \cdot V/T) \cdot 2^{(-40/T)}$; where $K = L_u/V_u$. (Elongation is assumed to be unipolar until cells reach length $2 \cdot L_u$, after which it is bipolar.)

Rule 5: cell division. Cell division is initiated when cell length reaches $2 \cdot L_u$ and requires a constant period of time (D) to complete. (This use of 'D' is different from its original use by Cooper & Helmstetter, 1968, but is indistinguishable under most conditions.)

Rule 6: integration of chromosome replication with cell growth and division. Initiation of chromosome replication is linked to cell volume, such that initiations of rounds of DNA replication take place at each successive doubling of the minimum (or unit) volume, V_u (Donachie, 1968). (Stating this more exactly, 2^n initiations of DNA replication take place at 2^n copies of the chromosomal origin of replication, *oriC*, when cell volume reaches $2^n \cdot 2 \cdot V_u$; where n is zero or any positive integer.) The time taken for replication forks to travel (bidirectionally) around the chromosome is a constant, C (Cooper & Helmstetter, 1968) which is independent of the growth rate of the cells (but is dependent upon other factors, such as temperature and the availability of DNA precursors).

Initiation of chromosome replication sets up a block to the initiation of septa (septa initiated before the start of the round of DNA replication are not affected). The reversal of this block requires the completion of the round of DNA replication, together with a period, P, of post-termination protein synthesis.

Rule 7: number and location of potential division sites. At each doubling of the minimum (or unit) length, L_u, one potential division site is formed. Each new potential division site arises at a point midway between a pair of pre-existing ones, including the cell poles (which may be considered as 'used' potential division sites) when the distance between them reaches $2 \cdot L_u$.

Rule 8: division potential. For each doubling in unit length, L_u, one quantum of 'division factor' (Teather, Collins & Donachie, 1974) is produced per $2 \cdot L_u$. Each quantum is sufficient for the formation of one septum and is assigned at random amongst all available potential division sites.

The above rules of growth for an ideal cell are not the only possible set. Some rules could be changed for others without much affecting the predicted behaviour. For example, rule 5 could be changed for a rule which states that cell division is initiated at each successive doubling of the unit *volume*, V_u, and that this process requires a different constant period of time to be completed. This period would equal $(40+D)$ minutes. This equivalence comes about because rules 2, 3 and 4 ensure that cell length will always be equal to $2 \cdot L_u$ at a time which is 40 min after cell volume has reached $2 \cdot V_u$. I prefer rule 5 as stated because it automatically includes a requirement for continuous cell growth during the 40 min period after each doubling in unit volume. This corresponds to the observed behaviour of *E. coli* (Pierucci & Helmstetter, 1969; Donachie, Begg & Vicente, 1976).

No values have been assigned to the various constants in the rules for the ideal cell because these may well vary with strains and conditions but we may give some approximate values for the strains of *E. coli* most often used and for the usual laboratory growth conditions. Thus, to a first approximation, $V_u = 0.2 \, \mu m^3$ (F. Trueba & C. L. Woldringh, personal communication), $L_u = 1.4 \, \mu m$ (Donachie *et al.*, 1976) $D = 20$ min and $C = 40$ min (Cooper & Helmstetter 1968), $P = 7$ min (Jones & Donachie, 1973) and T may have any value down to a minimum of 20 min (all for 37 °C).

The predicted cell cycles for an ideal cell growing relatively slowly ($T = 80$ min) or quite quickly ($T = 30$ min) are shown in Fig. 1. Similar self-consistent cycles can be drawn for any growth conditions, using the above rules.

Comparison between *E.coli* and the ideal cell

Not all of the predictions of the preceding model have yet been experimentally tested but there are numerous observations on the growth and division of *E.coli* which can be compared with the expected behaviour of the ideal cell.

(1) *Geometry*. Fig. 2 is a phase-contrast picture of some cells of *E. coli* from an asynchronous population in exponential growth. It can be seen that the cells are roughly cylindrical with hemispherical poles, that growth consists in elongation without much change in cell diameter, that nucleoid (light internal areas) duplication precedes cell division and that the septum forms between the two sister nucleoids. The exact shape of intact cells is hard to determine because they are so small (and because preparation for electron microscopy and perhaps even the procedure used to make Fig. 2 often involves distortion of the cells) but preparations of sacculi (consisting solely of the covalently linked peptidoglycan layer which is responsible for cell shape) show that these approximate closely to the

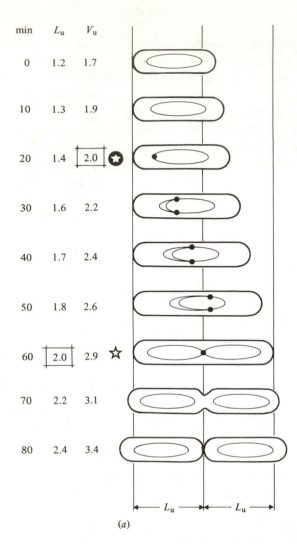

min	L_u	V_u
0	1.2	1.7
10	1.3	1.9
20	1.4	2.0
30	1.6	2.2
40	1.7	2.4
50	1.8	2.6
60	2.0	2.9
70	2.2	3.1
80	2.4	3.4

(a)

Fig. 1. Growth, chromosome replication and division of an 'ideal' *E. coli* cell at two generation times: 80 min (*a*) and 30 min (*b*). The same rules of growth hold at all growth rates (see text). Growth in cell volume is exponential. Growth in length is proportional to volume, times a growth rate related factor $(K \cdot 2^{(-40/T)}$, see text) such that cell proportions (length/width) are constant for any given fraction of the cell cycle for any growth rate and cell width is constant for any growth rate. Cell extension is unipolar until a cell reaches a length of $2 \cdot L_u$ (i.e. twice the minimum theoretical cell length). At this length septum formation is initiated at a point midway between the two cell ends and cell extension becomes bipolar. Septum formation takes 20 min ('*D*' min) to complete. (The process of septum formation is not well understood and the sequence shown is for illustrative purposes only.) Rounds of chromosome replication are initiated

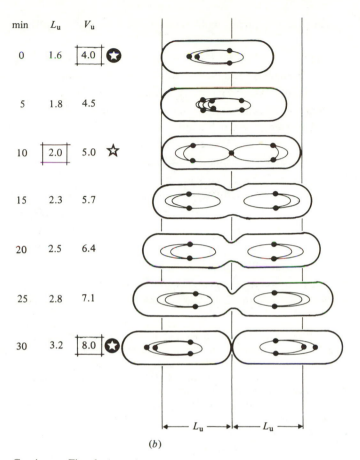

min	L_u	V_u
0	1.6	4.0
5	1.8	4.5
10	2.0	5.0
15	2.3	5.7
20	2.5	6.4
25	2.8	7.1
30	3.2	8.0

(b)

Caption to Fig. 1. (*cont.*)

at each doubling of the unit volume, V_u (the theoretical minimum cell volume) such that the ratio between cell volume and number of initiations is constant at the time of initiation ($V_i/n = 2 \cdot V_u$). Replication begins at a unique site (*oriC*) and proceeds at a constant rate, bidirectionally around the chromosome (which is not drawn to scale) to reach the terminus after 40 min ('*C*' min). Thus termination of rounds of replication always coincides with the time that cell length reaches $2 \cdot L_u$. When the generation time of the cell is less than *C* min, then new rounds are initiated before preceding rounds have been completed, to give dichotomously replicating chromosomes. When the generation time is greater than *C* min, then there is a 'G2-like' period of up to 20 min in duration during septum formation at the end of the cycle. When the generation time of the cell is greater than 60 min (*C* + *D* min), then cell volume at birth is less than $2 \cdot V_u$ and there is a 'G1-like' period at the beginning of the cycle during which no DNA synthesis takes place. The illustrated sequences are for cells in 'balanced growth' at constant growth rates but the same rules may be used to calculate cell cycles during periods of changing growth rates (with one possible additional rule; see comparison between *E. coli* and rule 6 in text).

ideal cell shape of a cylinder with hemispherical ends (Schwarz, Asmus & Frank, 1969).

(2) *Growth during the cell cycle*. Fig. 2 shows that cells do appear to grow only in length, without change in diameter, during culture at constant growth rate. Careful measurements (Marr, Harvey & Trentini, 1966; O. F. Trueba & C. L. Woldringh, personal communication) have confirmed that there is little if any change in cell diameter under such conditions. (A small change in diameter due to the change in the ratio of surface to volume during cell division may be expected and there is some evidence for this; O. F. Trueba & C. L. Woldringh, personal communication.) That growth takes place solely by elongation at constant growth rate is most easily seen when cell division is inhibited without change in volume growth rate (e.g. by cephalexin, low doses of penicillin G, ultraviolet irradiation, thymine starvation, etc.). Under such conditions, *E. coli* cells grow into long 'snakes' with little if any change in diameter. This behaviour is expected of the ideal cell as a consequence of rule 4.

(3) *Cell volume as a function of growth rate*. Rule 5, taken together with rule 4, implies that a cell of volume $2 \cdot V_u$ must grow for 60 min before dividing. To-

Fig. 2. The cell cycle of *Escherichia coli* B/r. The photograph is of living cells on agar from a log-phase asynchronous population. The agar contained chloramphenicol (200 μg ml^{-1}) to cause condensation of the nucleoids (light areas) (Zusman, Carbonell & Haga, 1973) and polyvinylpyrrolidone to increase the contrast between the cytoplasm and the nucleoids. The photograph was taken under phase-contrast. Cell growth is by elongation and the cycle ends with the formation of two new poles at the cell centre. Duplication of the nucleoid precedes cell duplication.

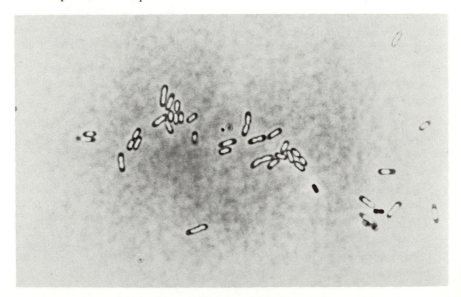

gether with rule 3, this gives the volume of the cell at division as $V_d = 2 \cdot V_u \cdot 2^{(60/T)}$. Using the age distribution of cells in an exponentially dividing population (Powell, 1956) and rule 3, this gives the relationship between average cell volume and growth rate ($R = 60/T$) as,

$$\bar{V} = \bar{V}_u \cdot 2^R \tag{1}$$

where $\bar{V}$ is mean cell volume and $\bar{V}_u$ is the mean volume of a population of cell with a growth rate approaching zero.

The growth rate of *E. coli* cells can be controlled by varying the composition of the growth medium. Fig. 3 shows the relationship between average cell volume and growth rate for some strains of *E. coli* in exponential phase growth in a number of synthetic media. In the same figure is the curve generated by equation (1). It can be seen that there is a good overall correspondence between the observed and predicted relationships between growth rate and average cell volume. It is also clear that cells growing in a particular growth medium may deviate to some extent from the ideal expectation: thus does the real world differ from the ideal! Nevertheless, when we consider how much these cells may differ in their metabolism in the different media, we should perhaps be impressed by how well they do correspond to such an abstraction as equation (1).

(4) *Cell length as a function of growth rate*. Rule 5 states that a cell initiates septum formation at length $2 \cdot L_u$ and that it completes division 20 min later. By a similar argument to that used for cell volume (above) this implies that

$$\bar{L} = \bar{L}_u \cdot 2^{(R/3)}. \tag{2}$$

Fig. 3 also shows the observed average cell lengths for the various growth rates, together with the relationship predicted by equation (2). Once again it is evident that the real cells behave impressively like their ideal abstractions. (N.B. The strains used had similar values for the parameter $\bar{V}_u$ but one had a higher value for $\bar{L}_u$ than the others. This strain is therefore relatively long and thin. It nevertheless behaves in a formally identical manner.)

(5) *Cell shape as a function of growth rate*. Equations (1) and (2) imply that the ratio between average cell length ($\bar{L}$) and average cell width ($2 \cdot \bar{r}$) is a constant at all growth rates. Such a constancy was first suggested by Zaritsky (1975). The data in Fig. 3 show that this must be approximately true (see also Fig. 4). Thus cells with different growth rates differ in absolute size but retain the same proportions, as required by the model.

(6) *Cell size and shape during changes in growth rate*. The above discussion has been concerned with 'balanced growth', in which cell size has reached some constant average value. We will now consider what the model predicts will happen when the growth rate is changed. In our model cell, the primary event following a change in growth conditions is considered to be the change in rate of

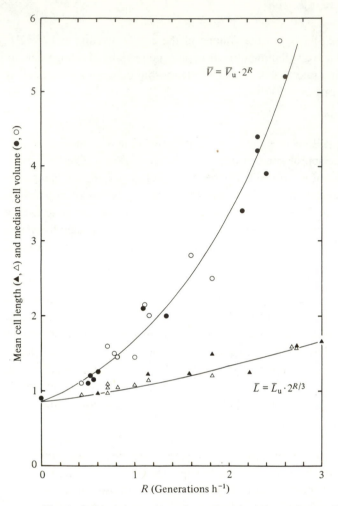

Fig. 3. Cell volume and length as a function of growth rate. Cells of two strains of *Escherichia coli* (B/rA, solid symbols; K12 CR34, open symbols) were grown at 37 °C in a number of different media (see Donachie *et al.* (1976) for examples of these media). The cells were maintained as far as possible in 'balanced growth' by frequent dilution into fresh medium. Cell volumes and lengths were measured on fixed samples from mid-log phase growth in each medium. Median volumes were estimated electronically using a Coulter Channelyzer (Begg & Donachie, 1978) and mean cell lengths were measured from projections of phase-contrast photographs (Donachie & Begg, 1970, Donachie *et al.*, 1976). The circles show median cell volumes and the triangles are mean cell lengths. The curves were drawn according to the two equations shown. The vertical scale is in arbitrary units, such that all curves share a common origin at $R = 0$. The point for cell volume at $R = 0$ was obtained from a stationary phase culture in rich medium (L-broth + glucose). The median volumes of the two different strains are much the same at comparable growth rates but CR34 cells are on average longer than B/rA cells (by a factor of about $\times$ 1.3). The data for mean cell lengths for the two strains have therefore been normalised to fit the same theoretical curve.

growth in volume. As a consequence of rule 5 it is expected that the rate of cell division will remain unchanged for approximately 60 min after an asynchronous population changes from one volume growth rate to another. Thus average cell volume will change continuously over this period and reach its new steady value by the end. Fig. 4 shows the course of change in cell volume and number following a shift of an *E. coli* population from a relatively poor to a relatively rich medium. As is well known (especially from the work of O. Maaløe and his colleagues; see Maaløe & Kjeldgaard, 1966) the behavior of real cells conforms well to this expectation.

Rule 4, concerning the relationship between the rate of cell elongation and cell volume and growth rate, can be used to predict the course of change in average cell length after a change in growth rate. Rule 4 predicts that mean cell length will begin to change at the same time as does cell volume but at a lower rate. In consequence, the cells will change their widths over the initial 60 min period after a change in growth rate but in such a way as to preserve (approximately) the constancy of average length to average width. Fig. 4 shows that this is so, within the accuracy of the experimental measurements. (In fact the approach to the new characteristic cell length for the new growth rate is expected to be asymptotic because the initial ratio of length to volume will be different from that for cells of the same volume at the new growth rate, because $\bar{L}/\bar{V} = (\bar{L}_u/\bar{V}_u) \cdot 2^{(-2/3)R}$, from rules 3, 4 and 5.)

So far, the rules of growth of the ideal cell have proved to predict behaviour which is closely similar to that which has been observed with asynchronous populations of *E. coli*. However, when synchronous populations have been examined in detail some differences from the predictions of the ideal cell model have been seen. Thus, rule 4, as stated, implies that cells at every stage in the cycle will respond similarly to a change in rate of volume increase. Consequently, if a population of, say, new-born cells were transferred from a poor to a rich medium, the rate of increase in cell volume would be expected to increase immediately and the rate of increase in cell length would be expected also to reflect immediately the changed rate of volume increase ($\times 2^{(-2/3)R}$, as in rule 4). The observed behaviour of such cells however does not correspond well to this prediction (Donachie, *et al.*, 1976). Instead, the rate of cell elongation appears to continue at more-or-less the pre-shift rate for an initial period, despite the fact that the rate of growth in cell volume has been immediately stimulated, after which it changes abruptly to the higher post-shift rate. We have therefore suggested that a cell is capable of changing its rate of elongation only at one stage in the cell cycle. We have proposed that cells can adjust their rate of elongation to match that of their rate of volume growth only when they reach cell lengths that are doublings of the unit length, L_u (Donachie *et al.*, 1976). Thereafter, the rate of elongation would be given by the equation in rule 4, as before.

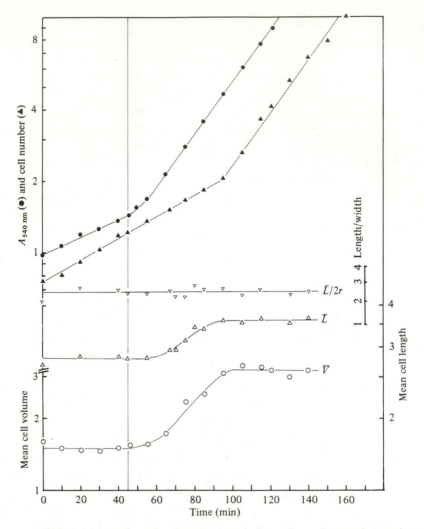

Fig. 4. Maintenance of cell proportions during a change in growth rate. An asynchronous culture of *E. coli* B/rA was grown to mid-log phase in a poor growth medium (M9 salts + glycerol) and shifted to a richer medium (Oxoid nutrient broth + glucose) at the time indicated by the vertical line. The culture was diluted periodically into fresh medium to keep cell density low; the points on the figure have been corrected for these dilutions. The absorbance of the culture (measured at 540 nm) showed an immediate stimulation after the shift in medium, reaching a new constant doubling time after about 15 min. In contrast, the rate of cell division remained at approximately its pre-shift value for about 52 min before shifting abruptly to parallel the rate of increase in absorbance. In consequence, mean cell volume (and absorbance per cell) increased continuously over this period (the measurements shown were made on fixed cells with a Channelyzer, as in Fig. 2) to reach the value characteristic of the new balanced growth rate in the rich medium. Measurements showed that mean cell length increased over the same initial period, although to a much smaller extent. Note

One of the attractive features of the above hypothesis is that it also fits in well with some observations on the mode of elongation of the cell envelope during the cell cycle at various growth rates (Begg & Donachie, 1977). Measurement of the distribution of receptor sites for bacteriophage T6 during cell growth has suggested that cells that are less than $2 \cdot L_u$ in length grow by elongation from one pole only but that growth from the second pole starts whenever length $2 \cdot L_u$ is reached. Thus the time of activation of a second growth site could correspond to the time at which the cell is able to adjust its rate of elongation.

Although such speculations are fascinating in themselves, they are very difficult to test experimentally (mostly because *E. coli* cells are so small and easy to distort) and they should not be allowed to obscure the fact that the simple rules of the 'ideal cell' give an excellent fit to most observations on the growth of the cells of *E. coli*.

(7) *Chromosome replication and growth rate.* Rule 6 accounts excellently for the observed relations between bacterial chromosome replication and the cell cycle, at least for generation times of about one hour or less (at 37 °C). The details of overlapping rounds of replication, etc. are well known and have often been described in review articles (e.g. Helmstetter, Cooper, Pierucci & Revelas, 1968; Donachie, Jones & Teather, 1973) and may be seen in operation in Fig. 1. Although there is general agreement about what happens to chromosome replication in fast-growing cells ($T = 60$, or less) there is as yet no consensus about what happens in more slowly-growing cells. The predictions of rule 6 are simple. At generation times greater than one hour it is predicted that DNA synthesis will begin 60 min before the end of the cycle and be completed 20 min before. Thus there would be no DNA synthesis during the remaining initial period of the cycle, which would therefore constitute a 'G1' period, in the terms used for eukaryotic cell cycles. Estimates of the value of C for growth rates from about 0.3 to nearly 3.0 generations h^{-1} have been made for two unrelated strains (K12 and B/rK) by Kubitschek & Freedman (1971) and Chandler, Bird & Caro (1975), using asynchronous log-phase cultures. Both groups concluded that C remained approximately constant at all growth rates. In addition, Kubitschek & Freedman were able to conclude that D also remained approximately constant over the

Caption to Fig. 4. (*cont.*)

that the vertical scale given for mean cell length is exaggerated in comparison with all the others, in order to show this change clearly. The relative changes in mean cell length and volume were such as to maintain the same average cell proportions throughout the period of observation. The ratio of mean cell length to mean cell width, $\bar{L}/2r$, was calculated from the values obtained from measurements of lengths and volumes on the assumption that the cells had the 'ideal' shape of cylinders with hemispherical poles. Because the volume measurements were not in absolute units, the calculated ratios are not exact. $\bar{L}$, mean cell length; $\bar{V}$, mean cell volume.

range of growth rates, thus implying the existence of a 'G1' period at slow growth rates. More recent studies on the pattern of DNA replication in synchronous cultures have directly demonstrated such a G1 period in slow-growing cultures of some strains (Helmstetter & Pierucci, 1976). However not all the results from synchronous cultures are consistent with the predictions of rule 6. Thus, Helmstetter & Pierucci claimed that the values of C increased to various extents with decreasing growth rate in various strains and that indeed in one strain (B/rA) it increased in proportion to the increase in the length of the cell cycle (for T greater than about 60 min), so that a G1 period was never observed. If this report is correct, then it would imply (in conjunction with the measurements of cell volume at different growth rates for B/rA, as in Fig. 3) that initiation of chromosome replication did not take place at $2 \cdot V_u$ in slow-growing cells of at least one strain of E. coli. We hope this problem will be resolved in the future. Thus there is clearly a need for a systematic study of the pattern of chromosome replication in relation to cell size in several strains at a variety of low growth rates. Meanwhile it is not possible to say whether our rule 6 does apply adequately under all conditions.

It is possible to slow down the rate of DNA synthesis per replication fork at higher growth rates, either by limiting the rate of supply of precursors, e.g. thymine (Pritchard & Zaritsky, 1970) or by using mutants (Lane & Denhardt, 1975). In such cases it seems that rule 6 applies very well. When the replication velocity is slowed down sufficiently, so that $C > T$, then multifork replication (due to the initiation of succeeding rounds in advance of the termination of earlier rounds) is observed and average cell volume is increased (Zaritsky & Pritchard, 1973). Because the growth rate of the cells is not much altered by such procedures, it would also be predicted from rule 4 that such enlarged cells would retain their normal diameter (for the particular growth rate) and be increased only in length. This has now been shown to be the case (Begg & Donachie, 1978; Pritchard, Meacock & Orr, 1978; Zaritsky & Woldringh, 1978).

Similarly, rule 6 predicts that the capacity to initiate rounds of chromosome replication will accumulate during cell growth under conditions which specifically inhibit DNA replication and this has now been amply demonstrated for a variety of situations (e.g. after thymine deprivation, ultraviolet irradiation and treatment with various antibiotics). After the release of such a block, very high numbers of replication forks per chromosome have been found (up to 15 pairs per chromosome; Donachie, 1969).

One of the conspicuous failures of the 'ideal cell' model has been to predict the pattern of replication of replicons other than the chromosome itself. Thus plasmid replication does not appear to be linked to cell size in any known case. In at least two cases, the plasmids F & R ldrd19 (Andresdottir & Masters, 1978; Gustafsson & Nordström, 1978) it seems reasonably clear that replication takes place at random times through the cycle. A major unsolved mystery is that such a pattern of replication is found in plasmids that are present in only one or two copies per cell and which nevertheless are lost from the cells at only extremely low frequencies.

(8) *Interactions between DNA replication and cell division.* Rule 6 states that cell division cannot be initiated until the completion of any ongoing rounds of chromosome replication. This prediction is fulfilled in wild-type strains of *E. coli*. However, there are certain mutant strains which do not appear to follow this rule. Inouye (1969, 1971) showed that *recA⁻* and some other strains were capable of continued cell division in the absence of DNA replication. It was partly the existence of such strains which first showed that the timing of cell division was not linked directly to the time of completion of chromosome replication, as had previously been the favoured hypothesis. Rule 6 states that it is the initiation of chromosome replication which sets up a potential block to further cell division. Therefore, if initiation could be prevented it would be predicted that cell division would proceed normally in the absence of chromosome replication. All *dnaA*(Ts) and some *dnaC*(Ts) mutants are classified as 'initiation' mutants, in the sense that they are able to complete any rounds of chromosome replication in progress at the time of a shift to the restrictive temperature but appear to be unable to start any new ones. Unfortunately, rule 6 is a purely formal statement which does not attempt to define the exact molecular meaning of 'initiation'. Thus we cannot predict whether a particular 'initiation' mutant will be blocked at a point before or after the stage at which the inhibition of septation is set up. Fortunately, at least some *dnaA* and *dnaC* mutants show the predicted behaviour. Hirota *et al.*, (1968) and Spratt & Rowbury (1971) have shown that such strains complete rounds of chromosome replication at the restrictive temperature, initiate no further DNA synthesis but continue to grow and divide to produce anucleate cells. Moreover, Hirota *et al.* (1968) have also shown that if such a *dnaA*(Ts) strain is shifted to the restrictive temperature and if, in addition, all further DNA synthesis is immediately prevented (by the addition of nalidixic acid) then division is blocked and the cells grow as long aseptate filaments, in exactly the same way as other classes of *dna*(Ts) mutants which are blocked in replication rather than initiation. Addition of the inhibitor of DNA synthesis after the time of completion of rounds of replication did not prevent subsequent cell division.

A fascinating but relatively little studied aspect of the interactions between DNA synthesis and cell division concerns the effect of inhibition of plasmid replication. Thus the introduction into a cell of ultraviolet-irradiated plasmids (of certain kinds) can result in a dramatic blocking of cell division. If the recipient cell is incapable of repairing the damage to DNA caused by the irradiation (*uvrA⁻*) then this block appears to be permanent, so that very long multinucleate filaments are produced (MacQueen & Donachie, 1977). Chromosomal replication is not grossly affected in such cells, although there do appear to be alterations in the segregation pattern of certain DNA strands as a result of the infection (MacQueen & Donachie, 1977). Autoradiography has shown that the damaged plasmid DNA is localised in the centres of the filamentous cells (MacQueen & Donachie, 1977). One hypothesis is that the existence of large numbers of thymine dimers in the plasmid DNA allows the initiation step of replication to take place (which perhaps involves binding to a membrane site) but prevents further replication (Donachie, 1974). The block to cell division would therefore be considered to be analogous in mechanism to that operating after a block to chromosomal DNA replication, as discussed above.

(9) *Localisation and number of division sites*. In the normal cell cycle of *E. coli*, a single septum is formed near the cell centre. The control of this process is precise, in as much as a single septum is formed almost exactly in the cell centre (although some strains show a low frequency of slightly asymmetric divisions; Cullum & Vicente, 1978) at a predictable cell length (Donachie, *et al.*, 1976; see rule 8). Cell division may be specifically inhibited without very much effect on cell growth, so that abnormally long cells are formed. When the block to division is reversed in most such cells, the numbers and positions of subsequent septa can be predicted from a knowledge of cell length alone (Donachie & Begg, 1970). The numbers of potential division sites increase according to the series 0, 1, 3, 7, etc., as cell length reaches successive doublings of an approximately constant length. This length is close to estimates of the unit length, L_u (Donachie & Begg, 1970).

An exception to the usual rules for determining septum positions is seen in certain mutants. One, which has been studied in this laboratory, is temperature-sensitive for septum location. At 30 °C this strain (which carries an amber mutation in a gene which we call '*ran*' (dom) in combination with a temperature-sensitive amber suppressor) forms septa in normal positions but, after a period of growth at 42 °C, septum localisation is grossly disturbed, so that cells of every possible length are produced, ranging from minute, spherical 'minicells', through every intermediate length to long filaments. A similar temperature-sensitive phenotype has been reported for a different mutant which has a temperature-sensitive DNA-gyrase (E. Orr, N. F. Fairweather & I. B. Holland, personal communication). Although the relevant mutations lie in different genes (unpublished experiments) these two strains have in common, apart from disturbed septum localisation, a block to the normal segregation of chromosomal DNA into separate nucleoids at the restrictive temperature. Thus the DNA becomes localised in a single, central mass in the longer cells, while the shorter cells contain no DNA. The existence of strains like these, in which both DNA segregation and septum localisation are disturbed, suggests that these two processes may share some common control.

One striking thing about the *'ran'* strain is that, although septum position appears to be random at the restrictive temperature, the *number* of septa per total cell mass, volume or length of the population is close to normal. Thus these strains still have good control over the number of septa formed per generation. However, the proof that there is a separate control system for septum number comes from measurements on a different mutant strain, the so-called 'minicell'-producing strain (P678–54) of Adler, Fisher, Cohen & Hardigree (1967). This strain behaves as if potential division sites, once formed, remain active indefinitely. This means that the poles of all cells are themselves potential division sites. If division occurs at a polar site (i.e. two new polar caps are formed there) then this results in the formation of a spherical minicell. (Minicells contain no chromosomal DNA and do not grow.) The mutant strain also appears to form normal numbers of potential division sites at normal positions along the cell length (as in rule 7). The very interesting observation is that this strain, despite having these 'extra' sites, has no more than the normal number of actual septa per cell length (Teather, *et al.,* 1974). Thus, a polar division, resulting in a minicell, prevents the formation of a septum at a normal-located potential division site, and vice versa. Analysis of septal frequencies and positions has shown clearly that a normal number of septa are formed per generation but that these are distributed at random amongst all the available potential division sites, including the polar ones. In wild-type cells the numbers of potential division sites and the numbers of septa formed are always equal, with the result that the two separate control systems for septal sites and septal number cannot be distinguished. Thus mutants such as *'ran'* and the minicell strains show the behaviour predicted by the formal 'rule 8' of the ideal cell.

How is the cell cycle controlled?

We have seen that the *E. coli* cell behaves remarkably like a well-regulated piece of machinery, with quite rigid rules of growth. Does all this 'natural history' help us to understand the mechanics of cell growth and regulation? The answer is that it does not. What it does do is to set out all the things that the cell's regulatory machinery must accomplish. To find out the nature of this machinery requires some molecular biology, rather than cell biology.

The most striking thing that we have seen about the *E. coli* cell cycle is that key events such as the initiation of chromosome replication, cell division and, perhaps, new surface growth sites, coincide with the times at which there is a doubling of a basic unit volume or of a basic unit length. The possibility therefore arises that the periodic events of the cycle are initiated as a direct consequence of the cell achieving certain critical dimensions. Reaching a critical size might set off a whole series of biochemical events but it would be the attainment of that cell size that would be the trigger for the initiation of the periodic events. An alternative, however, is that certain 'cell cycle control proteins' must reach critical intracellular amounts or concentrations in order to initiate the cycle events, and that the attainment of these critical quantities is somehow correlated with the attainment of particular cell dimensions. We have therefore attempted to find evidence for the existence of such a class of proteins.

Initially, we simply examined the pattern of synthesis of as many proteins as possible through the cell cycle (Lutkenhaus, Moore, Masters & Donachie, 1979). Asynchronous log-phase cultures were pulse-labelled with radioactive amino acids, chased, killed and separated according to cell size (and age) using a zonal rotor. Total labelled polypeptides were then prepared from each fraction and separated and autoradiogrammed using the two-dimensional electrophoretic method of O'Farrell (1975). About 750 polypeptides could be resolved but no systematic changes in the rates of synthesis of any of them could be detected. (Similar results were obtained using synchronous cultures.) This method of course has many limitations but it does show that the cell cycle does not involve the syntheses of any large number of specific proteins at well-defined times.

Our second approach was to construct mutant strains in which the synthesis of individual proteins of interest could be turned on or off at will. In *E. coli* this can be simply achieved by combining a nonsense mutation in a particular gene with a temperature-sensitive suppressor. Such a double mutant strain will produce a complete polypeptide at one temperature but only an inactive fragment at another. If a certain protein must reach a critical amount in order to initiate some cell cycle event, then it follows that its synthesis must be required anew between each such successive event. If we could turn off the synthesis of such a protein specifically in a population of cells then only those cells which already had a sufficient amount of this protein could complete the controlled cell cycle event. This expected behaviour for a control protein may be contrasted with the expectations for a re-utilisable protein, such as an enzyme which is required for a particular process to take place but not in any particular amount. If the synthesis of such a protein were switched off, the cell cycle process in which it was involved would be expected to continue at a constant rate during the subsequent growth of the population of cells. We have in fact now seen examples of both kinds of behaviour amongst our mutant strains. Thus strain OV42 (*ftsK*(Am) *supF*(Ts)) continues to grow exponentially after a shift to 42 °C (when further synthesis of the 90 kdalton *ftsK* protein is prevented) but cell division remains at a constant rate thereafter, so that cell length increases progressively with time (Lutkenhaus & Donachie, unpublished). Thus the mutant strain suggests that the *ftsK* protein is limiting for cell division but that its synthesis could not be responsible for the initiation of septation. This behaviour may be contrasted with that of strain OV16 (*ftsA*(Am) *supF*(Ts)). Fig. 5 shows that, when synthesis of the 50 kdalton *ftsA* protein is switched off by shifting an asynchronous population to 42 °C, cell division continues to parallel cell growth for about 20 min before stopping abruptly and completely. Cell growth, DNA replication and nucleoid segregation continue normally, so that long multinucleate filaments are produced (Fig. 6). If the cells are returned to 30°C, at which temperature we have shown that suppressor activity is restored almost immediately (Donachie *et al.*, 1979) then cell division resumes after a delay of about 20 min. This behaviour is consistent with the idea that the *ftsA* protein reaches a critical amount or concentration in the cell about 20 min before the completion of cell division. It will be

Fig. 5. Cell growth and division at 30°C and 42°C in mutant cells which are temperature-sensitive for the synthesis of the *ftsA*-protein (Lutkenhaus & Donachie, 1979). An asynchronous population of strain OV16 (*ftsA*(Am) *supF*(Ts)) was grown to mid-log-phase at 30°C and then shifted to 42°C at 0 min. The rate of increase in absorbance was immediately stimulated; cell division continued for the first 20 min and then stopped abruptly, so that median cell volume rose continuously during further growth at 42°C. After one hour at 42°C, the culture was returned to 30°C. The rate of increase in absorbance immediately decreased and cell division resumed after a delay of about 20 min. Median cell volume thereafter slowly returned to its 30°C value. Control experiments (Donachie *et al.*, 1979) have shown that suppression and therefore the synthesis of the *ftsA*-protein is switched off immediately after a shift to 42°C and switched on immediately after a return to 30°C. N, cell number (Coulter counts); V, median cell volume (Channelyzer). The values are expressed in arbitrary units.

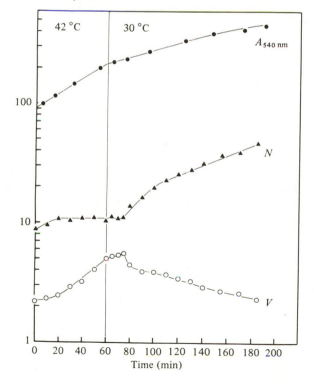

remembered that this is the approximate time at which cells become 'committed' to division, in the sense that after that time no further protein synthesis is required for septum formation and completion. Experiments with synchronous populations (Donachie *et al.*, 1979) have shown that synthesis of the *ftsA* protein is required for subsequent cell division only during a short (5–10 min) period immediately before the time of commitment to septation. Thus this protein does not require to be accumulated over the whole cycle in order for the cells to be able to divide. Such experiments have shown also that cells are 'competent' to respond to a period of permissive *ftsA* protein synthesis only during the short period immediately before septation. Shifts to 30 °C at any other times in the cycle do not permit cell division. There is thus one obligatory stage for *ftsA* protein synthesis in the cell cycle. There are several possible reasons for this (e.g. that the protein can be made only during this critical period, or that it is very unstable and that the cells are competent to respond to its synthesis only during that period) but it is clear that this protein has a very special role in cell division. Nevertheless, we remain unable to say whether it is the synthesis of the *ftsA* protein at the critical time which is responsible for the initiation of septation,

Fig. 6. Cells of strain OV16 (*ftsA*(Am) *supF*(Ts)) growing at 42°C. Phase-contrast photograph of living cells on polyvinylpyrolidone + chloramphenicol agar, as in Fig. 2.

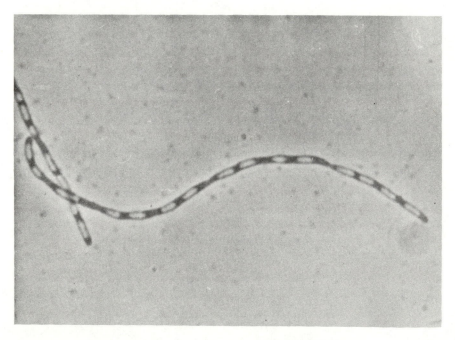

or whether some other event (such as the attainment of the critical length $2 \cdot L_u$) either results in the induction of this protein or makes the cell competent to respond to its synthesis.

Unpublished work from this laboratory has shown that mutants with amber mutations in genes other than *ftsA* give rather similar kinetics of cell division, but none of these have yet been studied in as much detail as OV16. In addition, we have one mutant in DNA synthesis, due to an amber mutation in an unknown *dna* gene ('*dna-37*') which behaves as if it can complete rounds of replication after a temperature shift but cannot initiate further rounds. If this strain is shifted to 42°C under conditions in which DNA synthesis is blocked by thymine deprivation then the capacity to complete rounds of replication remains intact even after prolonged periods at the restrictive temperature. Thus the *dna-37* protein behaves as if it is inactivated during use in DNA replication. This is certainly the behaviour expected of a cell cycle control protein.

One of the advantages of studying nonsense mutations in cell cycle genes is that it enables unambiguous identification of proteins that are specifically involved in the cycle. (This is done by cloning the amber mutation on a phage vector and comparing the set of polypeptides coded for by the cloned fragment in two types of ultraviolet-inactivated cells, which are either capable or incapable of suppressing the amber mutation; Lutkenhaus & Donachie, 1979.) In addition, specific penicillin-binding proteins have been identified, some of which appear to be intimately involved in cell elongation and division (Spratt, 1977). Thus the way is now open to begin to ask some specific questions about the molecular events which control the amazingly regular cell cycle of *E. coli*.

References

Adler, H. I., Fisher, W. D., Cohen, A. & Hardigree, A. A. (1967). Miniature *Escherichia coli* cells deficient in DNA. *Proceedings of the National Academy of Sciences, USA*, **57**, 321–6.

Andresdottir, V. & Masters, M. (1978). Evidence that F'*lac* replicates asynchronously during the cell cycle of *Escherichia coli* B/r. *Molecular and General Genetics*, **163**, 205–12.

Begg, K. J. & Donachie, W. D. (1977). The growth of the *E. coli* cell surface. *Journal of Bacteriology*, **129**, 1524–35.

Begg, K. J. & Donachie, W. D. (1978). Changes in cell size and shape in thymine-requiring *Escherichia coli* associated with growth in low concentrations of thymine. *Journal of Bacteriology*, **133**, 452–8.

Chandler, M., Bird, R. E. & Caro, L. (1975). The replication time of the *Escherichia coli* K12 chromosome as a function of cell doubling time. *Journal of Molecular Biology*, **94**, 127–32.

Cooper, S. & Helmstetter, C. E. (1968). Chromosome replication and the division cycle of *Escherichia coli* B/r. *Journal of Molecular Biology*, **31**, 519–40.

Cullum, J. & Vicente, M. (1978). Cell growth and length distribution in *Escherichia coli*. *Journal of Bacteriology*, **134**, 330–7.

Donachie, W. D. (1968). Relationship between cell size and time of initiation of DNA replication. *Nature, London*, **219**, 1077–9.

Donachie, W. D., (1969). Control of cell division in *Escherichia coli:* experiments with thymine starvation. *Journal of Bacteriology*, **100**, 260–8.

Donachie, W. D. (1974). Cell division in bacteria. In *Mechanism and Regulation of DNA Replication*, ed. A. R. Kolber & M. Kohiyama, pp. 431–45. New York: Plenum.

Donachie, W. D. & Begg, K. J. (1970). Growth of the bacterial cell. *Nature, London*, **227**, 1220–4.

Donachie, W. D., Begg, K. J., Lutkenhaus, J. F., Salmond, G. P. C., Martinez-Salas, E. & Vicente, M. (1979). Role of the *ftsA* gene product in control of *Escherichia coli* cell division. *Journal of Bacteriology*, **140**, 388–94.

Donachie, W. D., Begg, K. J. & Vicente, M. (1976). Cell length, cell growth and cell division. *Nature, London*, **264**, 328–33.

Donachie, W. D., Jones, N. C. & Teather, R. M., (1973). The bacterial cell cycle. *Symposium of the Society for General Microbiology*, **23**, 9–44.

Gustafsson, P. & Nordström, K., (1978). Selection and timing of replication of plasmids R1*drd 19* and F'*lac* in *Escherichia coli*. *Plasmid*, **1**, 187–203.

Helmstetter, C. E., Cooper, S., Pierucci, O. & Revelas, E. (1968). The bacterial life sequence. *Cold Spring Harbor Symposium on Quantitative Biology*, **33**, 809–22.

Helmstetter, C. E. & Pierucci, O. (1976). DNA synthesis during the division cycle of three substrains of *E. coli* B/r. *Journal of Molecular Biology*, **102**, 477–86.

Hirota, Y., Jacob, F., Ryter, A., Buttin, G. & Nakai, T. (1968). On the process of cellular division in *Escherichia coli*. *Journal of Molecular Biology*, **35**, 175–92.

Inouye, M. (1969). Unlinking of cell division from deoxyribonucleic acid replication in a temperature-sensitive deoxyribonucleic acid synthesis mutant of *Escherichia coli*. *Journal of Bacteriology*, **99**, 842–50.

Inouye, M. (1971). Pleiotropic effect of the *recA* gene of *Escherichia coli:* uncoupling of cell division from deoxyribonucleic acid replication. *Journal of Bacteriology*, **106**, 539–42.

Jones, N. C. & Donachie, W. D. (1973). Chromosome replication, transcription and the control of cell division. *Nature New Biology*, **243**, 100–3.

Kubitschek, H. E. & Freedman, M. L. (1971) Chromosome replication and the division cycle of *Escherichia coli* B/r. *Journal of Bacteriology*, **107**, 95–9.

Lane, H. E. D. & Denhardt, D. T. (1975). The *rep* mutation. IV. Slower movement of replication forks in *E. coli rep* strains. *Journal of Molecular Biology*, **97**, 99–112.

Lutkenhaus, J. F. & Donachie, W. D. (1979). Identification of the *ftsA* gene product. *Journal of Bacteriology*, **137**, 1088–94.

Lutkenhaus J. F., Moore, B. A., Masters, M. & Donachie, W. D. (1979). Individual proteins are synthesised continuously throughout the *Escherichia coli* cell cycle. *Journal of Bacteriology*, **138**, 352–60.

Maaløe, O. & Kjeldgaard, N. O. (1966). *Control of Macromolecular Synthesis*. New York: W. A. Benjamin.

MacQueen, H. A. & Donachie, W. D. (1977). Intracellular localisation and effects on cell division of a plasmid blocked in DNA replication. *Journal of Bacteriology*, **132**, 392–7.

Marr, A. G., Harvey, R. J. & Trentini, W. C. (1966). Growth and division of *Escherichia coli*. *Journal of Bacteriology*, **91**, 2388–9.

O'Farrell, P. H. (1975). High resolution two-dimensional electrophoresis of proteins. *Journal of Biological Chemistry*, **250**, 4007–21.

Pierucci, O. & Helmstetter, C. E. (1969). Chromosome replication, protein synthesis and cell division in *Escherichia coli*. *Federation Proceedings*, **28**, 1755–60.

Powell, E. O. (1956). Growth rate and generation time of bacteria, with special reference to continuous culture. *Journal of General Microbiology*, **15**, 402–51.

Pritchard, R. H., Meacock, P. A. & Orr, E. (1978.) Diameter of cells of a thermo-sensitive *dnaA* mutant of *Escherichia coli* cultivated at intermediate temperatures. *Journal of Bacteriology*, **135**, 575–80.

Pritchard, R. H. & Zaritsky, A. (1970). Effect of thymine concentration on the replication velocity of DNA in a thymineless mutant of *Escherichia coli*. *Nature, London*, **226**, 126–31.

Schwarz, U., Asmus, A. & Frank, H. (1969). Autolytic enzymes and cell division of *Escherichia coli*. *Journal of Molecular Biology*, **41**, 419–29.

Spratt, B. G. (1977). Temperature-sensitive cell division mutants of *Escherichia coli* with thermolabile penicillin-binding proteins. *Journal of Bacteriology*, **131**, 293–305.

Spratt, B. G. & Rowbury, R. J. (1971). Physiological and genetical studies on a mutant of *Salmonella typhimurium* which is temperature-sensitive for DNA synthesis. *Molecular and General Genetics*, **114**, 35–49.

Teather, R. M., Collins, J. F. & Donachie, W. D. (1974). Quantal behaviour of a division factor involved in the initiation of cell division at potential division sites in *E. coli*. *Journal of Bacteriology*, **118**, 407–13.

Zaritsky, A. (1975). On dimensional determination of rod-shaped bacteria. *Journal of Theoretical Biology*, **54**, 243–8.

Zaritsky, A. & Pritchard, R. H. (1973). Changes in cell size and shape associated with changes in the replication time of the chromosome of *Escherichia coli*. *Journal of Bacteriology*, **114**, 824–37.

Zaritsky, A. & Woldringh, C. L. (1978). Chromosome replication rate and cell shape in *Escherichia coli*: lack of coupling. *Journal of Bacteriology*, **135**, 581–7.

Zusman, D. R., Carbonell, A. & Haga, J. I., (1973). Nucleoid condensation and cell division in *Escherichia coli* MX74T2 *ts52* after inhibition of protein synthesis. *Journal of Bacteriology*, **115**, 1167–78.

P.NURSE & P.A.FANTES

Cell cycle controls in fission yeast: a genetic analysis

Introduction

The mitotic cycle of a growing cell is the period between the birth of a cell and the time when it undergoes division to form two daughters. It is made up of a series of events which are required for the cell to complete division successfully. The best known of these events are DNA replication, mitosis and cell division, all of which are concerned with the replication and partition of the hereditary material between the two daughter cells. Cell cycle controls are the cellular activities that regulate the progress of the cell through these various events leading to cell division. The roles of three of the controls which are clearly important for cell cycle regulation are as follows. (i) The transition of the cell from a non-proliferating state into the mitotic cell cycle. This control involves the initiation of cell cycle activities, and the commitment of the cell to that cell cycle as compared with other possible fates such as differentiation or entry into a non-growing state. (ii) The set of controls which ensure that the various cell cycle events occur in the correct sequence at the right times during the cell cycle, and are located in the correct region of the cell. (iii) The controls which determine the rate at which a cell can complete the cell cycle and undergo cell division. These define the major rate-controlling events which limit progress through the cell cycle.

In this chapter we shall describe some of the work that has been carried out in the simple eukaryote fission yeast for the purpose of investigating these cell cycle controls. The work has been primarily genetical in approach, and has concentrated on the controls that determine the rate at which a cell completes its cell cycle and that regulate the timing of DNA replication, mitosis and cell division.

Genes required for the cell cycle

The cell cycle of fission yeast or *Schizosaccharomyces pombe* is similar to most other eukaryotes in that it can be divided up into various phases: G1, S (DNA replication), G2, and M (mitosis) (Mitchison, 1970). G1 is short and occurs at the beginning of the cell cycle. It is followed by a short S phase and a long G2 which culminates in mitosis towards the end of the cell cycle. Mitosis is immediately followed by the formation of a septum dividing the cell, and then cell separation. *S. pombe* is a rod shaped organism (Fig. 1a), and during its cell cycle it grows mostly in length with little change in diameter.

The genes that are required for the cell to complete its cell cycle can be identified by isolating mutants which are specifically unable to undergo cell division. Such *cdc* (cell division cycle) mutants have to be conditionally defective in the mutant function since cells which could not divide would obviously be lethal. In *S. pombe, cdc* mutants have been isolated which grow normally like wild type at 25 °C but are unable to divide at 35 °C (Nurse, Thuriaux & Nasmyth, 1976). They can be recognised since at the higher temperature the cells cannot divide but continue to grow in mass and length producing long filamented cells (Fig. 1a). About 90 such mutants have now been isolated which have been found to define a total of 26 *cdc* genes. Physiological analysis of representative mutants in these genes has shown that nine genes are required for DNA replication, eight for mitosis and nine specifically for septation and cell division (Nurse *et al.*, 1976; Thuriaux, Sipiczki, & Fantes, 1980; Nasmyth & Nurse, 1981).

A rate-determining control in the cell cycle

Although *cdc* mutants can define those gene functions that are required for cell cycle completion, they cannot distinguish those genes which are important in determining the *rate* at which the cell cycle is completed. This point can be made clearer by making an analogy to control in a biochemical pathway. All the enzymes in the biochemical pathway are required before any flux through it is possible, but not all the enzymes are important in controlling the level of that flux. The pacemaker enzymes that are important are those that are rate-limiting for the flux. Any change in the level of their activity has a large effect on the level of flux through the entire pathway. Those *cdc* gene functions which are analogous to pacemaker enzymes are rate-limiting for progress through the cell cycle. If such a gene function was abolished, as could be the case in a *cdc* mutant, then cell division would be delayed or would not be possible. However, the same situation would also apply to any gene function required for cell division whether it was rate-limiting or not. Thus it is clear that *cdc* mutants cannot distinguish genes whose functions are rate-limiting for cell cycle progress from those which are not.

However, such a distinction is possible by isolating mutants that are *advanced* into cell division rather than delayed. If a gene function were not rate-limiting for cell cycle progress, then speeding that function up would have little effect on the timing of cell division. On the other hand, speeding up a function which was rate limiting would have an effect, as cells would be advanced into division. Cells advanced into division would divide at a smaller cell size than usual. This is because cells normally tend to divide at a particular size, and an advancement of cells into division would result in a shortened cell cycle. Since the rate at which cells increase their mass would not change, cell division would take place before the cell had grown to its normal size for division. In *S. pombe* such cells can be recognised as they undergo septation at a reduced cell length from normal (Fig. 1*b*). Fifty-two *wee* mutants showing this phenotype have now been isolated (*wee* because of their small size at division and Scottish origin). These mutants define two genes whose functions can be mutated to advance cells into division, and which therefore must be rate-limiting for progress through the cell cycle

Fig. 1. (*a*) Cells of wild type and *cdc* mutant. Cells have been stained with primulin and photographed using a fluorescence microscope. A, Wild-type cells: the bottom cell has a brightly fluorescing septum across its middle. B, *cdc* mutant cells: the mutant is defective in mitosis and has been incubated at 35 °C. The cells are elongated and have no septa. (*b*) Cells of wild type and *wee* mutant. Cells have been photographed using dark field optics. A, Wild-type cells: dividing cells have a septum across their middle. B, *wee* mutant cells: the dividing cells are much shorter than the wild types.

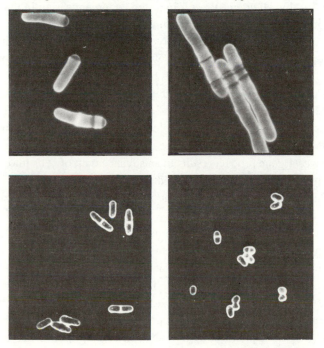

(Nurse, 1975; Thuriaux, Nurse & Carter, 1978; Nurse & Thuriaux, 1980.

The cell cycle event that the two *wee* genes control can be identified by using temperature-sensitive *wee* mutants of the two genes. These mutants divide at a larger size at 25 °C compared with 35 °C. On shift from 25 °C to 35 °C, cells are rapidly advanced into cell division at a smaller size. Analysis of the shift shows that the first cell cycle event to be advanced is mitosis. Rapidly after the shift there is a dramatic increase in the number of cells undergoing mitosis, indicating that cells which would have normally taken some time before undergoing mitosis do so almost immediately. Therefore the two *wee* genes must define rate-controlling functions which determine when mitosis is initiated. Furthermore, since cells are also advanced into cell division, the initiation of mitosis must act as a major rate-controlling event determining overall progress through the cell cycle.

Control initiating mitosis

Role of *wee genes*.

Genetic analysis of the *wee* mutants has given some idea how the two *wee* genes may function in the control initiating mitosis. Of the 52 *wee* mutants all but one map at the *wee* 1 locus. One of these mutant alleles of *wee* 1 has been shown to be suppressible by a nonsense suppressor tRNA (Nurse & Thuriaux, 1980). Therefore, this allele must result in a stop codon appearing in the middle of the *wee* 1 gene which prematurely terminates polypeptide synthesis. Premature termination results in a shortened polypeptide which is highly unlikely to have any biological activity. Consequently we can conclude that a lack of activity of the *wee* 1 gene product results in cells being advanced into mitosis and cell division. This conclusion is supported by the high frequency of mutations at the *wee* 1 locus. Most mutations result in loss of activity of gene products. Therefore a high frequency of mutation at a particular locus is suggestive that the phenotype produced by these mutations is due to a loss of gene product activity. If a lack of activity of the *wee* 1 gene product results in an advancement of cells into mitosis, then the wild type *wee* 1^+ gene product is likely to act as an inhibitor of mitosis. When this inhibitor loses activity as in the *wee* 1 mutants, cells are advanced into mitosis at a small size.

A different conclusion is reached concerning the function of the second gene *wee* 2. As implied above, only one *wee* mutant has been isolated at this locus. The rarity of *wee* 2 mutants suggests that its activity has to be modified rather than destroyed to yield the *wee* phenotype. This is supported by the observation that the *wee* 2 mutant is allelic with *cdc* 2, a gene which had been previously identified as one required for the completion of mitosis (Nurse & Thuriaux, 1980). A fine genetic map of this region showed that the *wee* 2 mutant allele mapped in the middle of the *cdc* 2 gene, and hence *wee* 2 and *cdc* 2 were the same gene. Confirmation for this was found when two of the original *cdc* 2 mutants were found to have a partial *wee* phenotype. At 25 °C they initiated mitosis at a small size and at 35 °C they were unable to initiate mitosis at all. Therefore, some mutant alleles in the *cdc* 2 gene result in cells being advanced into mitosis at one temperature and delayed from initiating mitosis at another temperature. These observations can be explained as follows. Gene *cdc* 2 codes for a product whose activity is required for the initiation of mitosis. In addition some aspect of this activity determines the cell cycle timing of mitosis. If the activity is lost cells cannot initiate mitosis, as was observed in the originally isolated *cdc* 2 mutants incubated at 35 °C. But if some aspect of the activity is modified in a particular way without loss of overall activity, cells can be advanced into mitosis as observed in the *wee* 2 mutant. The most likely interpretation of this is that the wild type *cdc* 2^+ gene codes for an activator in the mitotic control. Destroying the activator delays mitosis whilst some types of modification advance mitosis.

Influence of growth rate/nutritional status and cell size

The influence of two physiological cellular parameters on the rate at which a cell completes its cell cycle have also been investigated in *S. pombe*. One of these is the cell growth rate or nutritional status, and the other is cell size. Wild-type cells growing in different media supporting different growth rates undergo cell division over a range of sizes. Cells divide at a smaller size in media supporting slow growth and at a larger size under conditions of faster growth (Fantes & Nurse, 1977). If cells are shifted from media supporting fast growth to one supporting slow growth, cells are rapidly advanced into cell division at a small size. As with the temperature shifts of *wee* mutants, the first cell cycle event to be advanced is mitosis. This experiment demonstrates that there is something about shifting to a richer medium which delays mitosis. On transfer from richer medium to poorer medium the inhibition of mitosis is lessened and cells are advanced into mitosis. Therefore the growth medium has a direct effect on the rate-controlling functions which determine the cell cycle timing of mitosis. The cell monitors some parameter related to growth medium, like growth rate or nutritional status, and integrates this signal into the control initiating mitosis.

The second parameter investigated in *S. pombe* is cell size. Steady-state growing cells tend to divide when they have attained a certain cell size and appear to have some homeostatic mechanism for maintaining this cell size at division. If the cells are perturbed so that they divide at a different size than normal, then this is corrected for in the next cell cycle (Fantes, 1977). Larger cells at birth grow less before the next division, and smaller cells more. Consequently cell division takes place in cells closer to the normal size. The cell must have some mechanism for monitoring cell size and communicating this signal to the cell division cycle. Since attainment of a certain cell size is a normal requirement before cell division can take place, then the rate at which a cell accumulates mass is limiting for the rate at which a cell completes its cell cycle. It is very likely that the effect of cell size on limiting the rate at which cell division takes place is mediated by the control initiating mitosis. This control is also rate-limiting for cell cycle progress and in addition takes place just before cell division. If this is correct then the following model of the wild-type mitotic control can be made. Mitosis is initiated when the cell attains a critical size. In poorer media supporting slower growth, this critical size is modulated downwards so mitosis and cell division takes place in smaller cells. In richer media supporting faster growth, the critical size is modulated upwards so mitosis takes place in larger cells.

The involvement of the *wee* 1 and *cdc* 2 (*wee* 2) genes in the mediating of signals generated by the cell growth rate/nutritional status and cell size is not clear at the present time. It is possible that *wee* mutants are defective in the monitoring of cell size so that mitosis is not initiated at the correct size. Therefore the *wee* 1 and *cdc* 2 (*wee* 2) gene products may be elements in the mechanism by which the cell monitors its mass or feeds this signal into the nucleus. Another possibility is that the *wee* mutants are altered in the monitoring of the cell growth rate or nutritional status. The *wee* mutant cells may always 'consider' themselves growing in poor medium and so initiate mitosis at a small size even though the actual growth rate of *wee* mutant cells is unaffected. Further support for this interpretation of *wee* 1 comes from the observation that *wee* 1 mutant cells cannot respond to changes in the growth medium (Fantes & Nurse, 1978). A shift from media supporting fast growth to one supporting slow growth does not advance mitosis in *wee* 1 mutant cells, in contrast to the behaviour of wild-type cells. It may be that *wee* 1 mutant cells cannot detect the signal monitoring the cell growth rate or nutritional status and so initiate mitosis at a small size characteristic of very slow growth rates. Fantes (1979) discusses some of these possibilities.

Control initiating DNA replication

Study of *wee* mutants has also given some insight into the control which determines the rate at which DNA replication is initiated. Both *wee* 1 and *wee* 2 mutants have longer G1 periods with S phase being delayed by 0.25–0.30 of a cell cycle (where 1.00 is equivalent to the duration of one cell cycle) (Nurse, 1975; Nurse & Thuriaux, 1977). One possible reason for this delay would be that cells must attain a certain minimal size before DNA replication can be initiated. The *wee* mutant cells are beneath this size at the start of the cell cycle and so have to grow for a longer G1 period before they attain the critical minimal size. Wild-type cells are above this size even at the beginning of the cell cycle and so undergo S phase after a very short G1. Consequently, the minimal size requirement is 'cryptic' in wild-type cells since they are always too big (Fig. 2). However, the minimal cell size requirement would be expressed if the wild-type cells were made smaller. This prediction was tested in two ways (Nurse & Thu-

Fig. 2. Model of cell size controls acting over S phase and mitosis: (a) wild type; (b) wee mutant. The figure shows the increase in protein (as a measure of cell size) in individual cells during two cell cycles of each type of mutant. In the wild-type cell, mitosis (M) is initiated when the cell attains the critical cell size marked by the large open triangles. The cell size after cell division is higher than the minimum cell size required for the initiation of DNA replication (S) which is marked by the large closed triangles. In the wee mutant cell the critical size control over mitosis is not operative and as a consequence cell size after division is lower than the minimum cell size required for the initiation of DNA replication (S). The cells grow for a further quarter of a cell cycle before attaining this size marked by the large closed triangles.

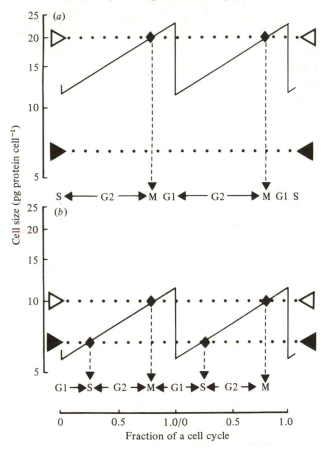

riaux, 1977). Firstly, wild-type cells were starved of nitrogen and so they accumulated in the G1 state and were much smaller than cells in the normal cell cycle. Secondly, diploid wild-type cells were sporulated when they formed small haploid spores also in the G1 state. Both types of small G1 cells were grown in fresh medium, and underwent DNA replication at a cell size very similar to the hypothetical critical minimal cell size. This was much smaller than the normal cell size at S phase in wild-type cells but similar to that observed in *wee* mutant cells. These results can be understood if all cells have a requirement to attain a critical minimal cell size before initiating DNA replication. Wild-type cells are normally too big for this to be expressed but in *wee* mutant cells it determines the timing of S phase.

Since the critical minimal cell size is cryptic in wild/type cells, S phase is initiated rapidly after mitosis is completed. A short time has to elapse between mitosis and S phase of about 0.1 of a cell cycle, which probably reflects a requirement for the cell to spend a short time in G1 before DNA replication can be initiated. Therefore, it appears that the timing of S phase in the wild-type cell cycle is determined by a dependency of the initiation of DNA replication upon the completion of mitosis, coupled with a requirement for the cell to spend a certain short minimum period in G1. The dependency of DNA replication upon the completion of mitosis has also been revealed by mutants blocked in mitosis which, as a consequence, are unable to undergo any further rounds of DNA replication. A change-over from this control operative in wild-type cells to the critical minimal cell size requirement in *wee* mutant cells would be expected as cells get smaller at mitosis. This change-over has been identified by measuring the timing of S phase and the cell size at S phase in a series of mutants undergoing mitosis and cell division over a range of cell sizes (Fig. 3) (Nasmyth, Nurse

Fig. 3. Cell size at S phase and length of G1 in various strains undergoing cell division over a range of sizes. (●) Cell size at S phase; (■) length of G1.

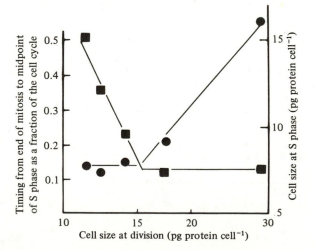

& Fraser, 1979). As cells become smaller at mitosis, G1 gets longer but cell size at S phase remains constant. As cells become larger at mitosis, G1 remains short and constant whilst cell size at S phase gets larger. This is exactly what would be expected if the two controls described above were determining the timing of S phase, one being operative in small cells and the other in large ones.

The existence of a minimal cell size requirement before DNA replication can also explain why *wee* mutant cells are viable and maintain a constant cell size at division. The normal mitotic control is defective in *wee* mutant cells and so the cell size control initiating mitosis which normally co-ordinates cell division with increase in cell size is not operative. However, the minimal cell size requirement for DNA replication is now exposed and can maintain a constant cell size at division. There appears to be a requirement for a minimum period in G2 before mitosis can be initiated. This requirement was established in experiments in which DNA replication was delayed (Fantes & Nurse, 1978). When the block on DNA replication was released, a certain period had to elapse before mitosis was possible. This minimum period in G2 was about 0.4 of a cell cycle, similar to the length of G2 in *wee* mutant cells. Therefore in *wee* mutant cells mitosis and division occur at a particular cell size since they take place at a fixed time after S phase which takes place at a constant cell size.

Before leaving the minimal cell size requirement for DNA replication it is of interest to note that a similar requirement has been detected in the budding yeast *Saccharomyces cerevisiae*. This control acts over 'start' a function that commits the cell to the mitotic cell cycle and determines the timing of S phase (Hartwell, 1974). Therefore it is possible that the control of S in *S. pombe* is also a point of commitment to division in the sense outlined in the Introduction (Nurse, 1981).

Control initiating septation and cell division

Under all growth conditions and in all mutants except those specifically blocked in septation, cell division takes place soon after mitosis is completed. Also mutants in eight of the nine *cdc* genes required for mitosis are unable to initiate septation when mitosis is blocked. These two observations suggest that the initiation of the cell division process is dependent upon the completion of mitosis and it is this dependency that determines the cell cycle timing of cell division. One mutant *cdc* 13.117 becomes blocked late in mitosis with condensed chromosomes visible and does make septa which cut acoss the mitotic nucleus (Nasmyth & Nurse, 1980). This can be understood if the signal initiating the cell division process is generated at a point in the middle of mitosis rather than at its completion. Mutant cells of *cdc* 13.117 are blocked after this point and so septation can be initiated. Since the septa cut across the dividing nucleus it is possible that the location of septation is determined by the position of the dividing nucleus.

Another mutant *cdc* 16.116 is also informative regarding the septation control. This mutant continues to accumulate septa between the divided nuclei and therefore appears to be defective in the control which ensures only one septum is made during each cell cycle (Minet, Nurse, Thuriaux & Mitchison, 1979). The continued initiation of septation in *cdc* 16.116 illustrates an important point regarding periodic events in the cell cycle. It is not sufficient to initiate a periodic event at a particular time in the cell cycle, there must be in addition a mechanism which prevents further initiations until another cell cycle is completed.

Molecular mechanisms for the controls

In our consideration of cell cycle controls in *S. pombe,* we have invoked a number of rather abstract notions such as attainment of critical cell sizes and requirements for minimum times in particular cell cycle phases. It is possible to propose molecular mechanisms to account for these notions and in this section we shall mention a few possibilities. The interested reader should follow up the references for more details and should also consult Fantes & Nurse in the present volume. The monitoring of cell size has attracted considerable theoretical attention and a number of molecular mechanisms have been proposed (see review by Fantes *et al.,* 1975). The requirement for a minimum time in say G1 or G2 may reflect nothing more than the time required to complete a series of biochemical reactions. One possibility would be the time required for decondensation and condensation of the chromatin which appears to undergo continuous changes in the cell cycle (Rao & Johnson, 1974). The monitoring of the growth rate/nutritional status of the cell may be analogous to the pleiotypic control in mammalian cells (Hershko, Mamont, Shields & Tomkins, 1971) or the relaxed-stringent control in bacteria (Cashel & Gallant, 1974). It is consistent with the latter possibility that Lusby & McLaughlin (1980) report that the level of a major acid soluble compound reflects the rate of protein synthesis in *Saccharomyces cervisiae*. Finally there are two types of explanation for the dependency of one cell cycle event upon the completion of a prior one. The first is a direct causal relationship between the two events. For example, if a nucleus has not completed mitosis the organisation of the chromatin may be quite inappropriate for initiation of DNA replication. Alternatively there may be a control pathway connecting the two events. An example of this may be the signal generated by the late mitotic nucleus which initiates septation.

Summary of controls, and conclusions

It will be useful at this stage to summarise our conclusions regarding cell cycle controls in *S. pombe*. Before a cell can initiate DNA replication it must satisfy two conditions: attainment of a minimal cell size and completion of the previous mitosis coupled with a minimum time in G1. Either of these two controls can determine the timing of S phase during the cell cycle depending on cell size at mitosis. Having completed S phase the cell proceeds through G2 to mitosis. The normal wild-type control initiating mitosis requires the cell to attain a critical size before initiation is possible. This size can be modulated up and down according to the growth rate or nutritional status of the cells. Two genes are involved in the mitotic control, *wee* 1 which codes for an inhibitor and *cdc* 2 (*wee* 2) which codes for an activator. If the mitotic control is defective as in *wee* mutant cells, then mitosis is initiated after a minimum time in G2. Towards the end of mitosis a signal is generated which initiates septation and cell division. A further control ensures only one septum is made during each cell cycle.

In this chapter we have described some of the cell cycle controls that appear to be operative in *S. pombe*. Our knowledge of these controls remains poor in two areas. The first is the controls involved in the transition of the cell from a non-proliferating state into the mitotic cell cycle, and in the correct spatial location of cell cycle events within the cell. The second concerns the molecular mechanisms which underlie the cell cycle controls.

Concerning transition probability

The genetic and physiological analysis described in this chapter has led us to propose that in *S. pombe* cell division takes place when the cell attains a critical size, but that this size can be modulated up or down according to whether the cells are growing in rich or poor media. An alternative view of division control called the revised transition probability model has been described by Brooks (this volume), in which there is no size control. Instead the control of cell division is brought about by two random transitions between cellular states coupled with two cell cycle timers. Since his analysis of the control is clearly at odds with our own we should like to consider his arguments in some detail to see how well they fit the *S. pombe* data.

Brooks first explains how cell size homeostasis at division may be maintained without resorting to a cell size control, as long as the rate of increase of mass during cellular growth is not proportional to mass. In the particular example he gives, he assumes cells grow linearly during the cell cycle. However, this mechanism of size homeostasis cannot apply to *S. pombe*. If it did apply, then on average, all cells regardless of their size at birth would grow the same amount during the succeeding cell cycle. This is not the case because large cells grow considerably less, since they divide sooner, than do small cells during their respective cell cycles (Fig. 3 in Fantes, 1977). As a consequence the progeny of a cell which is greater or smaller than the population mean at birth, will rapidly return to that mean, usually within one cell cycle. Such a rapid return would be expected by the cell size model, but not by the Brooks model.

A further argument against the Brooks model for cell size homeostasis is that it is very likely that the rate of increase of mass during cellular growth is in some way proportional to mass in *S. pombe*. A series of *cdc* 2 mutants are available which, when grown at the permissive temperature, undergo division at different cell sizes (Nasmyth, *et al.*, 1979). The cell size at division varies over a threefold range, some being smaller than wild type and others larger. Despite the differences in cell size at division their generation times are very similar. This means that during the time period of the cell cycle the cells which divide at the large size grow about three times as much as those that divide at the small size. Therefore, the rate of increase of mass of these cells is roughly proportional to the mass of the cells. The only other interpretation, that is *cdc* 2 mutants are directly altered in growth rates per cell as a consequence of the mutant lesion, is considered in the next paragraph.

Brooks then considers whether the *wee* mutant phenotype can be explained in terms of the revised transition probability model. He suggests that the *wee* mutant lesions have two effects, one which speeds up a cell cycle timer running from S through G2 to M and a second which slows down the growth rate per cell. To explain the behaviour of the series of *cdc* 2 mutants described above in a similar fashion, he would have to resort to the following arguments. As the mutants get larger, to about wild-type size, the mutant lesion results in the S + G2 + M getting longer whilst the growth rate per cell gets greater. In the mutants larger than wild type, the mutant lesion results in the growth rate per cell getting increasingly greater whilst the timer remains constant in length since S + G2 + M does not change. Finally, when incubated at the restrictive temperature, the timer becomes infinitely long as mitosis cannot take place, but there is little effect on the rate of growth of the cells. In other words this interpretation demands that the *cdc* 2 locus is directly involved in two rather different activities, one determining the length of the S + G2 + M timer and the other the rate of growth per cell, and furthermore, the different mutant lesions at *cdc* 2 alter these two activities in quite different ways as decribed above. This explanation should be compared to our own, in which the mutants are simply altered in their size at division, the differing growth rates per cell being a consequence of the fact that rate of increase in mass is roughly proportional to mass. In our opinion Brooks' interpretation demands too much of behaviour of mutant lesions at a single locus.

The final problem Brooks considers is the behaviour of cells on transfer between rich and poor media (Fantes & Nurse, 1977). He suggests that the data can be simply explained by a transient perturbation in the running of the cell cycle timer. Let us consider what these simple transient perturbations involve. When cells are grown in poor media supporting slow growth, then the cell cycle timer must be quite long to match the long generation time, and conversely when grown in rich media the cell cycle timer must be quite short. On transfer from poor media to rich, division is delayed and Brooks proposes that there is a transient set back in the progress of cells towards division. This means that on transfer the long timer is first transiently made longer and is then made much shorter as cells attain steady state. When considering a transfer from rich to poor media he would have to postulate that the short timer is first transiently made shorter and is then made much longer. In other words he has to argue that a transfer in this direction results in a transient acceleration of cells towards division even though their growth rate is less. This rather complex series of assumptions should be compared with our own model in which the cell size at division is just modulated up or down according to whether cells are growing in rich or poor medium. 'Occam's razor' must favour the latter model. In addition, for Brooks to explain the insensitivity of *wee* mutants to media shifts (Fantes & Nurse, 1978), he must propose that the mutant is not subject to the transient set backs and accelerations, in addition to his other proposals that the main effects of the mutant lesion are on the length of the cell cycle timer and the growth rate of the cells.

One final point concerns variability during the cell cycle of *S. pombe*. The transition probability model predicts that the most variable part of the cell cycle is the G1 period. This is not the case for rapidly growing wild-type *S. pombe* cells as the average difference in duration of the G1 phase between two sister cells was at most about 0.01 of a cell cycle, which is far smaller than the average difference in total cell cycle time between the sister cells which is about 0.1 of a cell cycle (Nasmyth, *et al.,* 1979). Although we have criticised this application of the transition probability model to the *S. pombe* data as we think it very contrived, this does not mean that we do not have sympathy with the model. Indeed we believe that the apparent conflicts between it and more deterministic models involving cell size may be eventually resolved to the benefit of both viewpoints. One example of how it may be resolved is given in the chapter by Fantes & Nurse in this volume and is further discussed by Nurse (1980).

It is a pleasure to acknowledge the various people who have participated in the work described in this chapter. Murdoch Mitchison provided a stimulating environment and much fruitful discussion, Pierre Thuriaux was involved in all stages of the work, Kim Nasmyth isolated many of the *cdc* mutants, Bruce Carter some of the *wee* mutants, and Michéle Minet worked on the control of septation.

References

Cashel, M. & Gallant, J. (1974). In *Ribosomes,* ed. M. Nomura, A. Tissieres & P. Lengyel, pp. 733–45. New York: Cold Spring Harbor Laboratory.

Fantes, P. (1977). Control of cell size and cycle time in *Schizosaccharomyces pombe. Journal of Cell Science,* **24,** 51–67.

Fantes, P. (1979). Epistatic gene interactions in the control of division in fission yeast. *Nature, London,* **279,** 428–30.

Fantes, P., Grant, W., Pritchard, R., Sudbery, P. & Wheals, A. (1975). The regulation of cell size and the control of mitosis. *Journal of Theoretical Biology*, **50**, 213–44.

Fantes, P. & Nurse, P. (1977). Control of cell size at division in fission yeast by a growth modulated size control over nuclear division. *Experimental Cell Research*, **107**, 377–86.

Fantes, P. & Nurse, P. (1978). Control of the timing of cell division in fission yeast. *Experimental Cell Research*, **115**, 317–29.

Hartwell, L. H. (1974). *Saccharomyces cerevisiae* cell cycle. *Bacteriological Reviews*, **38**, 164–98.

Hershko, A., Mamont, P., Shields, R. & Tomkins, G. (1971). Pleiotypic Response-hypothesis relating growth regulation in mammalian cells to stringent controls in bacteria. *Nature New Biology*, **232**, 206–11.

Lusby, E. W., Jr. & McLaughlin, C. S. (1980). The effect of amino acid starvation on a major, acid soluble compound in *Saccharomyces cerevisiae*. *Molecular and General Genetics*, **179**, 699–701.

Minet, M., Nurse, P., Thuriaux, P. & Mitchison, J. M. (1979). Uncontrolled septation in a cell division cycle mutant of the fission yeast *Schizosaccharomyces pombe*. *Journal of Bacteriology*, **137**, 440–6.

Mitchison, J. M. (1970). Physiological and cytological methods for *Schizosaccharomyces pombe*. *Methods in Cell Physiology*, ed. D. M. Prescott, vol. 4, pp. 131–65. New York & London, Academic Press.

Nasmyth, K., Nurse, P. & Fraser, R. (1979). The effect of cell mass on the cell cycle timing and duration of S phase in fission yeast. *Journal of Cell Science*, **39**, 215–33.

Nasmyth, K. & Nurse, P. (1981). Cell division cycle mutants altered in DNA replication and mitosis in the fission yeast *Schizosaccharomyces pombe*. *Molecular and General Genetics*, in press.

Nurse, P. (1975). Genetic control of cell size at cell division in yeast. *Nature, London*, **256**, 547–51.

Nurse, P. (1980). Cell cycle control – both deterministic and probabilistic. *Nature, London*, **286**, 9–10.

Nurse, P. (1981). Genetic analysis of the cell cycle. *Symposium of the Society for General Microbiology*, **31**, 291–315.

Nurse, P. & Thuriaux, P. (1977). Controls over the timing of DNA replication during the cell cycle of fission yeast. *Experimental Cell Research*, **107**, 365–75.

Nurse, P. & Thuriaux, P. (1980). Regulatory genes controlling mitosis in the fission yeast *Schizosaccharomyces pombe*. *Genetics*, in press.

Nurse, P., Thuriaux, P. & Nasmyth, K. (1976). Genetic control of the cell division cycle in the fission yeast *Schizosaccharomyces pombe*. *Molecular and General Genetics*, **146**, 167–78.

Rao, P. & Johnson, R. (1974). Induction of chromosome condensation in interphase cells. *Advances in Cell and Molecular Biology*, **3**, 135–89.

Thuriaux, P., Nurse, P. & Carter, B. (1978). Mutants altered in the control co-ordinating cell division with cell growth in the fission yeast *Schizosaccharomyces pombe*. *Molecular and General Genetics*, **161**, 215–20.

Thuriaux, P., Sipiczki, M. & Fantes, P. (1980). Genetical analysis of a sterile mutant by protoplast fusion in the fission yeast *Schizosaccharomyces pombe*. *Journal of General Microbiology*, **116**, 525–28.

B.L.A.CARTER

The control of cell division in
Saccharomyces cerevisiae

Introduction

The budding yeast *Saccharomyces cerevisiae* is a eukaryote which is amenable to genetical as well as biochemical and physiological analyses of the mitotic cell cycle. In the last decade numerous studies have been published concerning such analyses and the cell cycle of *S. cerevisiae* has been the subject of several reviews (Halvorson, Carter & Tauro, 1971; Hartwell, 1974; Hartwell, Culotti, Pringle & Reid, 1974; Hartwell, 1978; Simchen, 1978). I have attempted to concentrate here on areas that have not been reviewed extensively although inevitably some information that has been reviewed previously is discussed. My major concern has been to review information on controls affecting the rate of cell division and those which co-ordinate growth and cell division.

The cell cycle of a *S. cerevisiae* cell growing with a mass doubling time of two hours begins with an unbudded cell whose shape approximates a prolate spheroid. The cell grows in volume during this G1 phase of the cycle. The initiation of S phase coincides with the emergence of a bud. S phase lasts approximately 30 min and is followed by G2 phase. At the end of G2 phase the nucleus migrates to the junction of parent cell and bud. Nuclear division occurs without the nuclear membrane breaking down: the nucleus elongates such that part of it lies within the parent cell and part within the bud and then it divides. The time taken for nuclear division has been calculated as 15 min, from the percentage of cells which possess dividing nuclei in an exponential culture (Barford & Hall, 1976). Nuclear division is not immediately followed by cell division. Slater, Sharrow and Gart (1977) have calculated that in cells growing at 30 °C with a 2.2 h mass doubling time 0.52 h separates nuclear division and cell division.

Cell cycle landmarks

The most obvious landmarks dividing the cell cycle are DNA replication, bud emergence, nuclear division and cell division. The position of a cell

99

within the cycle can also be assessed by the ratio of bud size to parent cell size as the bud grows in size throughout the cell cycle (Mitchison, 1958; Sebastian, Carter & Halvorson, 1971; Shulman, Hartwell & Warner, 1973).

The ease of genetical analysis provided by the budding yeast has permitted the isolation of temperature-sensitive (ts) cell division cycle (*cdc*) mutants (Hartwell, Culotti & Reid, 1970). Such mutants divide normally at the permissive temperature (23 °C) but at the restrictive temperature (usually 36 °C) all cells in an asynchronous exponential culture arrest at a particular stage of the cycle. A number of these mutants have been shown to have a mutation in a single nuclear gene and genetic analysis of 148 such mutants showed that 32 different genes were represented (Hartwell, Mortimer, Culotti & Culotti, 1973). In recent years a few more *cdc* genes have been defined (Simchen, 1978; Hartwell, personal communication). Each of these mutants has a characteristic execution point (sometimes called transition point) within the cell cycle. Beyond this point in the cycle a ts *cdc* mutant can be shifted from the permissive to the restrictive temperature and still complete cell division: a shift to the restrictive conditions prior to this point in the cycle results in cell cycle arrest before division. If a ts cell cycle mutant is defective in gene product function at the restrictive temperature the execution point defines the time in the cycle when the gene product completes its function. Mutants defective in the synthesis of gene product but not its function at the restrictive temperature have execution points which reveal the time in the cycle when enough gene product is made to complete cell division.

Ts cell cycle mutants were recognised as mutant cultures in which almost all cells arrested with a uniform morphology with respect to the cell cycle (the terminal phenotype) at the restrictive temperature. The terminal phenotype of a particular cdc mutant has subsequently been termed the diagnostic landmark of a mutant by Hartwell (1974).

The *cdc* mutants have been used to determine the dependency relationships of various cell cycle events by Hartwell *et al.,* (1974). They reasoned that it is possible, for instance, to determine whether nuclear division is dependent on the prior completion of DNA synthesis by shifting to the restrictive temperature a mutant which cannot complete DNA synthesis at this temperature and noting whether nuclear division occurs in the absence of DNA synthesis. This approach has shown that the yeast cell cycle consists of two dependent pathways (Fig. 1). Events on the inner and outer pathways (Fig. 1) are independent of each other except that the event mediated by the *cdc* 28 gene product is necessary for both the initiation of DNA synthesis and bud emergence and in addition, cytokinesis and cell separation cannot occur in the absence of either or both bud emergence and nuclear division.

α-factor and the cell cycle

Haploid cells of *a* and α mating types will, in appropriate circumstances, fuse to form diploid cells. Mating is restricted to unbudded cells; budded cells of opposite mating type in a mating type mixture complete cell division before mating (Hartwell, 1973; Reid & Hartwell, 1977). When haploid cells of opposite mating type are mixed together the proportion of budded cells decreases rapidly and that of unbudded cells increases before mating (Hartwell, 1973). This appears to be brought about by the action of hormone-like molecules secreted by the two haploids. α cells produce α-factor, a small polypeptide of 13 amino acids (Stotzler & Duntze, 1976; Stotzler, Kiltz & Duntze, 1976) which arrests *a* cells in the G1 phase of the cell cycle (Bucking-Throm, Duntze, Hartwell & Manney, 1973). Similarly, *a* cells produce *a*-factor which transiently arrests cells of mating type α in G1 (Wilkinson & Pringle, 1974; Betz, MacKay & Duntze, 1977). *a*-factor is less well characterised than α-factor and in general has received less attention than α-factor.

The mutual arrest of cells of *a* and α mating types provides a mechanism whereby cells can be synchronised prior to mating. Whether the mating factors have an additional role in mating is not clear although such a role is suggested by the observations that α-factor-resistant mutants are often non-maters (Manney & Woods, 1976) and that mutants of *a* mating type selected as non-maters are

Fig. 1. Dependent pathway of landmarks in the cell cycle of *Saccharomyces cerevisiae*. Events connected by an arrow are proposed to be related such that the distal event is dependent for its occurrence upon the prior completion of the proximal event. Numbers refer to *cdc* genes that are required for progress from one event to the next (Hartwell, 1978).

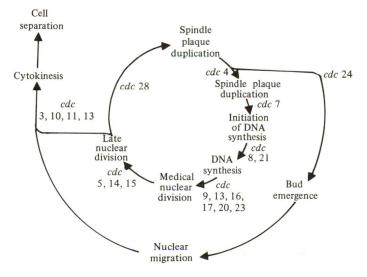

often α-factor-resistant (Mackay & Manney, 1974). If the sole function of mating factors was to synchronise cell cycles, mutants resistant to α-factor should mate albeit at reduced efficiency since only a small percentage of cells of either mating type in a mating mixture are unbudded.

The mating factors have been used as probes in studies on the control of cell division. When α-factor which can be extracted and purified from the culture filtrate of α cells (Bucking-Throm *et al.*, 1973; Stotzler & Duntze, 1976) is added to an asynchronous culture of a cells all the budded cells in the culture complete cell division. Most of the unbudded cells are arrested without forming

Fig. 2. Determination of the time in the cell cycle of the α-factor step for cells grown at different growth rates. Small cells were isolated from cultures of the haploid strain 2180-1A growing in (*a*) ethanol medium (mass doubling time 6.84 h) and (*b*) YEPD medium (mass doubling time 2.4 h) and were inoculated into similar media. The cultures were allowed to develop synchronously until the culture in ethanol medium contained 23% budded cells and the culture in YEPD medium contained 39% budded cells. Each culture was then subdivided and resuspended in two flasks and α-factor was added to one of them (■). The percentage of budded cells was determined at intervals for each flask. At least 1000 cells were counted for each sample and the standard error is given for each determination. The percentage of budded cells increased to a plateau value in cultures incubated with α-factor. This increase is due to a small percentage of cells being between the α-factor step and bud initiation at the time of α-factor

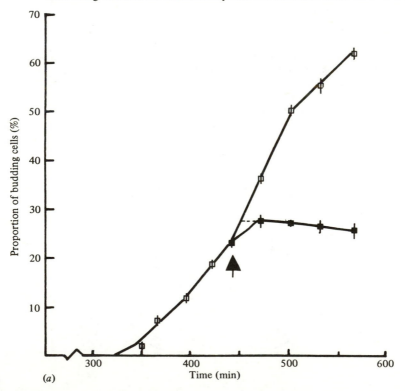

(*a*)

a bud although they continue to grow and cells become first pear-shaped and then dumbbell-shaped. These structures have been called shmoos (MacKay & Manney, 1974). A small proportion of unbudded cells produce a bud and divide in the presence of α-factor. This suggests that the last time in the cell cycle that cells are sensitive to α-factor (the α-factor step) is a short time before bud initiation. A. Lorincz & B. L. A. Carter (unpublished) place the α-factor step approximately 15 min before bud initiation in cells that are growing with mass doubling times of either 2.4 h or 6.8 h (Fig. 2). Hartwell and Unger (1977) also observed that the α-factor step occurred a constant time before bud initiation at

Caption to Fig. 2. (*cont.*)
addition. The length of time between the α-factor step and bud initiation (12.5 min in ethanol medium and 13 min in YEPD medium) is determined by calculating how long the control cultures take after the subdivision to reach the plateau value for percentage budded cells observed in the culture to which α-factor had been added. The arrow marks the time of the shift.

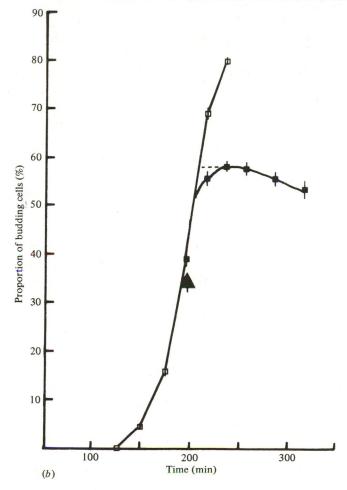

(*b*)

different growth rates. They placed the α-factor step approximately 25 min before bud initiation in cells growing on solid media where different growth rates were obtained by varying the concentration of the protein synthesis inhibitor cycloheximide.

A. Lorincz & B. L. A. Carter (unpublished) have examined whether cells are sensitive to α-factor for a large part of the unbudded phase of the cycle up to the α-factor step or whether there is only a discrete part of the unbudded phase during which cells are sensitive to α-factor. In cultures growing exponentially in rich medium (2.4 h mass doubling time) the unbudded phase of the cycle is very short but it takes at least 60 minutes to detect the morphological changes which follow α-factor treatment. Thus it is difficult to determine unambiguously whether all unbudded cells are susceptible to arrest by α-factor, or that cells only become affected by α-factor when they reach a certain size or stage of the unbudded phase of the cycle as all unbudded cells do in the 60 min period.

To distinguish between these two possibilities cells were grown to stationary phase on rich medium (YEPD) and small cells (12.5 μm^3) from the culture were selected by zonal centrifugation and placed in fresh medium. These cells were then exposed to α-factor and within 60 min cell shapes characteristic of α-factor arrest were observed. Even after 90 min in the presence of α-factor the volume of these cells (26.5 μm^3) is markedly less than that at which the α-factor step occurs (38.5 μm^3) in this medium. Thus cells are sensitive to α-factor for a considerable period and possibly all of the unbudded phase before the α-factor-sensitive step. It is likely that unbudded cells are sensitive to the action of α-factor until a particular cell cycle event has been completed late in the unbudded phase after which they are no longer sensitive to α-factor.

The α-factor step has been temporally sequenced in relation to the execution points of cdc 28, cdc 4 and cdc 7 (Hereford & Hartwell, 1974). Cells carrying the cdc 28 mutation were aligned at the α-factor step by incubation for 3 h in α-factor at 23 °C. The factor was then removed and cells were either shifted to 38 °C (the restrictive temperature for cdc 28) or maintained at 23 °C. The culture maintained at 23 °C underwent budding, DNA synthesis and cell division. Cells shifted to 38 °C remained unbudded. Thus the cdc 28 gene product did not complete its function while cells were arrested at the α-factor block. Cells were also aligned at the cdc 28 block at 38 °C, then shifted to 23 °C, the culture split and one sample given α-factor. Cells shifted to 23 °C without α-factor underwent DNA synthesis and bud initiation (although surprisingly no cell division was observed), whereas cells incubated with α-factor failed to produce a bud.

These results suggest that there is an event in the cell cycle at or before spindle plaque duplication (Fig. 1) that requires an adequate amount of the cdc 28 gene product for its completion and that in the presence of α-factor the event cannot be completed.

In other experiments it was shown that the products of two other genes cdc 4 and *cdc* 7 which are also necessary for the initiation of DNA synthesis are part of a dependent pathway and that the order of the steps mediated by these three genes products is

$$\underset{\alpha\text{-factor}}{\overset{cdc\ 28}{\longrightarrow}} \qquad \overset{cdc\ 4}{\longrightarrow} \qquad \overset{cdc\ 7}{\longrightarrow} \qquad \text{initiation of DNA synthesis.}$$

The action of α-factor on *a* cells indicates that the principle of specific cell–cell interactions mediated by hormones and chalones, that Prescott (1976) has suggested are used in the control of cell proliferation in multicellular organisms, is present in yeast. The peptide nature of serum hormones which modulate division in mammalian cells as described by Stiles, Cochran & Scher in this volume, illustrates the similarity. It is therefore attractive that genetic analysis of α-factor action is possible, and isolation of mutants resistant to the action of α-factor followed by complementation analysis will be useful in distinguishing and defining genetic lesions at each step in the pathway of hormone action. Thus yeast permits a genetic approach to the hormonal control of cell division in eukaryotes.

The cell cycle at different growth rates

Only comparatively recently has the cell cycle of yeast been examined at a variety of growth rates. The methods used to alter growth rate include nutrient limitation in continuous-flow culture (Meyenberg, 1968; Adams, 1977; Jagadish & Carter, 1977; Jagadish, Lorincz & Carter, 1977; Carter & Jagadish, 1978a,b; Johnston, Erhadt, Lorincz & Carter, 1979), variation in nutrient composition in batch media (Barford & Hall, 1976; Jagadish & Carter, 1977; Carter & Jagadish, 1978a,b; Jagadish *et al.*, 1977; Johnston *et al.*, 1979) and direct limitation of protein synthesis by growth in the presence of different concentrations of protein synthesis inhibitors (Hartwell & Unger, 1977).

These studies have shown that as the growth rate is slowed all phases of the cycle do not expand proportionately. The results of Meyenburg (1968) (although not quite correctly interpreted at the time) were the first to show that the lengthening of the cell cycle at slow growth rates is largely due to an expansion of the unbudded phase (G1) of the cell cycle although the time cells spend in the budded phase also increases slightly with increasing mass doubling time. Perhaps the most exhaustive analysis has been done by Hartwell & Unger (1977) who found that as cell cycle length increased at slow growth rate the major expansion in cycle time occurred in the time cells spent in the unbudded phase of the cycle but that the budded interval increased by 1.7 min for every 10 min increase in the mass doubling time of the culture. Jagadish & Carter (1977) have shown that the major expansion of cycle time at slow growth rates occurs in the unbudded interval that precedes the α-factor/*cdc* 28 step. The conclusion from these studies is that as culture doubling time increases, the increased cell cycle time can largely be accounted for by expansion of the time spent in the unbudded phase of the cycle prior to the α-factor step. The period from the α-factor step to division does however increase somewhat as mass doubling time increases.

Control of cell division at the *cdc* 28 event

Hartwell has suggested (Hartwell, 1974; Hartwell *et al.*, 1974) that overall control of cell division in the budding yeast is achieved in the G1 portion of the cell cycle at those steps that precede and include the step mediated by the *cdc* 28 gene product. It is envisaged that cells 'start' the cell cycle once they complete the *cdc* 28 event. A number of observations support this view. The *cdc* 28 step initiates and integrates two subsequently independent pathways namely DNA synthesis leading to nuclear division and bud emergence leading to nuclear migration (Hartwell, 1974; Hartwell *et al.*, 1974). When cultures are deprived of any one of several nutrients, cells arrest as unbudded cells (Johnston, Pringle & Hartwell, 1977). Byers & Goetsch (1975) have shown that such unbudded stationary phase cells have an unduplicated spindle plaque which is characteristic of cells arrested prior to the *cdc* 28/α-factor step. These results suggest that, as nutrients become limiting, cells only execute the *cdc* 28 step if they can subsequently complete cell division. This conclusion fits nicely with the observations described previously that only a slight increase in the interval from the α-factor step to division occurs over a wide range of growth rates. It is as if cells at slow growth rates take longer to prepare for 'start' but once this event is completed cells proceed to cell division in a time which varies little with growth rate.

Once cells progress beyond the *cdc* 28 step it has been claimed that they cannot mate (if they are haploids) or sporulate (if they are diploids) but are committed to cell division (Hartwell, 1974). Reid & Hartwell (1977) observed that cells arrested at the *cdc* 28 event were able to mate but cells arrested at the *cdc* 7 event and at various subsequent cdc mutant mediated events could not. Milne (quoted in Hartwell, 1974) observed that when budded diploid cells were challenged to undergo meiosis they complete cell division first. Similar shifts with unbudded cells revealed that a small percentage went on to divide but most cells initiated meiosis. The proportion of the latter was such that it was calculated that cells within ten minutes of bud initiation were committed to cell division. Hirschberg & Simchen (1977) found, however, that diploid cells arrested at the *cdc* 4 event could embark on meiosis without dividing first. Since the *cdc* 4 event is after the *cdc* 28 event this result indicates that the *cdc* 28 step is not the point of commitment to mitosis when cells are challenged to undergo meiosis.

It should be pointed out that although completing the event mediated by the *cdc* 28 gene has been claimed to 'start' the cell cycle it is possible that the *cdc* 28 event is simply the earliest event in the cycle in which so far a mutation has been isolated. 'Start' may occur at an earlier stage in the cycle at an event thus far not identified by a gene mutation. It is also important to note that commitment need not occur at 'start'. Indeed the results of Hirschberg & Simchen (1977) suggest that commitment to mitosis occurs after 'start'.

Co-ordination of growth and cell division

It has been suggested that cells require to attain a crucial size before a particular cell cycle event can be completed (Hartwell & Unger, 1977; Jagadish, et al., 1977; Johnston, Pringle & Hartwell, 1977; Carter, 1978; Carter & Jagadish 1978a). A number of observations point to the G1 phase as the site of this co-ordination. When very small cells isolated from a stationary phase culture were inoculated into batch media supporting different growth rates they initiated a bud at the size characteristic of cells in an exponentially growing culture in the same media (Johnston et al., 1977; Johnston, et al., 1979). Carter & Jagadish (1978b, and unpublished) separated cells grown on ethanol medium (mass doubling time 7.5 h) according to size and used two fractions of different size to inoculate two synchronous cultures in fresh ethanol medium. Cells of initial volume (13.5 μm^3) took 20 h for 50% of the cells to form a bud whereas cells of initial volume (16.8 μm^3) took 13 h. The mean cell volumes at bud initiation were 27.5 μm^3 and 26.9 μm^3 respectively. This is the size at bud initiation of age zero cells in an exponential culture growing in ethanol medium (Lorincz & Carter, 1979).

Although growth to a crucial size may be necessary for completion of a cell cycle event at or before bud emergence it is clear that cell size at bud initiation is dependent on growth rate (Johnston et al., 1979; Lorincz & Carter, 1979). Size at bud initiation in haploid cells varies with growth rate within the growth rate range 0.33 h^{-1} to 0.23 h^{-1} but at growth rates slower than 0.23 h^{-1} cells display a constant cell size at bud initiation independent of growth rate (Fig. 3). The same relationship between growth rate and size at bud initiation is observed in diploids: at all growth rates the ratio of diploid: haploid size at bud initiation is approximately 1.7 (Lorincz & Carter, 1979).

Regulation of cell size resides in the G1 phase of the cycle just prior to bud initiation (Lorincz & Carter, 1979). When unbudded cells from different stages of a synchronous culture in poor media (6.84 h mass doubling time – size at bud initiation 26.2 μm^3) were shifted to rich medium (2.4 h mass doubling time – size at bud initiation 41.6 μm^3) all cells except those very close to the size characteristic of bud production on poor medium go on to produce a bud at the size characteristic of the rich medium. Cells do not become committed to producing a bud at the size characteristic of poor medium until they are within 3 μm^3 of that size. All cells that are below the commitment size for poor medium at the time of shift to rich medium initiate their bud at the size characteristic of rich medium: no cells initiate a bud at a size intermediate between that of rich and poor medium. These results suggest that cells quickly sense their altered environment after a medium shift and that any progress cells make towards bud initiation on poor medium is irrelevant when they are shifted to rich medium unless they have reached the commitment point for bud initiation.

A. Lorincz & B. L. A. Carter (unpublished) have investigated whether the α-factor step precedes the commitment step, or vice versa, or occurs at the same

Fig. 3. Relationship between growth rate and size at bud initiation for a wild-type haploid strain (C 4,2) and a small mutant (*whi*-1). Growth rates were varied by glucose limitation in a chemostat. (▲), 0 bud scar class for C 4,2; (○), 1 bud scar class for C 4,2; (▼), 2 bud scar class for C 4,2; (□), 3 or more bud scar class for C 4,2; (■), 0 bud scar class for *whi*-1.

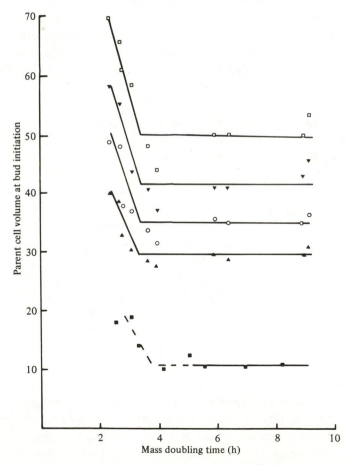

stage of the cycle. Synchronous cultures were grown in ethanol medium (a poor medium) until the proportion of budded cells reached 25% (Fig. 4). The culture was then divided into three parts: one part was replaced in fresh ethanol medium, another was placed in a rich medium (YEPD) and a third placed in YEPD medium containing α-factor (2 units ml^{-1}). The percentage of budded cells in ethanol medium increased in a manner consistent with synchronous growth in this medium. The proportion of budded cells in the culture shifted to YEPD medium increased from 25% at the time of the shift to 35% after which it remained constant before increasing again 30 min later. Our interpretation of these results is that cells which had passed the commitment point at the time of the shift continued to bud after the shift (at the pre-shift size). The plateau occurs

Fig. 4. Determination of the time in the cycle of the commitment step for size at bud initiation and the α-factor step. A synchronous culture of C 4,2 was allowed to grow in ethanol medium until the proportion of budded cells reached 25%. The culture was divided into three parts: one part was replaced in fresh ethanol medium (□), another was placed in YEPD medium (●) and a third placed in YEPD medium containing α-factor (2 units ml^{-1}) (■). The percentage of budded cells was determined at intervals for each flask. At least 1000 cells were counted for each sample and the standard error is given for each determination. The arrow marks the time of the division. See text for further details.

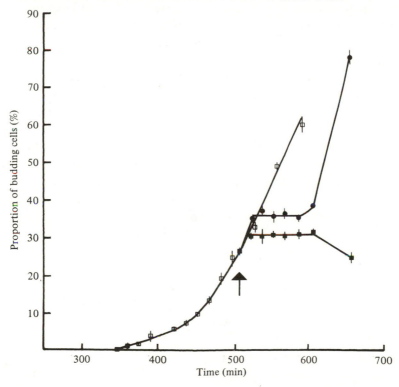

because unbudded cells which had not reached the commitment point when cells were shifted take time to reach the larger size at which bud initiation occurs in YEPD medium. When they reach this size the percentage of budded cells in the population increases again. In the culture shifted to YEPD medium plus α-factor the proportion of budded cells increased to 30% before decreasing. The increase is a result of those unbudded cells that are past the α-factor step at the time of the shift undergoing bud initiation after the shift. No cells that are before the α-factor step at the time of the shift initiate a bud thereafter and so the percentage of budded cells declines as such cells divide. The increase in budded cells after the shift to YEPD medium plus α-factor is 5%, which is less than the increase up to the plateau value (10%) in the culture incubated in YEPD medium alone. We conclude from these data that the commitment step occurs at an earlier stage of the cycle than the α-factor step. This conclusion is dependent on cells responding to the new medium with the same speed as they react to α-factor. If cells respond more slowly to a media shift then the commitment step will appear earlier in the cycle (relative to the α-factor step) than it should. We are investigating this further.

Studies on mutants that divide at a smaller size than normal have been important in establishing that controls exist to ensure the co-ordination of growth and cell division in the fission yeast (see Nurse & Fantes, this volume). Recently, B. L. A. Carter & P. E. Sudbery (unpublished) have devised a method which has been used to isolate such mutants in *S. cerevisiae*. Mutagenised cultures were grown to stationary phase, separated according to size by zonal centrifugation and small cells (approx. 15 μm^3) were inoculated into rich medium (YEPD) and allowed to develop into a synchronous culture at 37 °C. Previous experiments had shown that in this medium addition of α-factor prior to cells attaining a volume of 38.5 μm^3 resulted in cell cycle arrest with the consequent development of large aberrant cells. α-factor was added to the culture when the mean cell size reached 30 μm^3. It was reasoned that mutants which initiated a bud at a smaller size than normal, for instance at 25 μm^3, would have passed the α-factor step when the population reached 30 μm^3 whereas cell division in wild-type cells would be arrested. After 6 h incubation with α-factor we supposed that wild-type cells would become very large but small mutant cells would have divided once and be in the early stages of cell cycle arrest.

After such exposure to α-factor the culture was separated according to size and fractions containing the smallest cells were plated overnight at 37 °C. Colonies were then examined and those containing smaller than normal cells were reserved for further analysis. At the time of writing, nine mutants dividing at a smaller size than normal have been isolated but although the method was devised to permit isolation of ts small mutants all nine are small at 24 °C and 37 °C (Table 1). Another small mutant (*whi*-16) was isolated from a mutagenised culture grown at slow growth rates in the chemostat for three weeks. Genetic analysis of *whi*-1, *whi*-3, *whi*-7, *whi*-13, *whi*-16 has shown that all these segregate the small phenotype in a 2:2 manner when crossed to normal-sized cells. We conclude that in these mutants the small phenotype results from a single nuclear gene mutation. The heterozygous diploid (*whi*-1 × wild-type) is smaller than the diploid resulting from a cross between two wild-type sized strains showing *whi*-1 is codominant with the wild-type allele (Table 1). Two small segregants from a *whi*-1 × wild-type cross were mated to form a homozygous *whi*-1/*whi*-1 diploid. The size of this diploid was indistinguishable from the wild-type haploid (Table 1).

The size at bud initiation of *whi*-1 has been determined at various growth rates and it is evident that the relationship of size to growth rate in this mutant is similar to that obtaining in wild-type cells. The minimum cell size at slow growth rates is approx. 11 μm^3 whereas in wild-type cells it is approx. 30 μm^3 (Fig. 3). This indicates that the *whi*-1 mutation represents an alteration in one of the components of the co-ordination process rather than the elimination of such a component. Although the shape of the curve relating growth rate and size at bud

Table 1. *Cell size of parent cell portion of* S. cerevisiae *bearing buds.* (B. L. A. Carter & P. E. Sudbery, unpublished.)

Strain	Culture temperature (°C)	Cell volume (μm^3)	Cell mass (arbitrary units)
2180-1A	25	45.0 ± 3.0	11.5 ± 1.6
2180-1A	37	46.0 ± 7.0	—
whi-1	25	22.6 ± 1.3	5.4 ± 1.6
whi-1	37	21.5 ± 6.4	—
whi-3	25	19.5 ± 3.3	—
whi-7	25	21.7 ± 2.4	—
whi-13	25	23.5 ± 3.1	—
whi-16	25	23.1 ± 5.9	—
whi-21	25	32.6 ± 3.7	—
whi-51	25	26.0 ± 3.3	—
whi-52	25	30.5 ± 3.0	—
whi-59	25	27.9 ± 3.5	—
2180-1A × S30	25	75.0 ± 2.0	—
2180-1A × AN33	25	78.1 ± 8.3	—
whi-1 × S30	25	60.2 ± 7.6	—
whi-7 × AN33	25	62.4 ± 8.1	—
whi-1-1A × *whi*-1-1B	25	45.0 ± 2.7	—

initiation for wild-type haploids (Fig. 3) and diploids might suggest that different mechanisms are responsible for cell size at bud initiation in fast and slow-growing cells, the fact that a qualitatively similar relationship holds for the small mutant *whi*-1 indicates that only one mechanism is responsible for this relationship.

It is important to isolate more small mutants and to determine how many genes they define. It may be very few genes since three small mutants, which were isolated from separate cultures, have been subjected to genetic analysis by forming diploids between them and analysing at least twenty tetrads from each. No wild type progeny, which would result from recombination between the genes, were obtained, therefore the mutations are probably located in a single gene, *whi*-1 (Sudbery, Goodey & Carter, 1980).

We have also detected a type of mutant, *whi*-2, which affects the ability to control initiation of division under poor nutritional conditions. As starvation approaches, wild type cells do not initiate division and so they form stationary phase cultures of unbudded cells arrested in G1 phase. There may be parallels in the nutritional requirements to initiate division in mammalian cells (see Stiles, Cochran & Scher, this volume). Mutations of the gene *whi*-2 cause initiations of division to continue even under very poor growth conditions and cells then either just succeed in producing daughters, which are however smaller than under favourable growth conditions, or fail to complete the cycle, so remaining in the budded phase which has poor survival properties. These mutants are described by Sudbery *et al.* (1980).

Cell age and the cell cycle

In cells that divide by fission two identical cells result from cell division. *S. cerevisiae* has a budding mode of division and at cell separation the parent cell can be distinguished from the daughter (ex-bud) because the former retains a bud scar on its surface composed in part of chitin (Bacon, Davidson, Jones & Taylor, 1966; Cabib & Bowers, 1971) which is not present on daughter cells.

The age of a parent cell can be determined from the number of scars on its surface: each time a parent cell separates from a bud a scar is left on its surface. Bud scars can be visualised by fluorescent microscopy after staining with primulin (Streiblova & Beran, 1963) or calcofluor (Hayashibe & Katohda, 1973).

Studies on cells growing at different growth rates have revealed that there are marked differences in the cycle times of parents and daughters at slow growth rates (Adams, 1977; Hartwell & Unger, 1977; Jagadish *et al.*, 1977; Carter & Jagadish, 1978*a*). At fast growth rates the bud almost reaches the size of the parent at cell division and the parent and daughter resulting from division have similar cycle times. At slow growth rates the bud at division is much smaller than the parent cell and takes a much longer time to produce a bud than the parent in the subsequent cell cycle. Thus in a culture in which the population has a mass doubling time of 7 h, parents traverse the cycle in a considerably shorter period than 7 h and daughters take much longer than this to traverse the cycle.

These results can be attributed to the fact that the budded phase at slow growth rates is not much longer than at fast growth rates and therefore a bud can grow to a larger size in this time at fast growth rates than at slow growth rates. The long cycle time of daughters at slow growth rates would be expected if these small cells require to grow to a crucial size before they can complete a cell cycle event at or before bud initiation.

The disparity in size between daughter and parent cell at slow growth rates is not as great as would be expected from the above considerations alone. This is because size at bud initiation also declines at slow growth rates (see earlier). A diagrammatic representation of the cell cycle of parents and daughters at fast and slow growth rates is shown in Fig. 5.

If cells require to attain a crucial size before beginning a cell cycle then the

Fig. 5. Representation of the cell cycle of parents (P) and daughters (D) at fast (*a*) and slow (*b*) growth rates.

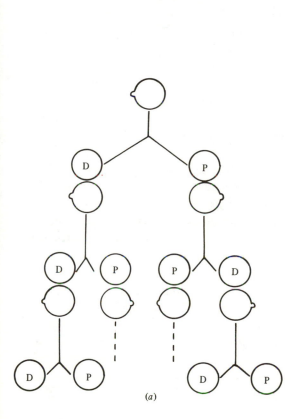

(*a*)

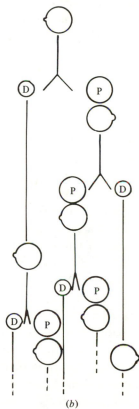

(*b*)

budding mode of division provides that once a cell has attained the crucial size it will in all subsequent cycles have fulfilled the size requirement for cell proliferation. Thus if size is the only requirement to start a cycle it seems reasonable to expect that after division a parent cell would produce its next bud without appreciable growth. This is not the case: the size of parent cells at bud initiation increases approximately 17% each generation (Hartwell & Unger, 1977; Johnston *et al.*, 1979), and since parent cells increase by only 7% during the budded period much of this increase must occur during the unbudded phase. Cell size at bud initiation increases by 17% with each generation at all growth rates investigated (Johnston *et al.*, 1979) which implies that parents spend longer in the unbudded phase of the cycle at slow growth rates than at fast growth rates.

Parent cells after division are not beyond the α-factor step. If they were, some unbudded cells (all the unbudded parents) in an exponential culture would be insensitive to α-factor and addition of α-factor to such a culture would not result in complete inhibition of division. A similar argument can be used to demonstrate that parent cells at division are not beyond the *cdc* 28 step. In addition A. Lorincz & B. L. A. Carter (unpublished) have observed that parents at division are not beyond the size commitment step. It is important to note that although it has been suggested that cells require to reach a crucial size it is unlikely that the cell monitors its own volume. It is however possible that the cell monitors, for instance, a specific protein whose synthesis is correlated with overall growth.

Further evidence that parent cells grow before the next bud initiation is provided by the observation that at all growth rates there is a significant percentage of unbudded parent cells in an exponential culture and that at slow growth rates parents spend a longer time unbudded than at fast growth rates although this trend is not as marked as that shown by daughters (Carter & Jagadish, 1978*a*; M. N. Jagadish & B. L. A. Carter, unpublished).

Recent results from this laboratory (J. R. Piggott & B. L. A. Carter, unpublished) indicate that the percentage of unbudded parents at any growth rate is independent of the generation age of parents. We have also observed that although at slow growth rates the ratio of daughters to parents increases (an indication that the generation time of daughters is longer than parents) the percentage of first generation parents is identical to the sum of subsequent generation parents at all growth rates indicating that the generation time of parents is independent of generation age. In other words the generation time of a parent that has produced one bud is the same as a parent cell that has produced two or more buds. Thus although there are major differences in the cycle time of parents and daughters at slow growth rates there are not differences between parents of different age. Similar conclusions have been reached by Hartwell & Unger (1977). Thus an exponential culture can be regarded as containing two sub-populations with respect to cell cycle behaviour, parents and daughters.

In summary, investigations of the cell cycle of cells of different ages support the hypothesis that cells require to grow to a crucial size before they complete a particular cell cycle event prior to bud initiation. These studies have also shown that earlier conclusions that at slow growth rates the unbudded phase of the cycle expands markedly are an over-simplification since there are two sub-populations under these growth conditions. At slow growth rates daughters resulting from division are small and have an extended unbudded phase. Parent cells resulting from division are capable of initiating a bud after a relatively short unbudded phase.

Summary

A number of observations suggest that completion of the *cdc* 28 event initiates the cell cycle programme and that once this programme is initiated cells can proceed to cell division in a time which varies little with growth rate. Cells can only complete the *cdc* 28 event if they can subsequently complete cell division in the environment in which they find themselves. Thus the major control over cell division operates at the time this process begins. Cells can only initiate this process (complete the *cdc* 28 event) if they grow to a crucial size. This crucial size varies with growth rate implying that cells are capable of monitoring their growth rate.

I would like to thank Peter Fantes and Paul Nurse for reading and criticizing this manuscript. A postal dispute during the period of writing this review has meant that recent articles by my colleagues have not been available to me. Any omission of their work is not due to caprice on my part but ignorance of their results. Research in my laboratory is supported by grants from the Medical Research Council of Ireland and the Irish Cancer Society.

References

Adams, J. (1977). The interrelationship of cell growth and division in haploid and diploid cells of *Saccharomyces cerevisiae*. *Experimental Cell Research*, **106**, 267–75.

Bacon, J. S. D., Davidson, E., Jones, D. & Taylor, I. F. (1966). The location of chitin in the yeast cell wall. *Biochemical Journal*, **101**, 36C–38C.

Barford, J. P. & Hall, R. J. (1976). Estimation of the length of cell cycle phases from asynchronous cultures of *Saccharomyces cerevisiae*. *Experimental Cell Research*, **102**, 276–84.

Betz, R., MacKay, V. L. & Duntze, W. (1977). *a* factor from *Saccharomyces cerevisiae:* partial characterization of a mating hormone produced by cells of mating type *a*. *Journal of Bacteriology*, **132**, 462–72.

Bucking-Throm, E., Duntze, W., Hartwell, L. H. & Manney, T. R. (1973). Reversible arrest of haploid yeast cells at the initiation of DNA synthesis by a diffusible sex factor. *Experimental Cell Research*, **76**, 99–110.

Byers, B. & Goetsch, L. (1975). Behaviour of spindles and spindle plaques in the cell cycle and conjugation of *Saccharomyces cerevisiae*. *Journal of Bacteriology*, **124**, 511–23.

Cabib, E. & Bowers, B. (1971). Chitin and yeast budding. Localization of chitin in yeast bud scars. *Journal of Biological Chemistry*, **246**, 152–9.

Carter, B. L. A. (1978). The yeast nucleus. *Advances in Microbial Physiology*, **17**, 244–97.

Carter, B. L. A. & Jagadish, M. N. (1978*a*). The relationship between cell size and cell division in the yeast *Saccharomyces cerevisiae*. *Experimental Cell Research*, **112**, 15–24.

Carter, B. L. A. & Jagadish, M. N. (1978*b*). Control of cell division in the yeast *Saccharomyces cerevisiae* cultured at different growth rates. *Experimental Cell Research*, **112**, 373–83.

Halvorson, H. O., Carter, B. L. A. & Tauro, P. (1971). Synthesis of enzymes during the cell cycle. *Advances in Microbial Physiology*, **6**, 47–106.

Hartwell, L. H. (1973). Synchronization of haploid yeast cell cycles, a prelude to conjugation. *Experimental Cell Research*, **76**, 111–17.

Hartwell, L. H. (1974). *Saccharomyces cerevisiae* cell cycle. *Bacteriological Reviews*, **38**, 164–98.

Hartwell, L. H. (1978). Cell division from a genetic perspective. *Journal of Cell Biology*, **77**, 627–37.

Hartwell, L. H., Culotti, J., Pringle, J. R. & Reid, B. J. (1974). Genetic control of the cell division cycle in yeast; a model. *Science*, **183**, 46–51.

Hartwell, L. H., Culotti, J. & Reid, B. (1970). Genetic control of the cell division cycle in yeast. I. Detection of mutants. *Proceedings of the National Academy of Sciences, USA*, **66**, 352–9.

Hartwell, L. H., Mortimer, R. K., Culotti, J. & Culotti, M. (1973). Genetic control of the cell division cycle in yeast. 5. Genetic analysis of cdc mutants. *Genetics*, **74**, 267–86.

Hartwell, L. H. & Unger, M. W. (1977). Unequal division in *Saccharomyces cerevisiae* and its implications for the control of cell division. *Journal of Cell Biology*, **75**, 422–35.

Hayashibe, M. & Katohda, S. (1973). Initiation of budding and chitin ring. *Journal of General and Applied Microbiology*, **19**, 23–39.

Hereford, L. M. & Hartwell, L. H. (1974). Sequential gene function in the initiation of *S. cerevisiae* DNA synthesis. *Journal of Molecular Biology*, **84**, 445–61.

Hirschberg, J. & Simchen, G. (1977). Commitment to the mitotic cell cycle in yeast in relation to meiosis. *Experimental Cell Research*, **106**, 245–52.

Jagadish, M. N. & Carter, B. L. A. (1977) Genetic control of cell division in yeast cultured at different growth rates. *Nature, London*, **269**, 145–47.

Jagadish, M. N., Lorincz, A. & Carter, B. L. A. (1977). Cell size and cell division in yeast cultured at different growth rates. *FEMS Microbiology Letters*, **2**, 235–7.

Johnston, G. C., Erhardt, C. W., Lorincz, A. & Carter, B. L. A. (1979). Regulation of cell size in the yeast *Saccharomyces cerevisiae*. *Journal of Bacteriology*, **137**, 1–5.

Johnston, G. C., Pringle, J. R. & Hartwell, L. H. (1977). Coordination of growth with cell division in the yeast *Saccharomyces cerevisiae*. *Experimental Cell Research*, **105**, 79–98.

Lorincz, A. & Carter, B. L. A. (1979). Control of cell size at bud initiation in *Saccharomyces cerevisiae*. *Journal of General Microbiology*, **113**, 287–95.

Manney, T. R. & Woods, V. (1976). Mutants of *Saccharomyces cerevisiae* resistant to the α mating factor. *Genetics*, **82**, 639–44.

MacKay, V. & Manney, T. R. (1974). Mutation affecting sexual conjugation and related processes in *Saccharomyces cerevisiae*. I. Isolation and phenotypic characterization of non-mating mutants. *Genetics*, **76**, 255–71.

Meyenburg, H. K. von (1968). The budding cycle of *Saccharomyces cerevisiae*. *Pathologia Microbiologica*, **31**, 117–27.

Mitchison, J. M. (1958). The growth of single cells. II. *Saccharomyces cerevisiae*. *Experimental Cell Research*, **15**, 214–21.

Prescott, D. M. (1976). *Reproduction of Eukaryotic Cells*, New York: Academic Press.

Reid, B. J. & Hartwell, L. A. (1977). Regulation of mating in the cell cycle of *Saccharomyces cerevisiae*. *Journal of Cell Biology*, **75**, 355–65.

Sebastian, J., Carter, B. L. A. & Halvorson, H. O. (1971). Use of yeast populations fractionated by zonal centrifugation to study the cell cycle. *Journal of Bacteriology*, **108**, 1045–50.

Shulman, R. W., Hartwell, L. H. & Warner, J. R. (1973). Synthesis of ribosomal proteins during the yeast cell cycle. *Journal of Molecular Biology*, **73**, 513–25.

Simchen, G. (1978). Cell cycle mutants. *Annual Review of Genetics*, **12**, 161–91.

Slater, M. L., Sharrow, S. O. & Gart, J. J. (1977). Cell cycle of *Saccharomyces cerevisiae* in populations growing at different rates. *Proceedings of the National Academy of Sciences, USA*, **74**, 3850–4.

Stotzler, D. & Duntze, W. (1976). Isolation and characterization of four related peptides exhibiting α-factor activity from *Saccharomyces cerevisiae*. *European Journal of Biochemistry*, **65**, 257–62.

Stotzler, D., Kiltz, H-H. & Duntze, W. (1976). Primary structure of α-factor peptides from *Saccharomyces cerevisiae*. *European Journal of Biochemistry*, **69**, 397–400.

Streiblova, E. & Beran, K. (1963). Demonstration of yeast scars by fluorescent microscopy. *Experimental Cell Research*, **30**, 603–5.

Sudbery, P. E., Goodey, A. R. & Carter, B. L. A. (1980). Genes which control cell proliferation in the yeast *Saccharomyces cerevisiae*. *Nature, London*, **288**, 401–4.

Wilkinson, L. E. & Pringle, J. R. (1974). Transient G1 arrest of *S. cerevisiae* cells of mating type α by a factor produced by cells of mating type *a*. *Experimental Cell Research*, **89**, 175–87.

C.D.STILES, B.H.COCHRAN & C.D.SCHER

Regulation of the mammalian cell cycle by hormones

Introduction

The goal of this chapter is to present a broad overview of the mammalian cell cycle. Where possible, we have attempted to compare and contrast the growth cycle of mammalian cells to that of lower organisms, both eukaryotic and prokaryotic. As we will demonstrate, the general features of the mammalian cell cycle are qualitatively similar to those described for lower eukaryotes. For this reason, we have focused on two aspects of the cell cycle that are unique to mammalian cell biology — namely, the regulation of the cycle by hormones and the effects of malignant transformation on regulation of the mammalian cell cycle.

The most detailed analyses of the mammalian cell cycle have been conducted with tissue culture systems. Fibroblast cells have been particularly well-studied because (i) the role of serum in control of fibroblast growth is well-documented, (ii) a family of well-characterized RNA and DNA tumor viruses are available, which serve as model carcinogens for fibroblasts, (iii) fibroblast cells are readily cultured *in vitro*. For these reasons, this review of the mammalian cell cycle emphasizes experimental findings drawn from the analysis of fibroblast cell growth. Where possible, however, we have cited parallel studies conducted with hematopoietic and epithelioid cells both *in vitro* and *in vivo*.

The cell cycle during proliferative growth: M, G1, S and G2 phases

During proliferative growth, mammalian cells replicate their cytoplasmic components and their DNA at separate times. The process of mammalian cell division has been classically described as a cycle (Howard & Pelc, 1953; Baserga, 1968; Mueller, 1969; Pardee, Dubrow, Hamlin & Kletzien, 1978). Using the physical separation of two daughter cells (mitosis) as a point of reference, the first stage in the cell cycle is seen as an interval of protein and ribonucleic acid synthesis termed 'G1'. The G1 phase of the cell cycle is followed by a period of preparation for mitosis termed 'G2'. During G2, proteins are synthesized, including tubulin, a component of the mitotic apparatus (Snyder & McIntosh, 1976). Following G2, the cell (containing what is now a tetraploid content of DNA) undergoes mitosis ('M' phase), followed by cytokinesis and the cycle begins anew. There is considerable variability in the transit time of individual cells through the cycle. As is the case with some lower eukaryotes such as budding yeast, most of the variability in the mammalian cell cycle is confined to the G1 component (see chapter by Brooks, in this volume).

These four components of proliferative mammalian cell growth (M, G1, G2 and S) together with the intrinsic variability in cell cycle transit times, have been documented both for whole mammalian tissues (liver, bone marrow) as well as for fibroblast cells and epithelioid cells in tissue culture (Lamerton & Fry, 1963; Lajtha, Gilbert, Porteus & Alexander, 1964; Post & Hoffman, 1965; Baserga, 1976). These features of mammalian cell proliferation are not qualitatively different from the cell cycle of other eukaryotic organisms including plants, protozoans and yeast (Mitchison, 1971; Hartwell, Culotti, Pringle & Reid, 1974; Shilo, Simchen & Pardee, 1978). The growth cycle of mammalian cells, and in eukaryotes in general, does differ qualitatively from the proliferative process of prokaryotic cells. Prokaryotic cells seem to replicate all of their cellular components co-ordinately during growth rather than at temporally distinct intervals as noted with most eukaryotic cells (Kuempel & Pardee, 1963; Helmstetter, 1969; Lark, 1969; Nierlich, 1978).

Resting cells: the G0 state
Mammalian cells display a much wider range of chemical and physical prerequisites for growth than lower eukaryotes and prokaryotic cells. These growth requirements can be most readily discerned with tissue culture cell systems where components can be added to, or withdrawn from the growth medium. Early passage embryo fibroblast cultures from chickens, rodents, and humans, as well as density-inhibited fibroblast cell lines of the 3T3 genre display three categories of growth requirements *in vitro:* nutritional, physical and hormonal (Brooks, 1975; Holley, 1975; Rothblat & Cristofalo, 1972). Nutritional growth requirements manifest themselves as a need for amino acids such as leucine and isoleucine (Ley & Tobey, 1970; Pohjanpelto & Raine, 1972), trace metals (zinc, selenium) (Rubin, 1972; McKeehan, Hamilton & Ham, 1976) and also fatty acids (Holley, Baldwin & Kiernan, 1974). Physical growth requirements include a need for 'anchorage' to a solid flat substrate (Macpherson & Montaignier, 1964; Stoker, 1968). Hormonal growth requirements are manifested as a need for a variety of small polypeptide growth factors which can be isolated from serum and from other animal tissues (See reviews by Brooks, 1975; Holley, 1975; Gospodarowicz & Moran, 1976).

When normal fibroblast cells are suddenly deprived of some component which is necessary for growth *in vitro* (as for example by reduction of the serum content in the culture medium) the proliferation of the cell population comes to a halt. However, unless the deprivation is drastic, the cells do not immediately cease growth-related activities; rather those cells that are in late G1 phase, S phase and G2 will proceed about their business until they have passed through mitosis. After mitosis, division-related activities are slowed considerably and a growth-arrested population with a diploid (G1) content of DNA is attained (Rothblat & Cristofalo, 1972; Nilhausen & Green, 1965).

The growth of serum or nutrient-starved fibroblast cells may be initiated by re-addition of the missing components to the culture medium. Curiously, the minimum lag time between re-addition of the missing components and the entry of the first cell into the S phase exceeds the length of time required for traversal of G1 during proliferative growth (Temin, 1971). Therefore, it has been necessary to propose that under conditions which are sub-optimal for growth, normal fibroblast cells enter a special resting state which lies outside the proliferative region of the cell cycle. This resting state has been variously termed 'quiescence', 'G0' or the 'A' state (Lajtha, 1963; Epifanova & Terskikh, 1969; Smith & Martin, 1973). In this volume Brooks suggests that cells can leave this quiescent state randomly and then embark on a process L, after which they may begin division processes. For purposes of this review, we will use the term G0 to describe the quiescent state of the cell cycle. The entire mammalian cell cycle including this 'G0' state is illustrated schematically in Fig. 1.

Fig. 1. Schematic view of the mammalian cell cycle. The time intervals indicated for G1, S, G2 and M represent the average duration of these phases in exponentially growing cultures of Balb/c-3T3 cells as measured by the method of Yen & Pardee (1978). When density-arrested G0 cultures of Balb/c-3T3 cells are stimulated to divide by the addition of fresh serum, the minimum lag time until the onset of DNA synthesis is 12 h. (See Pledger *et al.*, 1977; also Fig. 6.)

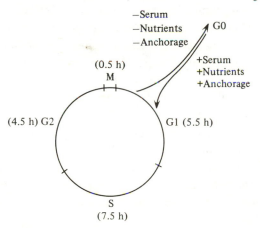

The G0 state can be examined from a phylogenetic point of view. The cells of lower eukaryotes behave similarly to mammalian cells under conditions that are restrictive for growth; plants, yeasts and protozoans enter a 'G0' state under restrictive conditions which can be distinguished kinetically and (in some cases chemically) from G1. Curiously, bacteria, which do not exhibit a temporally distinct G1, S or G2 phase in their cell cycle, are capable of entering a quiescent state under conditions which are restrictive for growth. Bacterial quiescence is superficially quite similar to the G0 state in mammalian cells (Keumpel & Pardee, 1963; Schaechter, 1968; Mitchison, 1971).

The biochemical basis of bacterial quiescence has been well-defined in contrast to the situation with eukaryotic cells. When bacteria which are autotrophic for an amino acid are deprived of that essential nutrient, there is a rapid, co-ordinated repression of protein synthesis, DNA synthesis and ribosomal RNA synthesis. This slowing of synthesis in bacteria is controlled by raised levels of guanosine tetraphosphate. 'Relaxed' mutants of bacteria do not synthesize guanosine tetraphosphate in response to nutrient starvation and do not demonstrate the stringent repression of metabolic activity (Lazzarini, Cashel & Gallant, 1971; Block & Haseltine, 1974). Hershko, Mamont, Shields & Tomkins (1971) observed that entry into (and exit from) the G0 state in mammalian cell growth regulation resembles the stringent control response of bacteria. However, neither mammalian cells nor any other eukaryotic cell have been shown to contain guanidine tetraphosphate (Stanners & Thompson, 1974; Thammana, Buerk & Gordon, 1976).

Control of cell growth by serum hormones

As the technology of animal cell culture was evolving in the forties and fifties, a major goal was to develop a chemically defined growth medium for animal cells. However, the early workers soon discovered a puzzling phenomenon: no matter how many amino acids, vitamins and other low molecular weight nutrients they included in their formulations, mammalian cells refused to proliferate unless the culture medium was supplemented with animal serum (Eagle, 1955). Animal cells which were derived from tumors had a lower serum growth requirement than cells taken from normal tissue. Moreover, when normal animal cells were exposed *in vitro* to oncogenic viruses or chemicals, the serum requirement for growth was substantially reduced (see Brooks, 1975 for review). In subsequent years, it has been shown that serum contains many components which function in different ways to support cell growth. Some low molecular weight components including amino acids, lipids, and metals are required in trace amounts for cellular growth and may be provided by the serum supplement to tissue culture medium. The transmembrane transport of low molecular weight nutrients may be facilitated by carrier proteins in serum such as transferrin, which is an iron carrier (Cedarblad, 1979). Finally serum contains factors which are not required in any nutritional sense for cell growth, but rather play a regulatory role in cell proliferation. These growth regulatory factors contained in serum constitute a newly recognized class of hormones (see review by Gospodarowicz & Moran, 1976). This point was driven home dramatically by the work of Sato and his colleagues (Hayashi & Sato, 1976) who cultured animal cells in a chemically defined medium using combinations of pure hormones in lieu of animal sera to supplement the tissue culture medium.

In recent years the serum hormones which modulate fibroblast growth have been relatively well-defined. A critical growth regulatory hormone for fibroblasts is the newly discovered platelet-derived growth factor (PDGF) (Ross, Glomset, Kariya & Harker, 1974; Kohler & Lipton, 1974; Westermark & Wasteson, 1976), and which has recently been purified to homogeneity (Antoniades, Scher & Stiles, 1979). PDGF is a heat stable, cationic polypeptide hormone with a molecular weight in the vicinity of 37 000 (by the criterion of SDS gel electrophoresis under non-reducing conditions). Pure preparations of PDGF stimulate division of fibroblast cells at concentrations as low as 10^{-10} M (Antoniades *et al.*, 1979). Among the polypeptide hormones, PDGF is unique in that it does not travel through the blood in a soluble form; rather, PDGF is sequestered within the α-granules of circulating blood platelets (Kaplan *et al.*, 1979). When blood platelets come into contact with traumatized tissue, the contents of the α-granules are secreted. Thus, PDGF may function as a 'wound hormone' *in vivo* which is delivered specifically to the site of damaged tissue and released by blood platelets (see Scher, Shepard, Antoniades & Stiles, 1979 for review).

Animal serum, which is obtained from clotted blood, is a rich source of PDGF. Normal animal fibroblasts grow very efficiently in culture medium which is supplemented with animal serum. Platelet-poor plasma can be prepared from unclotted animal blood by centrifuging away the cellular components. Platelet-poor plasma is chemically almost identical to clotted platelet-rich serum except that it lacks PDGF. Normal animal fibroblasts do not proliferate in culture medium which has been supplemented with platelet-poor plasma (Fig. 2) (Balk, Whitfield, Youdale & Braun, 1973; Ross *et al.*, 1974; Kohler & Lipton, 1974, Scher *et al.*, 1978). When pure or partially pure preparations of PDGF are added to a culture medium containing platelet-poor plasma, normal animal fibroblasts can resume growth. Thus serum, which is required for the growth of normal fibroblast cells, may be conceived as the summation of PDGF plus platelet-poor plasma (Fig. 3).

Purified PDGF is not the only hormone needed for the growth of normal animal fibroblast cells. Pledger, Stiles, Antoniades & Scher (1977) found that platelet-poor plasma component of serum plays an active role in fibroblast cell growth and contains still other hormones which function synergistically with PDGF to promote an optimum growth response. The synergistic relationship between PDGF and plasma in promotion of fibroblast growth is illustrated in Fig. 4. Stiles *et al.* (1979*a*) found that plasma from hypophysectomized donors was deficient in its ability to complement PDGF (Fig. 5). These workers found that somato-medin C, an insulin-like polypeptide hormone whose concentration in blood is modulated largely by pituitary growth hormone, was a key factor in co-mitogenesis with PDGF. Somatomedin C belongs to a family of closely related insulin-like polypeptides, which has been postulated to mediate the actions of

Fig. 2. Growth of normal and SV40-transformed mouse fibroblast cells in serum or in plasma-supplemented culture medium. Early passage cultures of Balb/c-3T3 embryo fibroblasts (*a*), the density-inhibited Balb/c-3T3 cell line (*b*), and the SVT2 line of SV40-transformed Balb/c-3T3 embryo fibroblasts (*c*) were cultured in medium supplemented with serum (●) or an equivalent concentration of plasma (○). (From Scher *et al*, 1978.)

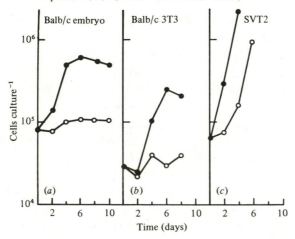

Fig. 3. Schematic illustration of the relationship between clotted blood serum, platelet-poor plasma and the platelet-derived growth factor (PDGF). Serum, which is customarily used as a culture medium supplement, supports the growth of normal fibroblast cells while plasma does not (see Fig. 2); this is because serum contains PDGF, a polypeptide growth factor, which is absent in platelet-poor plasma. PDGF is contained within the α-granules of blood platelets and is discharged into serum when blood clots.

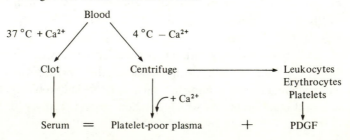

pituitary growth hormone on developmental growth (Salmon & Daughaday, 1957; Van Wyk & Underwood, 1978).

The role of hormones in growth control of mammalian cells is a feature that sets them qualitatively apart from lower eukaryotes such as yeast and from pro-karyotic cells. In lower organisms, cell proliferation is modulated by the avail-able nutrient supply (Schaechter, 1968; Mitchison, 1971; Nierlich, 1978). Mam-mals have homeostatic mechanisms which maintain the nutritional content of blood at a relatively constant level. For this reason a sophisticated set of chemical controls (hormones) has evolved which regulate cell proliferation although they are not required in any nutritional sense for cell growth. The role of hormones in modulation of animal cell growth has been outlined here in detail for fibroblast cells; however, hemopoietic and epithelioid tissues in mammals are also under hormonal growth control (Gospodarowicz & Moran, 1976).

Control of the mammalian cell cycle by serum hormones

In the previous section, we reviewed the role of serum in the growth regulation of normal fibroblast cells. A major role of serum is to provide hor-mones which modulate cell division. Studies on the fibroblast cell cycle by many

Fig. 4. Both PDGF and plasma are needed for normal fibroblast cells to undergo an optimum mitogenic response. Quiescent G0 cultures of Balb/c-3T3 cells were exposed to $100\,\mu g$ ($\bullet$), $10\,\mu g$ ($\bigcirc$) $5\,\mu g$ ($\triangle$) or $0\,\mu g$ (x) of partially purified PDGF together with the indicated concentration of platelet-poor plasma. The culture medium contained $5\,\mu$ Ci ml^{-1} of [^{3}H]dThd. After 36h the cultures were fixed and the percentage of the cells to have entered the S phase was determined by autoradiography. The percentage of radio-labeled nuclei is increased some-what by plasma alone or by the higher concentrations of PDGF in the absence of plasma; however, PDGF and plasma together have a synergistic action in pro-moting DNA synthesis. (From Pledger et al., 1977).

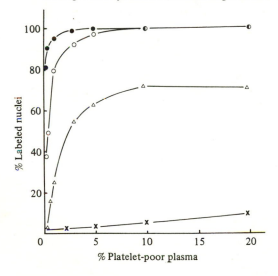

investigators have shown that serum seems to modulate events in the G0 or G1 phase of the cell cycle. For this reason, it seemed logical to examine this portion of the fibroblast cell cycle in an analysis of the action of serum growth hormones. Some of the most detailed studies of the cellular growth response to serum hormones such as PDGF have been conducted with mouse 3T3 cells *in vitro*.

When 3T3 cells are grown to high density without a change of culture medium, the serum is depleted of growth regulatory factors and the cells become growth arrested in G0 (Todaro, Lazar & Green, 1965; Jainchill & Todaro, 1970). Growth can be re-initiated by the addition of fresh culture medium containing whole serum (which contains both PDGF and plasma). However, neither PDGF nor plasma alone can efficiently promote exit from the G0 phase of the cell cycle and subsequent DNA synthesis (Pledger *et al.*, 1977; Pledger, Stiles, Antoniades & Scher, 1978).

When confluent G0 arrested cultures of Balb/c-3T3 cells are stimulated to divide again by the addition of fresh serum to the growth medium, the first cells do not begin to replicate their DNA until after a minimum lag time of 12 h. Following this, the cells begin to enter the S phase with kinetics that approximate

Fig. 5. Plasma prepared from hypophysectomized rats is deficient in 'progression activity'. Quiescent, G0-arrested Balb/c-3T3 cells were exposed to PDGF and then incubated in medium containing the indicated concentration of plasma from either normal (○) or from hypophysectomized (●) rats. 5 μCi ml^{-1} of [^{3}H]dThd were included in the culture medium. After a 24 h incubation at 37 °C the cultures were fixed and the percentage of cells to enter the S phase was determined by autoradiography. (From Stiles *et al.*, 1979*a*.)

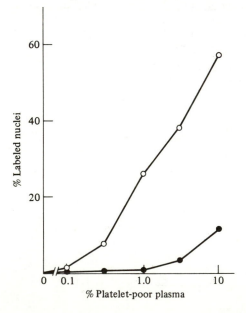

a first order process. By separating serum into its component parts (PDGF and plasma) it has been possible to examine the temporal relationship of hormone modulated growth regulatory events. Pledger *et al.* (1977) found that a transient exposure to the PDGF component of whole serum was sufficient to evoke replicative DNA synthesis in confluent, G0 arrested Balb/c-3T3 cells. 'Pulses' of PDGF as brief as 30 min were sufficient to induce replicative DNA synthesis following a 12 h lag (Fig. 6); however, PDGF-induced DNA synthesis was contingent upon subsequent addition of platelet-poor plasma to the cell culture medium. Whereas PDGF was only needed transiently, the platelet-poor plasma components of whole serum were needed continually for an optimum mitogenic response (Pledger *et al.*, 1978). PDGF appears to induce the first event in the cell cycle of rodent fibroblasts. Treatment of G0 arrested Balb/c-3T3 cells in plasma prior to the addition of PDGF does not shorten the lag phase until DNA synthesis. Such a shortening might be expected if plasma controlled the first replicative events.

When 3T3 cells were treated with PDGF and transferred to medium which lacked plasma, they did not enter the S phase. However, PDGF-treated cells remained 'competent' to replicate their DNA even after withdrawal of PDGF. This state of competence persisted for at least 13 h after removal of PDGF. When platelet-poor plasma was re-added to PDGF-treated competent cells, progression through the G0 and G1 phases was initiated and cells began to enter the S phase after a lag of 12 h (Fig. 6) (Pledger *et al.*, 1977).

Distinct growth regulatory points have been noted in the plasma mediated 'progression sequence' of G1. PDGF-treated cells were treated with plasma for various lengths of time and then transferred back to a culture medium that lacked plasma. After intervals, plasma was then added back to the culture medium and the cells entered the S phase. The lag period between the final addition of plasma and the onset of replicative DNA synthesis in these experiments was determined by the length of the initial exposure to plasma. (*a*) PDGF-treated competent cells that were incubated with plasma for 5 h during the initial exposure, remained competent, but did not progress through G0/G1. These cells began DNA synthesis 12 h after the final addition of plasma. (*b*) Cells treated with plasma for 10 h before withdrawal, initiated DNA synthesis within 6 h following the final addition of plasma; evidently, these cells became growth arrested at a point midway through G0/G1 – 6 h before S phase. (*c*) Cells treated with plasma for 15 h during the first period began replicative DNA synthesis immediately following the final addition of plasma; evidently, these cells were growth-arrested at a point immediately prior to the G1/S phase boundary. Thus, these studies indicated that there are at least four growth regulatory points in the G0/G1 phase of the Balb/c-3T3 cell cycle. (1) G0 'incompetent' cells; growth arrested 12 h prior to the G1/S phase boundary. (2) PDGF-treated 'competent' cells; growth arrested 12 h prior

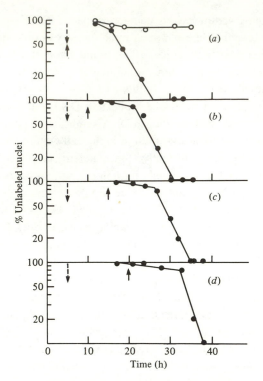

Fig. 6. The state of 'competence' induced by PDGF is stable. (a) Quiescent, G0-arrested cultures of Balb/c-3T3 cells were incubated with 50 μg of a partially purified PDGF preparation in culture medium containing 5% human plasma and 5 μCi ml^{-1} of [^{3}H]dThd. At the indicated times, cultures were fixed and processed for autoradiography. The data are plotted as the logarithm of cells remaining in G1 versus time after the fashion of Smith & Martin (1973). A minimum lag time of 12 h transpires before the first cells enter the S phase following treatment with PDGF and plasma together. (b–e) As above except the cultures were treated with 50 μg of the PDGF preparations only for 5 h. At the end of 5 h (↓) the PDGF medium was removed. The cell monolayers were washed twice with saline and returned to fresh culture medium containing 5 μCi ml^{-1} of [^{3}H]dThd but no PDGF or plasma. At the times indicated (↑) the medium was supplemented with an optimal (5%) (●) or a suboptimal (0.25%) (○) concentration of plasma. The cultures were fixed and processed for autoradiography at the times indicated. The experiments show (i) that PDGF-treated cells require an optimal concentration (5%) of plasma to leave G0/G1 and enter the S phase (see also Fig. 5), (ii) that PDGF is not required continually during the 12 h lag time that precedes the onset of DNA synthesis and (iii) delaying the addition of plasma to PDGF-treated cells delays the onset of DNA synthesis by a corresponding time. PDGF-treated cells remain 'competent' to replicate their DNA for as long as 13 h after exposure to PDGF has been terminated (e). (From Pledger et al., 1977.)

to the G1/S phase boundary. (3) PDGF and plasma-treated cells; growth arrested 6 h prior to the G1/S phase boundary. (4) PDGF and plasma-treated cells; growth arrested at the G1/S phase boundary. These growth arrest points for Balb/c-3T3 cells are illustrated schematically in Fig. 7.

Relationship between nutrients and serum factors in control of the cell cycle

Normal animal fibroblasts require both serum and low molecular weight nutrients for growth in culture. When either serum or essential nutrients are withdrawn, normal animal fibroblasts respond in the same way – by becoming growth-arrested in the region of the cell cycle with G1 DNA content. Because the growth arrest mediated by serum or nutrient withdrawal is similar, Holley (1972) speculated that the growth factors found in serum function to promote the transport of nutrients into the cell. Pardee (1974) described a 'restriction control point' in the G0/G1 phase of the cell cycle in hamster fibroblasts, which is sensitive to both nutrients and calf serum.

Fig. 7. Model depicting several sequential events in the G0/G1 phases of the cell cycle that precede DNA synthesis in Balb/c-3T3 cells. Under conditions of high cell density or serum starvation, Balb/c-3T3 cells become growth arrested in G0, a point at least 12 h prior to the G1/S phase boundary. Exposure to PDGF renders G0 cells 'competent' to replicate their DNA; however, PDGF-treated competent cells do not begin to progress through the 12 h chain of events leading to DNA synthesis until they are exposed to other growth factors contained in plasma. The 'progression' factors contained in plasma control at least two regulatory events denoted V and W located, respectively, 6 h prior to and immediately prior to the G1/S phase boundary. As cells traverse the W control point they next pass through a transition $(\frac{T}{c})$ to a state 'committed' to DNA replication in which they are relatively independent of exogenous growth factors until after mitosis (from Pledger et al., 1978). Further studies have established a linkage between the V and W control events and the nutritional content of the tissue culture medium (Stiles et al., 1979b).

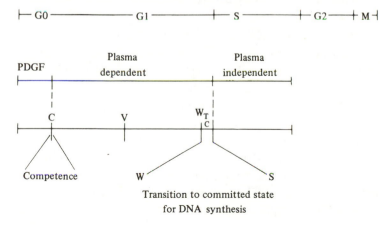

⊢ G0 ─────────── G1 ───────── ⊢─ S ───────── ⊢ G2 ──── ⊢ M ⊣

PDGF   Plasma   Plasma
                dependent   independent

        C   V W_T
                                  $\frac{}{c}$

  Competence   W   S

Transition to committed state
for DNA synthesis

In accord with the restriction control point model of growth regulation, fibroblast cells from baby hamster kidney (BHK) (Pardee, 1974) or Chinese hamster lung (CHL), (Martin & Stein, 1976) transferred from an amino-acid-deficient medium to a serum-deficient medium, do not traverse a single S phase. However, some experimental data have been inconsistent with the concept that serum and nutrients regulate a competent event in the mammalian fibroblast cell cycle; thus, after the addition of complete medium containing both serum and amino acids to BHK, CHL, or 3T3 fibroblast cells, the time interval until DNA synthesis depends on whether these cells were initially starved with respect to serum or amino acids (Burstin, Meiss & Basillico, 1974; Martin & Stein, 1976; Yen & Pardee, 1978).

Much controversy regarding the relationship between nutrients and serum factors in the control of the cell cycle has been reconciled by recognizing that serum contains multiple factors which regulate different events in the cell cycle. For fibroblast cells, serum can be conceived as a mixture of a platelet-derived growth factor (PDGF) and plasma. Studies with Balb/c-3T3 cells have demonstrated that nutrients are not required for the cellular response to PDGF; however, amino acids are required for plasma to promote the entry of PDGF-treated, competent cells into the S phase (Stiles *et al.*, 1979*b*). PDGF-treated, Balb/c-3T3 cells cultured in amino acid deficient growth medium become growth retarded at two points in the plasma-dependent portion of the cell cycle located six hours prior to and immediately prior to the G1/S phase boundary. The amino acid regulated growth arrest points correspond in time to the plasma-dependent control points noted in the study by Pledger *et al.* (1978) (see Fig. 7). Thus, these two growth regulatory events appear to be 'restriction points' in the sense defined by Pardee because cells become growth-retarded at these times after withdrawal of either amino acids or of plasma. Furthermore, re-addition of amino acids to amino acid starved cultures of Balb/c-3T3 cells does not allow DNA synthesis unless an optimal concentration of plasma is also present.

Malignant transformation and the cell cycle

In neoplastic disease, the regulation of growth and/or differentiation is perturbed at the cellular level. The growth of mammalian cells appears to be modulated at discrete control points within G0 and G1 phases of the cell cycle; for this reason, it is of interest to examine cell cycle parameters within malignantly transformed animal cells. Once again, tissue culture studies with fibroblast model systems have yielded the bulk of our data.

Fibroblast cells from rodents, birds and humans can be infected *in vitro* with a variety of tumor viruses. Following infection, the morphological and growth characteristics of these cells *in vitro* are altered in characteristic ways. The serum requirement for growth may be reduced. Fibroblast cells which could not form colonies in agarose suspension may acquire the ability to do so. The property of density dependent growth arrest may be abolished and the cell cultures will grow to much higher final cell densities. Finally, fibroblast cells which could not form tumors when inoculated into animal hosts may acquire the ability to do so. This set of growth regulatory modifications is referred to as 'transformation'. Some virus-transformed cells exhibit changes in all of their in-vitro growth regulatory properties (anchorage dependence, serum requirement, density dependent growth inhibition) in culture, while other transformed cell lines show changes only in one or a few of these growth regulatory parameters (see Sanford, 1974 for review of growth behavior *in vitro* and transformation).

Transformation by the Papova group of viruses appears to perturb the ability of normal fibroblast cells to enter G0 when transferred to culture medium containing low serum concentrations or when grown to the confluent monolayer state in culture medium supplemented with 10% calf serum (Holley & Kiernan, 1974). SV40-transformed 3T3 cells (SV-3T3) do not stop traversing the cell cycle under these non-permissive growth conditions; rather, SV-3T3 cells continue to divide (at a slower rate) in either low serum medium or in high density cultures (Bartholomew, Yokota & Ross, 1976).

The SV-3T3 cell line exhibits the full spectrum of the transformed phenotype relative to the parental 3T3 fibroblast line; SV-3T3 cells grow in low serum concentrations, grow to high densities, form colonies in soft agar suspension, and form tumors in mice (Paul, Lipton & Klinger, 1971). As previously mentioned, not all transformed cell lines exhibit such a completely aberrant growth phenotype in culture. SV40 transformants of 3T3 cells can be isolated which maintain the phenotype of low saturation density (Scher & Nelson-Rees, 1971). Vogel and colleagues selected revertants of SV-3T3 cells that have lost the ability to grow in low serum concentrations and will only proliferate in medium containing high serum supplements (Vogel & Pollack, 1974; Vogel, Ley & Pollack, 1974). Although the consequences of SV40 transformation on cell phenotype in culture appear to be variable, the loss of G0 growth regulation seems to be a constant feature since neither flat transformants nor serum revertants of SV-3T3 are able to enter a state of G0 growth arrest under non-permissive conditions (Scher & Nelson-Rees, 1971; Risser & Pollack, 1974; Dubrow, Pardee & Pollack, 1978). Whereas normal 3T3 cell cultures cease to replicate their DNA at high cell densities, SV40 'flat transformants' of 3T3 cells continue to replicate their DNA at high cell densities; the extra cells are sloughed into the medium thus maintaining a steady state cell population on the surface of the culture dish. Similarly, normal Balb/c-3T3 cells cease to replicate their DNA in low serum medium. The SV40 'serum revertants' continue to replicate their DNA in low serum containing medium, but the extra cells are sloughed off the surface of the tissue culture dish (Vogel & Pollack, 1974). A variety of studies with rodent fibroblasts demonstrates that under conditions which result in G0 growth arrest of normal cells, the SV40 virus transformed counterparts do not arrest in G0. Rather each phase of the cell cycle is proportionately lengthened (see Table 1). Under extreme conditions of serum starvation, however, some SV40 transformed rodent embryo fibroblasts may accumulate in the G2 phase of the cell cycle (Paul, Henahan & Walter, 1974).

Rodent fibroblasts can also be transformed by a number of RNA tumor viruses (retroviruses). The effect of retrovirus transformation on fibroblast cell cycle parameters has not been as thoroughly studied as the Papova viruses. As is the case with Papova virus transformation, rodent fibroblast cells transformed by retroviruses may acquire the ability to grow in low serum concentration, form colonies in agarose suspension and grow to high cell densities on monolayers. With respect to cell cycle control, however, retrovirus transformed cells behave differently from Papova virus transformed cells in several important respects. In

Table 1. *Effect of serum starvation on the cell cycle in normal and SV-40 virus-transformed 3T3 cells* [a]

	Serum conc. (%)	G1		S		G2 + M	
		Percent of population	Average residence time (h)	Percent population	Average residence time (h)	Percent of population	Average residence time (h)
3T3 cells	0.4	96.9	—	0	—	3.1	—
	1.0	93.7	—	1.7	—	4.6	—
	2.0	95.6	—	0	—	4.4	—
	4.0	80.1	—	6.1	—	12.8	—
	6.0	59.5	—	18.0	—	22.4	—
	8.0	50.9	6.8	24.1	4.0	25.0	5.2
	10.0	48.6	6.4	28.9	4.9	22.5	4.7
	16.0	42.8	5.6	31.0	5.0	26.2	5.4
	20.0	38.6	5.0	38.2	6.2	23.2	4.8
SV40-3T3 cells	0.1	34.6	—	9.6	—	55.9	—
	0.4	49.8	—	10.0	—	40.2	—
	1.0	49.6	21.0	22.6	12.0	27.8	18.0
	2.0	47.7	11.4	28.1	8.5	24.2	9.1
	4.0	56.6	9.6	22.3	4.9	22.2	5.5
	6.0	45.3	5.9	25.9	4.2	28.8	5.9
	8.0	49.1	6.6	27.5	4.5	23.5	4.9
	10.0	50.0	6.6	24.7	4.2	25.3	5.2
	16.0	46.8	6.2	27.1	4.5	26.1	5.3

[a] adapted from Bartholomew *et al.* (1976).

general, retrovirus-transformed fibroblasts do not grow to as high cell densities as do DNA virus transformants; moreover, at confluence, cells transformed by retroviruses become selectively growth retarded in the G1 phase of the cell cycle (O'Neill, 1978; Magun, Thompson & Gernor, 1979).

Animal cells may also be transformed *in vitro* by chemical carcinogens. The study of chemical transformation is probably most relevant to tumorigenesis in humans since most human neoplasms are thought to be chemically induced. Mouse 3T3 cells transformed by chemicals, form colonies in agarose suspension, grow to higher than normal cell densities and are tumorigenic animal hosts. These chemically transformed 3T3 cells show a serum requirement for growth that is lower than normal, but the serum growth requirement is higher than that displayed by SV-3T3 cells. It is of interest that the chemically transformed 3T3 cells will exhibit G0 growth arrest under conditions of serum starvation. The growth arrested cells can be induced to divide by the addition of serum or fibroblast growth factor (Holley, Baldwin, Kiernan & Messmer, 1976). Fibroblast growth factor is a basic peptide which is similar in some of its chemical and biologic properties to PDGF (Gospodarowicz & Moran, 1976; Scher *et al.*, 1979). Thus, the cell cycle controls of chemically transformed 3T3 cells seem to be less perturbed than those of SV40-transformed 3T3 cells. In this respect, chemical transformants of rodent embryo fibroblasts are similar to retrovirus transformants.

Summary
The proliferation of mammalian cells can be described as a cycle containing the M, G1, S, and G2 phases. Under conditions which are nonpermissive for the growth of normal mammalian cells, a state of quiescent growth arrest termed 'G0' is entered. Cells may be stimulated to leave the G0 state and enter the proliferative portion of the cell cycle by a variety of mitogenic agents. There is substantial variability in the transit times of individual mammalian cells through the cell cycle; most of this variability can be ascribed to the time that individual mammalian cells dwell in the G0 and G1 phases of the cycle (see chapter by Brooks in this volume). In all of these respects, the control of the mammalian cell cycle is not qualitatively different from the cell cycle that has been described for lower eukaryotes including yeast and single cell protozoans.

The mammalian cell cycle has been most thoroughly studied with tissue culture systems. Animal serum has been shown to contain a variety of hormones and growth factors which modulate the growth of normal animal cells. For fibroblast cells, some of these growth factors have been identified. Serum growth hormones appear to modulate several distinct events in the G0 and G1 phases of the mammalian cell cycle. The G1 phase of the cell cycle is also most sensitive to the nutritional content of the culture medium. Under most conditions *in vivo*, however, it is hormones, rather than nutrients *per se* which modulate the proliferation of animal cells; the regulation of the mammalian cell cycle may thus differ qualitatively from cell cycle regulation of lower eukaryotes; in these organisms, cell cycle traverse appears to be more directly coupled to the nutritional *milieu*, although the mating hormones of yeast cause arrest in G1 (Carter, this volume).

Neoplastic disease is a growth regulatory problem unique to multicellular organisms. The effects of oncogenic agents on cell cycle control parameters can be studied using mammalian cells in culture. Some tumor viruses interfere with the ability of normal animal fibroblasts to enter the G0 state. The G0 control mechanisms may not be the only target for oncogenic agents, however; cells transformed by chemicals and by other tumor viruses may retain their control of G0 while still exhibiting a malignant phenotype in culture and in animals.

Portions of the work reviewed here were supported by grants CA22042, CA22427 and CA18662 from the NIH. C. D. Scher is a Scholar of the American Leukemia Society. We are grateful to Ms. D. Krepcio for assistance with the manuscript.

References

Antoniades, H. N., Scher, C. D. & Stiles, C. D. (1979). Purification of the human platelet-derived growth factor. *Proceedings of the National Academy of Sciences, USA,* **76,** 1809–13.

Balk, S. D., Whitfield, J. F., Youdale, T. & Braun, A. C. (1973). Roles of calcium, serum, plasma, and folic acid in the control of proliferation of normal and rous sarcoma virus-infected chicken fibroblasts. *Proceedings of the National Academy of Science, USA,* **70,** 675–9.

Bartholomew, J. C., Yokota, H. & Ross, P. (1976). Effect of serum on the growth of Balb 3T3 A31 mouse fibroblasts and an SV40-transformed derivative. *Journal of Cellular Physiology,* **88,** 277–86.

Baserga, R. (1976). *Multiplication and Division in Mammalian Cells,* New York: Marcel Dekker.

Baserga, R. (1968). Biochemistry of the cell cycle: a review. *Cell and Tissue Kinetics,* **1,** 167–91.

Block, R. & Haseltine, W. A. (1974). In-vitro synthesis of ppGpp and ppCpp. In *Ribosomes,* ed. M. Nomora, A. Tissieres & P. Lengyel, pp. 733–45. New York: Cold Spring Harbor Laboratory.

Brooks, R. F. (1975). Growth regulation *in vitro* and the role of serum. In *Structure and Function of Plasma Proteins,* ed. A. C. Allison, vol. 2, pp. 239–89. New York: Plenum Press.

Burstin, S. J., Meiss, H. K. & Basillico, C. (1974). A temperature sensitive cell cycle mutant of the BHK cell line. *Journal of Cellular Physiology,* **84,** 397–407.

Cedarblad, G. (1979). Plasma proteins involved in haem metabolism and in transport of metals, hormones and vitamins. In *Plasma Proteins,* ed. B. Blomback & L. Hanson, vol. 1, pp. 99–102. Chichester: John Wiley & Sons.

Dubrow, R., Pardee, A. B. & Pollack, R. (1978). 2-amino-isobutyric acid and 3-O-methyl-D-glucose transport in 3T3 SV40-transformed 3T3 and revertant cell lines. *Journal of Cellular Physiology,* **95,** 203–11.

Eagle, H. (1955). Nutrition needs of mammalian cells in culture. *Science,* **122,** 501–4.

Epifanova, O. I. & Terskikh, V. V. (1969). On the resting periods in the cell life cycle. *Cell and Tissue Kinetics,* **2,** 75–93.

Gospodarowicz, D. & Moran, J. S. (1976). Growth factors in mammalian cell culture. *Annual Review of Biochemistry,* **45,** 531–58.

Hartwell, L. H., Culotti, J., Pringle, J. R. & Reid, B. J. (1974). Genetic control of the cell division cycle in yeast. *Science,* **183,** 46–51.

Hayashi, I. & Sato, G. (1976). Replacement of serum by hormones permits growth of cells in defined medium. *Nature, London,* **259,** 132–4.

Helmstetter, C. E. (1969). Sequence of bacterial reproduction. *Annual Review of Microbiology,* **23,** 223–38.

Hershko, A., Mamont, P., Shields, R. & Tomkins, G. M. (1971). Pleiotypic response – hypothesis relating growth regulation in mammalian cells to stringent controls in bacteria. *Nature New Biology,* **232,** 206–11.

Holley, R. W. (1972). A unifying hypothesis concerning the nature of malignant growth. *Proceedings of the National Academy of Sciences, USA,* **69,** 2840–1.

Holley, R. W. (1975). Control of growth of mammalian cells in culture. *Nature, London,* **258,** 487–90.

Holley, R. W., Baldwin, J. H. & Kiernan, J. A. (1974). Control of growth of a tumor cell by linoleic acid. *Proceedings of the National Academy of Sciences, USA,* **71,** 3976–8.

Holley, R. W., Baldwin, J. H., Kiernan, J. A. & Messmer, T. O. (1976). Control of growth of benzo(a)pyrene-transformed 3T3 cells. *Proceedings of the National Academy of Sciences, USA,* **73,** 3229–32.

Holley, R. W. & Kiernan, J. A. (1974). Control of the initiation of DNA synthesis in 3T3 cells: serum factors. *Proceedings of the National Academy of Sciences, USA,* **71,** 2908–11.

Howard, A. & Pelc, S. R. (1953). Synthesis of desoxyribonucleic acid in normal and irradiated cells and its relation to chromosome breakage. *Heredity, London, (Suppl.),* **6,** 261–73.

Jainchill, J. A. & Todaro, G. J. (1970). Stimulation of cell growth in vitro by serum with and without growth factor. *Experimental Cell Research,* **59,** 137–46.

Kaplan, D. R., Chao, F. C., Stiles, C. D., Antoniades, H. N. & Scher, C. D. (1979). Platelet α-granules contain a growth factor for fibroblasts. *Blood,* **53,** 1043–52.

Kohler, N. & Lipton, A. (1974). Platelets as a source of fibroblast growth-promoting activity. *Experimental Cell Research,* **87,** 297–301.

Kuempel, P. L. & Pardee, A. B. (1963). The cycle of bacterial duplication. *Journal of Cellular and Comparative Physiology,* **62,** 15–30.

Lajtha, L. G. (1963). On the concept of the cell cycle. *Journal of Cellular and Comparative Physiology,* **62,** 143–5.

Lajtha, L. G., Gilbert, C. W., Porteus, D. D. & Alexander, R. (1964). Kinetics of a bone marrow stem cell population. *Annals of the New York Academy of Sciences,* **113,** 742–52.

Lamerton, L. F. & Fry, R. J. M. (1963). *Cell Proliferation*. Oxford: Blackwell.

Lark, K. G. (1969). Initiation and control of DNA synthesis. *Annual Review of Biochemistry*, **38**, 569–604.

Lazzarini, R. A., Cashel, M. & Gallant, J. (1971). On the regulation of guanosine tetraphosphate levels in stringent and relaxed strains of *Escherichia coli*. *Journal of Biological Chemistry*, **246**, 4381–5.

Ley, K. D. & Tobey, R. A. (1970). Regulation of initiation of DNA synthesis in Chinese hamster cells. *Journal of Cell Biology*, **47**, 453–9.

Macpherson, I. & Montagnier, L. (1964). Agar suspension culture for the army of cells transformed by polyoma virus. *Virology*, **23**, 291–4.

Magun, B. E., Thompson, R. L. & Gernor, E. W. (1979). Regulation of DNA replication by serum and the transforming function in cultured rat fibroblasts transformed by rous sarcoma virus. *Journal of Cellular Physiology*, **99**, 207–16.

Martin, R. G. & Stein, S. (1976). Resting state in normal and simian virus 40 transformed Chinese hamster lung cells. *Proceedings of the National Academy of Sciences, USA*, **73**, 1655–9.

McKeehan, W. L., Hamilton, W. C. & Ham, R. G. (1976). Selonium is an essential trace nutrient for growth of WI-38 diploid human fibroblasts. *Proceedings of the National Academy of Sciences, USA*, **73**, 2023–7.

Mitchison, J. M. (1971). *The Biology of the Cell Cycle*. Cambridge University Press.

Mueller, G. C. (1969). Biochemical events in the animal cell cycle. *Federation Proceedings*, **28**, 1780–9.

Nierlich, D. P. (1978). Regulation of bacterial growth, RNA, and protein synthesis. *Annual Review of Microbiology*, **32**, 393–432.

Nilhausen, K. & Green, H. (1965). Reversible arrest of growth in G1 of an established fibroblast line (3T3). *Experimental Cell Research*, **40**, 166–8.

O'Neill, F. J. (1978). Differential in-vitro growth properties of cells transformed by DNA and RNA tumour viruses. *Experimental Cell Research*, **117**, 393–401.

Pardee, A. B. (1974). A restriction point for control of normal animal cell proliferation. *Proceedings of the National Academy of Sciences, USA*, **71**, 1286–90.

Pardee, A. B., Dubrow, R., Hamlin, J. L. & Kletzien, R. F. (1978). Animal cell cycle. *Annual Review of Biochemistry*, **47**, 715–50.

Paul, D., Henahan, M. & Walter, S. (1974). Changes in growth control and growth requirements associated with neoplastic transformation in vitro. *Journal of the National Cancer Institute*, **53**, 1499–503.

Paul, D., Lipton, A. & Klinger, I. (1971). Serum factor requirements of normal and simian virus 40-transformed 3T3 mouse fibroblasts. *Proceedings of the National Academy of Sciences, USA*, **68**, 645–48.

Pledger, W. J., Stiles, C. D., Antoniades, H. N. & Scher, C. D. (1977). Induction of DNA synthesis in Balb/c-3T3 cells by serum components: reevaluation of the commitment process. *Proceedings of the National Academy of Sciences, USA*, **74**, 4481–5.

Pledger, W. J., Stiles, C. D., Antoniades, H. N. & Scher, C. D. (1978). An ordered sequence of events is required before Balb/c-3T3 become committed to DNA synthesis. *Proceedings of the National Academy of Sciences, USA*, **75**, 2839–43.

Pohjanpelto, P. & Raine, A. (1972). Identification of a growth factor produced by human fibroblasts in vitro as putrescine. *Nature New Biology,* **235,** 247–8.

Post, J. & Hoffman, J. (1965). Further studies on the replication of rat liver cells in vivo. *Experimental Cell Research,* **40,** 333–9.

Risser, R. & Pollack, R. E. (1974). Biological analysis of clones of SV40 infected mouse 3T3 cells. In *Control of Proliferation in Animal Cells,* ed. B. Clarkson & R. Baserga, pp. 139–50. New York: Cold Spring Harbor Laboratory.

Ross, R., Glomset, J., Kariya, B. & Harker, L. (1974). A platelet dependent serum factor that stimulates the proliferation of arterial smooth muscle cells. *Proceedings of the National Academy of Sciences, USA,* **71,** 1208–10.

Rothblat, G. H. & Cristofalo, V. J. (eds.) (1972). *Growth nutrition, and metabolism of cells in culture.* New York: Academic Press.

Rubin, H. (1972). Inhibition of DNA synthesis in animal cells by ethylene diamine tetraacetate and its reversal by zinc. *Proceedings of the National Academy of Sciences, USA,* **69,** 712–6.

Salmon, W. D., Jr. & Daughaday, W. H. (1957). A hormonally controlled serum factor which stimulates sulfate incorporation by cartilage in vitro. *Journal of Laboratory and Clinical Medicine,* **57,** 825–36.

Sanford, K. K. (1974). Biological manifestations of oncogenesis in vitro: a critique. *Journal of the National Cancer Institute,* **53,** 1481–5.

Schaechter, M. (1968). Growth of cells and populations. In *Biochemistry of Bacterial Growth,* ed. J. Mandelstam & K. McQuillen, pp. 136–62. New York: Wiley.

Scher, C. D. & Nelson-Rees, W. A. (1971). Direct isolation and characterization of 'flat' SV40-transformed cells. *Nature New Biology,* **233,** 263–5.

Scher, C. D., Pledger, W. J., Martin, P., Antoniades, H. N. & Stiles, C. D. (1978). Transforming viruses directly reduce the cellular growth requirement for a platelet derived growth factor. *Journal of Cellular Physiology,* **97,** 371–80.

Scher, C. D., Shepard, R. C., Antoniades, H. N. & Stiles, C. D. (1979). Platelet-derived growth factor and the regulation of the mammalian fibroblast cell cycle. *Biochimica et Biophysica Acta,* **560,** 217–41.

Shilo, B., Simchen, G. & Pardee, A. B. (1978). Regulation of cell cycle initiation in yeast by nutrients and protein synthesis. *Journal of Cellular Physiology,* **97,** 177–88.

Smith, J. A. & Martin, L. (1973). Do cells cycle? *Proceedings of the National Academy of Sciences, USA,* **70,** 1263–7.

Snyder, J. A. & McIntosh, J. R. (1976). Biochemistry and physiology of microtubules. *Annual Review of Biochemistry,* **45,** 699–720.

Stanners, C. P. & Thompson, L. H. (1974). Studies on a mammalian cell mutant with a temperature-sensitive leucyl-tRNA synthetase. In *Control of Proliferation in Animal Cells,* ed. E. Clarkson & R. Baserga, pp. 191–203. New York: Cold Spring Harbor Laboratory.

Stiles, C. D., Capone, G. T., Scher, C. D., Antoniades, H. N., Van Wyk, J. J. & Pledger, W. J. (1979a). Dual control of cell growth by somatomedin c and platelet-derived growth factor. *Proceedings of the National Academy of Sciences, USA,* **76,** 1279–83.

Stiles, C. D., Isberg, R., Pledger, W. J., Antoniades, H. N. & Scher, C. D. (1979b). Control of the Balb/c-3T3 cell cycle by nutrients and serum factors: analysis using platelet-derived growth factor and platelet-poor plasma. *Journal of Cellular Physiology,* **99,** 395–406.

Stoker, M. (1968). Abortive transformation by polyoma virus. *Nature, London*, **218,** 234–8.

Temin, H. A. (1971). Stimulation by serum of multiplication of stationary chicken cells. *Journal of Cellular Physiology,* **78,** 161–70.

Thammana, P., Buerk, R. R. & Gordon, J. (1976). Absence of ppGpp production in synchronized Balb/c-3T3 cells on isoleucine starvation. *FEBS letters,* **68,** 187–90.

Todaro, G. J., Lazar, G. K. & Green, H. (1965). The initiation of cell division in a contact-inhibited mammalian cell line. *Journal of Cellular and Comparative Physiology,* **66,** 325–34.

Van Wyk, J. J. & Underwood, L. E. (1978). The somatomedins and their actions. In *Biochemical Actions of Hormones,* ed. G. Litwack, pp. 101–47. New York: Academic Press.

Vogel, A. & Pollack, R. (1974). Isolation and characterization of revertant cell lines VII. DNA synthesis and mitotic rate of serum sensitive revertants in non-permissive growth conditions. *Journal of Cellular Physiology,* **85,** 151–62.

Vogel, A., Ley, J. & Pollack, R. E. (1974). Two classes of revertants isolated from SV40-transformed 3T3 mouse cells. In *Control of Proliferation in Animal Cells,* ed. B. Clarkson & R. Baserga, pp. 125–38. New York: Cold Spring Harbor Laboratory.

Westermark, B. & Wasteson, A. (1976). A platelet factor stimulating human normal glial cells. *Experimental Cell Research,* **98,** 170–4.

Yen, A. & Pardee, A. B. (1978). Arrested states produced by isoleucine deprivation and their relationship to the low serum produced arrested state in Swiss 3T3 cells. *Experimental Cell Research,* **114,** 389–94.

W. SACHSENMAIER

The mitotic cycle in *Physarum*

Introduction

The myxomycete *Physarum polycephalum,* in its vegetative stage, forms multinuclear plasmodia which may be cultivated under axenic conditions on a semi-defined liquid medium (Guttes, Guttes & Rusch, 1961; Daniel & Baldwin, 1964). Suspension cultures consist of numerous microplasmodia containing up to several hundreds of nuclei. Stationary macroplasmodia are prepared by fusion of pelleted microplasmodia which subsequently grow as giant disc-shaped plasmodia on a sheet of filter paper or millipore filter (Fig. 1). A single macroplasmodium with a diameter of 6 cm contains about 10^8 nuclei, 20 mg protein, 2 mg RNA and 0.2 mg DNA. All nuclei within a single plasmodium divide synchronously every 8–10 h at 26 °C. This natural synchrony is amazingly sharp: more than 99% of the nuclei pass through mitosis within less than 5 min under optimal culture conditions.

The synchronous behaviour and the convenient techniques available for cultivating large plasmodia qualify this organism as an ideal tool for studies of molecular events related to the control of nuclear division. A brief survey of some cell cycle dependent metabolic features will be given in the first part of this contribution. In the second part, models will be discussed describing the possible nature of the mitotic clock in *Physarum.*

Fig. 1. Surface culture of macroplasmodia of *Physarum polycephalum.* (Nygaard *et al., 1960; Daniel & Baldwin, 1964).*

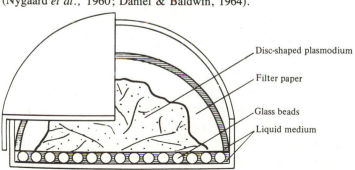

Disc-shaped plasmodium

Filter paper

Glass beads

Liquid medium

Nucleic acid metabolism

DNA synthesis in *Physarum* nuclei, as measured by the uptake of radio-active precursors (Nygaard, Guttes & Rusch, 1960; Sachsenmaier, 1964) or by isotope dilution techniques (Hall & Turnock, 1976) commences immediately after telophase and proceeds at a high rate during the first third of the cycle (~ 3 h). Incorporation of labelled thymidine continues at a low rate throughout the rest of the cycle which at least partially reflects 'maturation' of chromosomal DNA, i.e. joining of individual replicons (Funderud & Haugli, 1975; Funderud, Andreassen & Haugli, 1978). Some of the DNA labelling in G2 may also be attributed to a few asynchronous nuclei with aberrant characteristics as judged from autioradiographic studies (Guttes, E. & Guttes, S., 1969). These nuclei, however, appear to be abnormal in the sense that their DNA is not replicated again in the following cycle (Vogt & Braun, 1977). Perhaps they represent a population of dying nuclei.

The bulk of nuclear DNA, however, replicates during the first third of the cycle which is generally regarded as S phase. No G1-period has yet been demonstrated in normal plasmodia, even under various experimental conditions that affect the timing of mitosis, e.g. treatment with ultraviolet light, X-rays, anti-metabolites, temperature shocks. In all these cases, onset of a high rate of DNA replication remains strongly linked to the onset of nuclear mitosis. Under certain conditions, i.e. after heat-shock treatment in early prophase, nuclei enter mitosis but cannot complete the whole sequence through telophase; DNA replication nevertheless may start on time and lead to the formation of polyploid nuclei (Brewer & Rusch, 1968). Thus the signal for initiation of DNA replication is given by some yet unknown event related to the initiation of mitosis but not depending on the completion of the entire sequence of this complex morphological process. Perhaps decondensation of the chromatin, which normally occurs after telophase but may occur also in arrested mitotic nuclei, actually triggers the onset of DNA synthesis.

Replication of the entire genome proceeds in an ordered sequence (Braun, Mittermayer & Rusch, 1965; Braun & Wili, 1969) and requires concomitant protein synthesis (Muldoon, Evans, Nygaard & Evans, 1971). There is some suggestive evidence that the sequential activation of replicons is controlled by specific proteins appearing sequentially at particular stages during the S period (Wille & Kauffman, 1975).

RNA synthesis in synchronous *Physarum* plasmodia also exhibits some cycle dependent alterations (reviews by Grant, 1973; Melera, 1980). The overall rate drops during mitosis as indicated by the incorporation of radioactive precursors (Sachsenmaier & Becker, 1965; Kessler, 1967). A biphasic pattern of label uptake was reported by Mittermayer, Braun & Rusch (1964), but this is not seen with the isotope dilution technique (Hall & Turnock, 1976). The majority of

RNA synthesised at all stages of the cycle is ribosomal RNA, however, there are subtle differences in the proportion of certain RNA types formed at different times: poly(A)-containing RNA comprises a greater proportion of newly synthesised RNA during S-phase as compared to G2 phase (Fouquet, Bierweiler & Sauer, 1974; Fouquet *et al.*, 1974). Also, the complexity of RNA synthesized during S is higher than in G2 as demonstrated by RNA/DNA hybridisation under conditions which permit participation of low frequency DNA sequences (Fouquet & Braun, 1974). Nuclear poly(A)-containing RNA contains more redundant sequences when isolated during G2 than during S (Fouquet *et al.*, 1974; Fouquet & Sauer, 1975).

Protein synthesis

Discontinuous production of proteins, i.e. enzymes, during the cell cycle has long been suggested to reflect the sequential transcription of chromosomal genes (linear reading) or oscillatory repression (Mitchison, 1971). However, its possible significance in the control of the cell cycle has not been demonstrated yet. In most cases, the variations observed refer to enzyme activities which do not necessarily indicate changes of enzyme synthesis but may result from discontinuous activation and inhibition of continuously synthesised enzyme proteins. *Physarum,* unlike other systems, has revealed relatively few examples of true cycle specific enzymes. Among the enzymes studied, mostly those related to nucleic acid metabolism exhibit cell cycle dependent changes of activity: thymidine kinase (Sachsenmaier & Ives, 1965), deoxycytidine kinase (Sachsenmaier *et al.*, 1973; Sachsenmaier & Dworzak, 1976), deoxyadenosine kinase (Woertz & Sachsenmaier, 1979), DNA-polymerase (Brewer & Rusch, 1965, 1966; Schiebel *et al.* 1976), RNAase (Braun & Behrens, 1969), NAD-pyrophosphorylase (Solao & Shall, 1971), H1-histone phosphokinase (Bradbury, Inglis & Matthews 1974; Bradbury, Inglis, Matthews & Langan, 1974) ornithine decarboxylase (Mitchell, Campbell & Carter, 1976; Mitchell & Carter, 1977; Sedory & Mitchell, 1977; Mitchell, Carter & Rybski, 1978). A moderate but significant periodic behaviour during the mitotic cycle was found also for mitochondrial activities, e.g. respiration capacity and cytochrome oxidase activity (Forde & Sachsenmaier, 1979). None of these enzymes, except perhaps histone phosphokinase, appears to play a regulatory role on the cycle itself, but in contrast may be influenced secondarily by the clock mechanism of the cycle. Nevertheless, investigation of the basis of cycle specific fluctuations of enzyme production may help to elucidate the nature of the molecular timing mechanism governing the cell cycle.

One of these enzymes, thymidine kinase (EC 2.7.1.75), has been studied in more detail in *Physarum*. Specific activity of this enzyme increases sharply at the end of the G2-period, continues to increase during mitosis and reaches a maximum in early S phase (Fig. 2). Various findings from studies with inhibitors (Sachsenmaier, Fournier & Gürtler, 1967; Wright & Tollon, 1979a) and [²H] – labelled amino acids (Oleinick, 1972) suggest that the stepwise increase of enzyme activity reflects true discontinuous synthesis of enzyme protein. Thymidine kinase production appears to be under transcriptional and possibly also translational control; this follows from the observation that actinomycin inhibits the increase of enzyme activity when administered up to about one hour prior to the expected onset of the enzyme step; addition of the drug is ineffective after this transition point (Sachsenmaier et al., 1967).

Thymidine kinase induction is strongly linked to the onset of nuclear mitosis in *Physarum*. This is observed in normal unperturbed plasmodia as well as in plasmodia subjected to various experimental treatments which affect the timing of mitosis, e.g. ultraviolet- and X-irradiation (Fig. 3) (Sachsenmaier, Bohnert, Clausnizer & Nygaard, 1970a; Sachsenmaier, Dönges, Rupff & Czihak, 1970b; Oleinick, 1972; Sachsenmaier & Hansen, 1973; Sachsenmaier & Dworzak, 1976). This suggests that both events, initiation of enzyme synthesis and the

Fig. 2. Continuous and discontinuous enzyme production during the nuclear cycle of *Physarum polycephalum*. Specific enzyme activities were measured in the high speed supernatant of plasmodial homogenates obtained at various stages of the nuclear cycle. (▼), Glucose-6-phosphate dehydrogenase; (●), thymidine kinase; (▲), thymidylate kinase; (■), deoxycytidine kinase, control; (□) deoxycytidine kinase, after addition of 3×10^{-3} M hydroxyurea at the time indicated by the vertical arrow: Abscissae. Total length of one cycle was 10–12 h. Ordinates: one scale unit corresponds to 4×10^{-2} (▼), 2×10^{-5} (●), 1.5×10^{-4} (▲), 1×10^{-5} (■,□) enzyme units (μmol substrate min^{-1} mg^{-1} protein), respectively. M, mitosis. (Sachsenmaier et al., 1973).

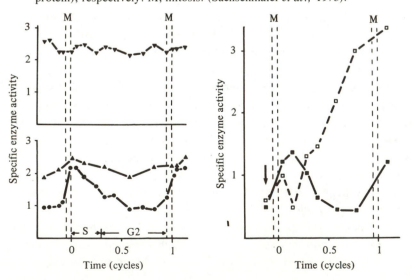

synchronous entrance of nuclei into division are controlled by the same metabolic trigger event. Recent findings by Wright & Tollon (1979*b*) indicate that a certain regimen of heat shock treatment may dissociate to some extent thymidine kinase induction from mitosis, however, this is explained as a differential response of these two events to the same trigger mechanism.

The duration of the induction period and the amount of enzyme produced can be altered by experimental conditions much more easily than can the position in the cycle of the onset of enhanced enzyme synthesis. Treatment of plasmodia with inhibitors of DNA replication during the S period prolongs the period of production of thymidine kinase, leading to an excess accumulation of enzyme activity (Sachsenmaier *et al.*, 1973; Sachsenmaier & Hansen, 1973). A similar effect is seen with deoxycytidine kinase (EC 2.7.1.74) (Fig. 2). This may indicate that DNA synthesis must proceed normally in order to 'shut off' enzyme synthesis at the end of the induction period. Perhaps the structural genes for thymidine kinase and deoxycytidine kinase are no longer accessible for transcription by RNA polymerase after replication of the corresponding part of the genome.

Fig. 3. Effects of ionizing radiation and ultraviolet (UV) light on thymidine kinase in *Physarum polycephalum*. (●), Control plasmodia; (○), irradiated plasmodia. (*a*) X-rays (1 krad, 29 kV, 25 mA, 0.3 mm Al filter, Siemens 'Dermopan') administered at metaphase of mitosis M−1, (not shown) which occurred at 10 h prior to mitosis (M). (*b*) UV light (260 J m⁻², λ max 254 nm, germicidal lamp 'Mineralight') administered during G2, i.e. at 3 h prior to mitosis (M). Ordinates: arbitrary units of enzyme activity in high speed supernatants of plasmodial homogenates. M^x, delayed mitosis in irradiated plasmodia. (Sachsenmaier & Hansen, 1973; Sachsenmaier & Dworzak, 1976).

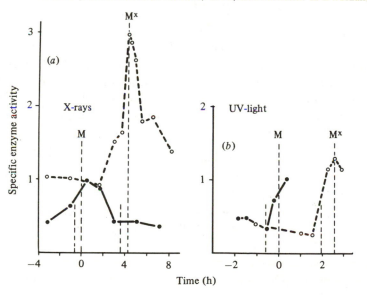

X-irradiation too, enhances the production of thymidine kinase (Fig. 3) (Sachsenmaier *et al.*, 1970a; Sachsenmaier & Dworzak, 1976) and of deoxycytidine kinase (Sachsenmaier & Dworzak, 1976), although probably by a different mechanism, since the stimulatory effect is observed without significantly inhibiting DNA synthesis; also the effect is most pronounced after a delay of a complete cycle following irradiation. Interestingly, the periods of maximum sensitivity in the cycle for X-ray induced mitotic delay (Fig. 6) and for thymidine kinase stimulation are identical and coincide with metaphase of the mitosis $(M - 1)$ that precedes the delayed mitosis (M) (Sachsenmaier *et al.*, 1970a; Sachsenmaier *et al.*, 1973). This coincidence suggests that mitotic delay as well as the superinduction of thymidine kinase result from the same primary radiation effect, i.e. from X-ray-induced single strand breaks in nuclear DNA. Whereas most of the single strand breaks are rapidly repaired, some of them may be 'fixed' during early S phase when the replication fork passes the gap before it has a chance to be healed. Alkaline sucrose gradient studies by Brewer & Nygaard (1972) have shown indeed that more DNA breaks remain unrepaired after X-irradiation of *Physarum* plasmodia in S phase compared with a similar treatment in G2-phase.

Regulation of thymidine kinase during the course of the mitotic cycle thus involves the following aspects.

(1) A steep increase in the rate of enzyme production at the end of the G2-period which is probably triggered by a yet unknown endogenous signal which also triggers the onset of nuclear division. This process cannot be provoked so far at other stages of the cycle by experimental treatments, e.g. by addition of substrate to the medium (W. Sachsenmaier, unpublished).

(2) Termination of enzyme synthesis occurs during early S phase and depends on normal DNA replication. Perhaps, changes of the chromatin structure associated with DNA replication and/or histone modification prevents further transcription of the corresponding genome. Inhibition of DNA synthesis prevents the termination of transcription causing excess production of enzyme protein.

(3) The amount of enzyme synthesised during the induction period appears to be largely influenced by a process occurring much earlier in the cycle which is particularly sensitive to X-irradiation. Perhaps an early replicating gene provides the information of a repressor which partially blocks enzyme synthesis at the translation level. X-irradiation at the beginning of the S period interferes with the formation of this factor thus causing unrestricted enzyme synthesis during the subsequent induction period. Alternatively, X-irradiation may induce additional starting points for RNA polymerase which increases the supply of mRNA

specific for thymidine kinase. However, this assumption cannot explain why the production of most other enzymes remains unaffected by X-irradiation.

Several distinct variants of thymidine kinase can be demonstrated in plasmodial extracts by isoelectric focussing (Gröbner & Sachsenmaier, 1976). The pattern of these variants changes characteristically during the mitotic cycle. Variant (C) and a satellite (C_1) with isoelectric points of pH 6.3 and 5.9 respectively appear first during the induction period followed later on by variants (A) and (A_1) with isoelectric points at pH 7.8 and 7.2. Species (A) dominates in the second half of the cycle whereas (C) and (C_1) disappear almost completely in G2 (Fig. 4). Variants ($C + C_1$) can be converted into variant (A) *in vitro* either spontaneously during storage of crude plasmodial extracts or by treating of partially purified preparations of ($C + C_1$) with alkaline phosphatase (Gröbner & Sachsenmaier, 1979). This behaviour and the sequential appearance of these variants *in vivo* suggest that a single nuclear gene codes for thymidine kinase which is transcribed at the end of the G2 period and during mitosis. The most acidic

Fig. 4. Thymidine kinase enzyme variants of *Physarum polycephalum* separated by isoelectric focussing of crude plasmodial extracts in 5% polyacrylamide gels. (*a*) Extract obtained in G2-period, i.e. 1.5 h prior to the expected onset of the second synchronous mitosis (M_2) after the inoculation of the culture; (*b*) extract obtained at telophase of mitosis M_2. See text for further explanation. (Gröbner & Sachsenmaier, 1976).

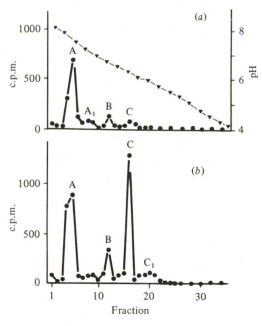

variant (C$_1$) probably reflects the primary translation product which is modified subsequently by stepwise dephosphorylation, giving rise to less acidic enzyme species. Variant (A) ultimately accumulates in the second half of the cycle as the most stable species.

Although thymidine kinase exhibits interesting features as a true cycle dependent step enzyme in *Physarum*, it cannot be regarded as a control element in the cycle itself as pointed out above. This follows also from the existence of several thymidine kinase deficient mutants (Haugli & Dove, 1972) which exhibit quite normal cycle characteristics including DNA replication (Sachsenmaier & Dworzak, 1976; Dworzak *et al.*, 1976*a*; Dworzak, Mohberg, Sachsenmaier & Czihak, 1976*b*; Mohberg, Dworzak & Sachsenmaier, 1980; Mohberg, Dworzak, Sachsenmaier & Haugli, 1980). The strong linkage between the onset of mitosis and the acceleration of enzyme production in wild-type plasmodia, however, focusses attention on the nature of the critical event triggering both processes.

Control of the timing of nuclear mitosis
Effect of inhibitors

Onset of nuclear mitosis can be inhibited or delayed by various chemical or physical treatments (Fig. 5). Reversible inhibition of DNA synthesis with 5-fluoro-2'-deoxyuridine delays the following mitosis by a period of time com-

Fig. 5. Sensitivity of synchronous plasmodia of *Physarum polycephalum* during the nuclear cycle to various mitotic inhibitors. The horizontal lines indicate the portion of the nuclear cycle during which the application of each agent prevents or strongly delays the onset of mitosis (M). Extension of some of these lines into the early part of mitosis means that application of the inhibitor during prophase blocks further progression of mitosis. Dotted lines indicate portions of the cycle with reduced radiation sensitivity as shown in more detail in Fig. 6. FUDR, 5-fluoro-2'-deoxyuridine; AM, actinomycin D or C; CH, cycloheximide; DNP, dinitrophenol; N$_2$, 99.96% nitrogen atmosphere; UV, ultraviolet light, λ_{max} 254 nm; X, X-rays.

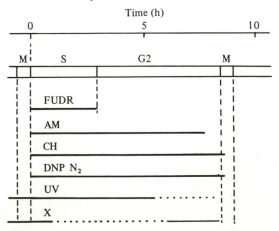

parable to the period of DNA inhibition (Sachsenmaier & Rusch, 1964). This suggests that progression of the cycle is halted during DNA blockade although overall protein synthesis continues almost normally for several hours (unbalanced growth). Inhibition of RNA synthesis with actinomycin also prevents or delays the onset of mitosis when applied prior to a transition point in late G2 period which precedes prophase by 0.5 h (~ 10% of the cycle) (Sachsenmaier & Becker, 1965; Sachsenmaier, 1966; Sachsenmaier *et al.*, 1967). Protein synthesis is required throughout the cycle as indicated by inhibitor studies with cycloheximide. This drug immediately stops progression of the cycle at any stage up to early prophase (Cummins, Brewer & Rusch, 1965; Cummins, Blomquist & Rusch 1966; Sachsenmaier *et al.*, 1967; Tyson, Garcia-Herdugo & Sachsenmaier, 1979). It appears from these studies that preparation for and initiation of mitosis requires the continuous synthesis of one or more proteins during most of the cycle which depends on the prior completion of DNA replication and synthesis of short-lived mRNA.

Physical treatments like temperature shocks or irradiation with ultraviolet light and X-rays interfere with the timing of mitosis apparently by affecting different cycle-specific processes. Heat shocks are most effective in delaying the onset of the next mitosis when applied during late G2 period, i.e. 2–3 h prior to the expected prophase (Brewer & Rusch, 1968). By analogy with the classical studies on *Tetrahymena* (review by Zeuthen & Rasmussen, 1971) this suggests that a temperature labile 'division protein' accumulates during the cycle which provides the signal for the initiation of mitosis.

The sensitivity of *Physarum* plasmodia to ultraviolet light and X-irradition

Fig. 6. Cycle dependent variation of radiation-induced mitotic delay in synchronous plasmodia of *Physarum polycephalum*. UV, ultraviolet light, 260 J m^{-2}, λ_{max} 254 nm; X, X-rays. Abscissa: time of irradiation. Ordinate: delay of the onset of mitosis (M), in fractions of the total cycle time (10 h), as compared to non-irradiated controls. (Sachsenmaier *et al.*, 1970*a*,*b*).

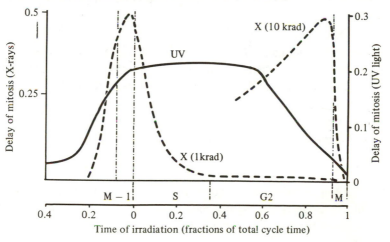

changes drastically during the cycle (Fig. 6). Ultraviolet light causes delays of the next mitosis (M) which are of equal magnitude when it is applied at any time during the first two-thirds of the cycle including mitosis $(M-1)$, S period and early G2 period (Sachsenmaier, 1966; Sachsenmaier *et al., 1970b*). The delaying effect declines steadily during the last third of the cycle approaching zero immediately prior to the onset of prophase. The nature of the division-specific process affected by ultraviolet light is not yet clear. The shape of the sensitivity curve is not compatible with the assumption that inhibition of DNA synthesis primarily causes the antimitotic effect, although ultraviolet light does inhibit DNA synthesis when applied during the S period. On the other hand, the sensitivity curve would agree with the idea that ultraviolet light interferes with the production of an initiator of mitosis which has to be formed in a certain quantity in order to trigger mitosis. The retarding effect of a single dose of ultraviolet light on mitosis therefore should become ever smaller the more initiator has already accumulated during the time prior to irradiation. This interpretation would also agree nicely with fusion experiments: fusion of an irradiated plasmodium with a control piece of equal size reduces the radiation-induced mitotic delay by one half, as predicted by the above hypothesis (Gnamusch & Sachsenmaier, 1979).

Ionizing radiation is most effective in retarding the onset of mitosis (M) when applied during metaphase of the preceding mitosis $(M-1)$; the sensitivity drops rapidly during early S phase and a second peak is detected with higher doses (10 krad) at the end of the G2 period (Sachsenmaier *et al., 1970a*). The shape of the sensitivity curve suggests that the primary radiosensitive processes are quite different from those affected by ultraviolet light. Most likely the sharp peak of X-ray sensitivity in metaphase of $M-1$, i.e. immediately prior to the onset of DNA replication, results from the induction of unrepaired DNA single-strand breaks which become 'fixed' during the ensuing S period (Brewer & Nygaard, 1972). These DNA lesions may interfere with the production of the mitotic initiator at the transcription level. On the other hand, the steady increase of radiation sensitivity during the G2 period and its sudden drop just prior to prophase suggests a 'set back' by ionizing radiation of the plasmodium in its progress toward mitosis, similar to the effect of heat shocks (Brewer & Rusch, 1968; Zeuthen & Rasmussen, 1971) or protein inhibitors (Rasmussen & Zeuthen, 1962; Tyson *et al.,* 1979). Again, this would agree with the assumption that a mitotic initiator accumulates during the cycle; high doses of X-rays destroy the initiator already formed at the time of irradiation.

Synchronisation by plasmodial fusion

The amazing spontaneous synchrony of nuclear divisions in *Physarum* plasmodia suggests that the initiation of mitosis is triggered by an extranuclear signal or by factors which serve to communicate between all nuclei of the syncytium. The synchronising effect of this system operates quite fast as demonstrated in fusion experiments (Fig. 7). Plasmodial pieces coalesce spontaneously when placed on top of each other. Mixed plasmodia can be prepared by this simple technique containing different sets of nuclei which originally represented different stages of the cycle. Under these conditions both sets of nuclei become synchronised within a fraction of the normal cycle time and enter the next mitosis

Fig. 7. Synchronisation of heterophasic nuclei in *Physarum polycephalum* by plasmodial fusion. Plasmodial pieces (A, B) with various phase differences were fused at the time indicated by double arrows, i.e. shortly after mitosis in the parent plasmodium A. Solid bars show mitosis in parent cultures and in self-fused control explants (A/A and B/B). Open bars show mitosis in mixed plasmodia; (●) cycle phase of A and B at the time of fusion; (○) cycle phase to which A and B nuclei are synchronized in the mixed plasmodium. In experiment III, many non-synchronous (n.s.) mitotic B-nuclei were observed shortly after fusion during a period of 1–2 h as indicated by the dotted line; the time interval after the start of the fusion (0.5 h) was obviously too short to achieve complete mixing of the plasmodial contents prior to the expected onset of mitosis in B-nuclei. (Sachsenmaier, 1976).

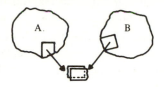

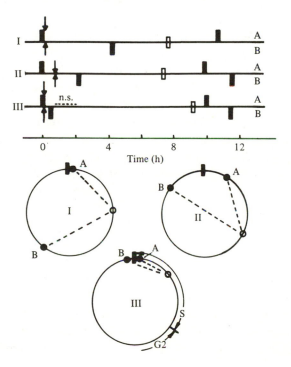

simultaneously. The timing of this first synchronous mitosis reflects an arithmetic compromise between the expected division times of the unfused parent cultures (Rusch, Sachsenmaier, Behrens & Gruter, 1966; Sachsenmaier, Remy & Plattner-Schobel, 1972; Chin, Friedrich & Bernstein, 1972).

Experiments of this type have shown that synchronisation of heterophasic nuclei in a fused pair of plasmodia always occurs in that part of the phase circle that does not include mitosis (Sachsenmaier, 1976). For example, in experiment II of Fig. 7, plasmodium (A) represents early interphase and plasmodium (B) late interphase at the time of fusion. Both sets of nuclei in the mixed plasmodium (X) enter mitosis simultaneously at a time later than (B) but earlier than (A). The sequence of mitotic divisions therefore is B⟶ X⟶ A⟶ B. Alternatively, if the synchronisation would occur in the shorter arc of the phase circle (including mitosis), the sequence would be A⟶ X⟶ B⟶ A.

Models of the mitotic clock

The synchronising behaviour of heterophasic nuclei in mixed plasmodia strongly suggests that the timing of nuclear mitosis is controlled by an extranuclear factor. This factor appears to accumulate during the cycle up to a threshold at which it initiates mitosis. Subsequently the system reverts to its original state in an oscillatory manner; the initiator is removed or inactivated and new initiator molecules are formed during the next cycle. Fusion of plasmodial pairs with various phase differences indicate that the mitotic clock in *Physarum* does not resemble a sinusoidal oscillator but rather a relaxation oscillator. Also, it appears that onset of mitosis itself is an obligatory component of the oscillating system. In other words, the mitotic clock is reset like an 'hourglass' each time the nuclei enter mitosis (Sachsenmaier *et al.*, 1972; Sachsenmaier, 1976).

This contrasts with the concept of a continuous limit cycle oscillator which visualises mitosis as a secondary event merely driven by the oscillator but not constituting an obligatory part of the oscillating system (Kauffman & Wille, 1976a,b). The latter concept implies that two (or more) interacting components fluctuate in an oscillatory manner autonomously; mitosis is triggered if one component (initiator) increases beyond a threshold level. Under certain conditions, the system may continue to oscillate even if mitosis does not occur, e.g. if the amplitudes of the oscillations are too small to push the initiator level beyond the threshold.

The limit cycle oscillator concept, unlike the hourglass model, would permit synchronisation of fused heterophasic plasmodia to the short arc of the phase circle (including mitosis) in experiments like those described in Fig. 7. Kauffman & Wille (1976a,b,) have reported some apparent cases of synchronisation to the short arc (including mitosis) from fusion experiments with *Physarum* plasmodia. However, these may be explained by shortcomings of their fusion technique (abutting edges of cut plasmodial pieces) which does not permit quick homogenous mixing of the plasmodial contents like the overlaying or 'sandwich' technique. Thorough studies on the synchronisation behaviour of nuclei from plasmodia with various phase differences using the 'sandwich' technique have shown that even under extreme conditions, e.g. fusion of a very late G2 plasmodium with a very early S phase plasmodium, synchronisation occurs to the long arc (excluding mitosis) of the phase circle (Sachsenmaier, 1976; Tyson & Sachsenmaier, 1978; Loidl, Linortner & Sachsenmaier, 1979). These observations are best explained by assuming that the initiation of a new cycle of the mitotic clock is strongly linked to the onset of mitosis.

Another important aspect of the control mechanism of mitosis in *Physarum* is the tendency of growing plasmodia to adjust the frequency of nuclear divisions to its overall growth rate. This is reflected by the observation that the ratio of total plasmodial mass per chromosome set, i.e. total protein : DNA, is relatively constant at the time of each nuclear division. Certain experimental treatments (ultraviolet light and X-irradiation, heat shocks, cycloheximide pulses) may transiently increase this ratio (Fig. 9) by inducing a delay of mitosis. The plasmodium, however, tends to revert to the normal ratio by speeding up subsequent nuclear mitoses (Fig. 10; Sachsenmaier *et al.*, 1970*b*; Sudbery & Grant, 1975; Sachsenmaier & Dworzak, 1976; Tyson *et al.*, 1979). Evidence for a cell size control of key events in the cell cycle, i.e. DNA synthesis and cell or nuclear division, have been reported also for prokaryotes (Donachie & Masters, 1969; Helmstetter, 1969), fission yeast (Nurse & Thuriaux, 1977) and other systems (for general discussion see chapters by Donachie and by Fantes & Nurse, this volume).

Taking into account these findings, the following model for the timing mechanism of mitosis in *Physarum* is proposed (Sachsenmaier *et al.*, 1972; Sachsenmaier, 1976; Tyson *et al.*, 1979) which may apply to other systems as well (Fig. 8). An extranuclear initiator of mitosis (I) is formed during the cycle and accumulates at a rate proportional to the increase of total cell mass. This factor, presumably a protein, reacts in a stoichiometric way with a given number of nuclear components, i.e. receptor sites (N). Mitosis is initiated as soon as all nuclear sites are saturated and/or the concentration of free initiator passes a threshold. During mitosis or during the ensuing S phase the number of nuclear sites increases by a factor of 2 and the new set of nuclear sites has to be saturated by newly synthesised initiator before the next division is triggered. The effects of pulse treatments with inhibitors of protein synthesis (Fig. 9, 10) suggest that the initiator is an unstable protein with a half-life of 1–2 h. Its accumulation may be described by the following simple equation (Tyson *et al.*, 1979):

$$dI/dt = kV - lI \tag{1}$$

where I is the total number of initiator molecules, V the volume (total mass) of the cell and l the rate constant for first order decay of the unstable initiator. Onset of mitosis occurs if the number of initiator molecules equals the number of nuclear sites:

$$I = N. \tag{2}$$

An additional assumption has to be made in order to account for the observation that plasmodia treated with 5- fluoro-2′-deoxyuridine (FUDR), an inhibitor of DNA synthesis, lose their capacity to advance 'young' nuclei in fusion experiments (Sachsenmaier *et al.*, 1972). This result suggests that the mitotic initiator is not synthesised during the S period and inhibition of DNA replication prevents the induction of initiator synthesis.

This model agrees well with all currently known experimental results concerning the timing of mitosis in normal and perturbed *Physarum* plasmodia. In particular, the striking advancement of post-irradiation mitoses following ultraviolet treatment (Fig. 11) is readily explained. As mentioned before, ultraviolet-treatment causes an initial delay of mitosis (Fig. 6) which is followed by one or more shortened cycles, depending on the dose of ultraviolet light applied (Sachsenmaier *et al.*, 1970*b*; Sudbery & Grant, 1975). Mitoses in the ultraviolet-treated plasmodia eventually overtake the control schedule resulting in a net gain of time. This contrasts with the effects seen following mitotic delays induced by

Fig. 8. Model of the mitotic clock in *Physarum*. Kinetics of total cell mass (——) and number of nuclei (– – –) in exponentially growing synchronous plasmodia. *I*, initiator, synthesised in proportion to the increase of total cell mass; *N*, nuclear binding sites, doubling stepwise concomitantly with the number of nuclei; *M*, synchronous nuclear mitoses. *I/N*, ratio of number of initiator molecules per number of nuclear sites; (– · –) ratio of *I/N* assuming that initiator is not synthesised during S. Ordinate: cell mass and number of nuclei in arbitrary units.

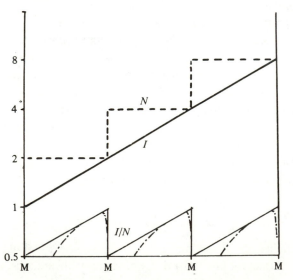

Fig. 9. Effects of cycloheximide pulses during the cycle on the timing of the next mitosis in *Physarum* at 26°C. Abscissa: start of 1.5 h pulses with 10μg ml⁻¹ cycloheximide expressed as fraction of the cycle between the second and third synchronous mitoses (M₂, M₃) after the inoculation of the cultures; total cycle time (M₂–M₃), 7h at 26°C. Standard errors of the mean are of five to eight replicates. Horizontal lines indicate protein : DNA ratio in controls (100 ± 5). The greater delay of mitosis caused by later inhibitions indicates synthesis of a mitosis initiator subject to turnover. (Tyson *et al.*, 1979).

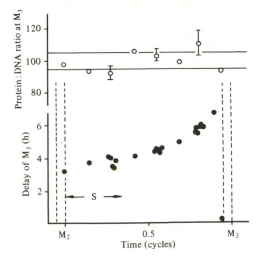

Fig. 10. Effects of cycloheximide pulses during the cycle on the timing of two consecutive mitoses in *Physarum polycephalum* at 19°C. Abscissa: start of 2-h pulses with 10 μg ml⁻¹ cycloheximide, expressed as fraction of the cycle, between the second and the third synchronous mitosis (M₂, M₃) after the inoculation of the cultures. Uninhibited controls performed M₃ at 14.2 ± 0.3 h and M₄ at 31.8 ± 1.0 h after M₂, as indicated by the horizontal lines. A later M₃ caused by inhibition is followed by a shorter cycle, so that M₄ in all treated cultures is close to 38h. (Tyson *et al.*, 1979).

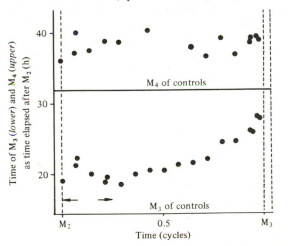

other means, e.g. X-irradiation, heat shocks, cycloheximide pulses. In these cases, the subsequent mitoses merely catch up with the control. The additional advance of post-irradiation mitoses in ultraviolet-treated plasmodia most likely results from the fact that many irradiated nuclei (20–30% after 200–500 J m^{-2}) disintegrate during the first post-irradiation S-phase (Sachsenmaier *et al*, 1970*b*). The protein : DNA ratio increases substantially above normal by this loss of nuclei. Assuming that mitotic initiator is still formed at an almost normal rate, i.e. proportional to the total cell mass (equation 1), less time is required to saturate the reduced number of nuclear sites. The interdivision time of subsequent cycles is shortened until the mass : DNA ratio reverts to its normal value, resulting in a permanent advance of the mitotic rhythm (Tyson *et al.*, 1979). The opposite effect – delaying mitosis by decreasing the mass : DNA ratio – was demonstrated long ago by Hartmann (1928) in his classical amputation experiments with *Amoeba proteus*. He was able to prevent growing amoebae from entering mitosis by repeated partial amputations of the cytoplasm. This effect also agrees well with our 'titration' model, provided that the presumptive initiator is indeed unstable.

Attempts to isolate and characterise the mitotic initiator from *Physarum* have not yet given conclusive results. Advancing effects of crude plasmodial extracts or homogenates on recipient plasmodia have been observed (Morhenn-Gruter, 1966; Oppenheim & Katzir, 1971); however, these effects were relatively small and difficult to reproduce for unknown reasons (Sachsenmaier & Dworzak, unpublished). On the other hand, significant advances of nuclear mitosis in *Physarum* were achieved with partially purified preparations of histone phosphoki-

Fig. 11. Effect of ultraviolet irradiation on the schedule of synchronous nuclear divisions in *Physarum polycephalum*. A single macroplasmodium was dissected into halves and one moiety was irradiated immediately (750 J m^{-2}, λ_{max} 254 nm) on the arrow, i.e. 3 h prior to the second synchronous mitosis (M_2) after the inoculation of the culture. Five consecutive mitoses (M_2–M_6) were observed in explants of the control and treated moieties over a period of 50 h. Solid bars show control mitoses; open bars show mitoses in the irradiated explant. (Sachsenmaier & Dworzak, 1976).

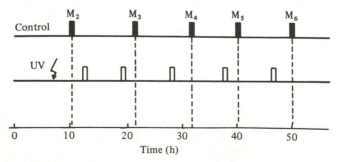

nase from mouse ascites tumour cells (Bradbury *et al.*, 1974*a,b*; illustrated in the chapter by Matthews, this volume). These findings support the assumption, discussed by Matthews in this volume, that H_1-histone phosphorylation is a key event in the triggering of chromosome condensation. In terms of our 'titration' model, histone phosphokinase and H_1-histone phosphorylation sites might well be candidates for the mitotic initiator (I) and the nuclear sites (N) respectively. However, there are still some question marks. For instance, histone phosphokinase of *Physarum* is strongly bound to the nuclear fraction whereas the mitotic initiator most likely is a cytoplasmic component. Also, the observed increase of histone phosphokinase activity during the cycle with a maximum in G_2 mostly reflects an activation process rather than synthesis of new enzyme protein (Mitchelson, Chambers, Bradbury & Matthews, 1978) which would not correspond well with the cycloheximide pulse experiments (Tyson *et al.*, 1979). Perhaps, the rate-limiting factor is a cytoplasmic activator of the nuclear histone phosphokinase. In any case, further experimental data are needed to prove or disprove the existence of the presumptive mitotic initiator.

The model proposed for the timing mechanism of nuclear mitoses in *Physarum* conforms with various observations in other systems too. The concept of an unstable 'division protein' was first developed by Zeuthen from studies with *Tetrahymena* (Rasmussen & Zeuthen, 1962; Zeuthen & Rasmussen, 1971). Fusion experiments with mammalian cells have shown rapid synchronisation of DNA synthesis and mitosis in multinuclear cells suggesting the important role of cytoplasmic factors in the timing of key events of the cell cycle (Johnson & Rao, 1971). The basic features of the 'titration' model remain unchanged if certain variations are applied (Fantes *et al.*, 1975; Sudbery & Grant, 1975), for example if the progressive removal during the cycle of an inhibitor is proposed instead of the accumulation of an initiator. Genetic evidence in fission yeast appears to favour the idea that the size control of the division cycle operates by the dilution of an inhibitor (Thuriaux, Nurse & Carter, 1978). On the other hand, fusion studies with *Physarum* plasmodia pretreated with an inhibitor of DNA synthesis (FUDR) tend to rule out the inhibitor dilution concept (Sachsenmaier *et al.*, 1972). None of these findings are absolutely decisive, however, as discussed in detail by Tyson *et al.* (1979).

References

Bradbury, E. M., Inglis, R. J. & Matthews, H. R. (1974*a*). Control of cell division by very lysine rich histone (F_1) phosphorylation. *Nature, London,* **247,** 257–61.

Bradbury, E. M., Inglis, R. J., Matthews, H. R. & Langan T. A. (1974*b*). Molecular basis of control of mitotic cell division in eukaryotes. *Nature, London,* **249,** 553–6.

Braun, R. & Behrens, K. (1969). A ribonuclease from *Physarum;* biochemical properties and synthesis in the mitotic cycle. *Biochemical and Biophysical Research Communications,* **195,** 87–98.

Braun, R., Mittermayer, C. & Rusch, H. P. (1965). Sequential temporal replication of DNA in *Physarum polycephalum. Proceedings of the National Academy of Sciences, USA,* **53,** 924–31.

Braun, R. & Wili, H. (1969). Time sequence of DNA replication in *Physarum. Biochimica et Biophysica Acta,* **174,** 246–52.

Brewer, E. N. & Nygaard, O. F. (1972). Correlation between unrepaired radiation induced DNA strand breaks and mitotic delay in *Physarum polycephalum. Nature New Biology,* **239,** 108–10.

Brewer, E. N. & Rusch, H. P. (1965). DNA synthesis by isolated nuclei of *Physarum polycephalum. Biochemical and Biophysical Research Communications,* **21,** 235–41.

Brewer, E. N. & Rusch, H. P. (1966). Control of DNA replication: Effect of spermine on DNA polymerase activity in nuclei isolated from *Physarum polycephalum. Biochemical and Biophysical Research Communications,* **25,** 579–84.

Brewer, E. N. & Rusch, H. P. (1968). Effect of elevated temperature shocks on mitosis and on the initiation of DNA replication in *Physarum polycephalum. Experimental Cell Research,* **49,** 79–86.

Chin, B., Friedrich, P. D. & Bernstein, I. A. (1972). Stimulation of mitosis following fusion of plasmodia in the myxomycete *Physarum polycephalum. Journal of General Microbiology,* **71,** 93–101.

Cummins, J. E., Blomquist, J. C. & Rusch, H. P. (1966). Anaphase delay after inhibition of protein synthesis between late prophase and prometaphase. *Science,* **154,** 1343–4.

Cummins, J. E., Brewer, E. N. & Rusch, H. P. (1965). The effect of actidione on mitosis in the slime mold *Physarum polycephalum. Journal of Cell Biology,* **27,** 337–41.

Daniel, J. W. & Baldwin, H. H. (1964). Methods of culture for plasmodial myxomycetes. In *Methods in Cell Physiology,* ed. D. M. Prescott, vol. 1, pp. 6–41. New York: Academic Press.

Donachie, W. D. & Masters, M. (1969). Temporal control of gene expression in bacteria. In *The Cell Cycle,* ed. G. M. Padilla, G. L. Whitson & I. L. Cameron, pp. 37–76. New York: Academic Press.

Dworzak, E., Mohberg, J., Forde, J., Sachsenmaier, W. & Czihak, G. (1976*a*). Characterization of BUDR-resistant mutants of *Physarum polycephalum. 3rd European Physarum Workshop, Rüttihubelbad (Bern),* Abstr. p. 26. University of Bern.

Dworzak, E., Mohberg, J., Sachsenmaier, W. & Czihak, G. (1976*b*). Charakterisierung von BUDR-resistenten Mutanten von *Physarum polycephalum.* In *Proceedings of the 3rd Annual Meeting of the Austrian Biochemical Society, Innsbruck 1976,* pp. 32–3. Publications of the University of Innsbruck 108. Innsbruck University Press.

Fantes, P. A., Grant, W. D., Pritchard, R. H., Sudbery, P. E. & Wheals, A. E. (1975). The regulation of cell size and the control of mitosis. *Journal of Theoretical Biology,* **50,** 213–44.

Forde, B. G. & Sachsenmaier, W. (1979). Oxygen uptake and mitochondrial enzyme activities in the mitotic cycle of *Physarum polycephalum. Journal of General Microbiology,* **115,** 135–43.

Fouquet, H., Bierweiler, B. & Sauer, H. W. (1974). Reassociation kinetics of nuclear DNA from *Physarum polycephalum. European Journal of Biochemistry,* **44,** 407–10.

Fouquet, H., Böhme, R., Wick, R., Sauer, H. W. & Braun R. (1974). Isolation of adenylate-rich RNA from *Physarum polycephalum*. *Biochimica et Biophysica Acta*, **353**, 913–22.

Fouquet, H. & Braun, R. (1974). Differential RNA synthesis in the mitotic cycle of *Physarum polycephalum*. *FEBS Letters*, **38**, 184–6.

Fouquet, H. & Sauer, H. W. (1975). Variable redundancy in RNA transcripts isolated in S and G_2 phase of the cell cycle of *Physarum*. *Nature, London*, **255**, 253–5.

Funderud, S., Andreassen, R. & Haugli F. (1978). Size distribution and maturation of newly replicated DNA through the S and G_2 phases of *Physarum polycephalum*. *Cell*, **15**, 1519–26.

Funderud, S. & Haugli, F. (1975). DNA replication in *Physarum polycephalum*: characterization of replication products *in vivo*. *Nucleic Acids Research*, **2**, 1381–90.

Gnamusch, A. & Sachsenmaier, W. (1979). Influence of plasmodial fusion on radiation induced mitotic delay in *Physarum polycephalum*. In *Current Research on Physarum. Proceedings of the 4th European Physarum Workshop, Innsbruck-Seefeld 1979*, ed. W. Sachsenmaier, pp. 151–6. Publications of the University of Innsbruck 120. Innsbruck University Press.

Grant, W. D. (1973). RNA synthesis during the cell cycle in *Physarum polycephalum*. In *The Cell Cycle in Development and Differentiation*, ed. M. Balls & F. S. Billett, pp. 77–109. Cambridge University Press.

Gröbner, P. & Sachsenmaier, W. (1976). Thymidine kinase enzyme variants in *Physarum polycephalum;* change of pattern during the synchronous mitotic cycle. *FEBS Letters*, **71**, 181–4.

Gröbner, P. & Sachsenmaier, W. (1979). Cycle dependent enzyme variants of thymidine kinase in *Physarum polycephalum*. In *Current Research on Physarum. Proceedings of the 4th European Physarum Workshop, Innsbruck-Seefeld 1979*, ed. W. Sachsenmaier, pp. 136–42. Publications of the University of Innsbruck 120. Innsbruck University Press.

Guttes, E. & Guttes, S. (1969). Replication of nucleolus-associated DNA during 'G_2-phase' in *Physarum polycephalum*. *Journal of Cell Biology*, **43**, 229–36.

Guttes, E., Guttes, S. & Rusch H. P. (1961). The morphology and development of *Physarum polycephalum* grown in pure culture. *Developmental Biology*, **3**, 588–614.

Hall, L. & Turnock, G. (1976). Synthesis of ribosomal RNA during the mitotic cycle in the slime mould *Physarum polycephalum*. *European Journal of Biochemistry*, **62**, 471–7.

Haugli, F. B. & Dove, W. F. (1972). Mutagenesis and mutant selection in *Physarum polycephalum*. *Molecular and General Genetics*, **118**, 109–24.

Hartmann, M. (1928). Über experimentelle Unsterblichkeit von Protozoen-Individuen. Ersatz der Fortpflanzung von *Amoeba proteus* durch fortgesetzte Regeneration. *Zoologischer Jahrbücher*, **45**, 973–87.

Helmstetter, C. E. (1969). Regulation of chromosome replication and cell division in *E.coli*. In *The Cell Cycle*, ed. G. M. Padilla, G. L. Whitson, I. L. Cameron, pp. 15–35. New York: Academic Press.

Johnson, R. T. & Rao, P. N. (1971). Nucleo-cytoplasmic interactions in the achievement of nuclear synchrony in DNA synthesis and mitosis in multinucleate cells. *Biological Reviews*, **46**, 97–155.

Kauffman, S. A. & Wille, J. J. (1976*a*). Evidence that the mitotic 'clock' in *Physarum polycephalum* is a limit cycle oscillator. In *The Molecular Basis of Circadian Rhythms*, ed. J. W. Hastings & H. G. Schweiger, pp. 421–31. Life Science Research Report 1. Berlin: Dahlem Konferenzen.

Kauffman, S. A. & Wille, J. J. (1976b). The mitotic oscillator in *Physarum polycephalum. Journal of Theoretical Biology*, **55**, 47–54.

Kessler, D. (1967). Nucleic acid synthesis during and after mitosis in the slime mold *Physarum polycephalum. Experimental Cell Research*, **45**, 676–80.

Loidl, P., Linortner, C. & Sachsenmaier, W. (1979). Timing of mitosis and DNA replication in mixed heterophasic plasmodia of *Physarum polycephalum*. In *Current Research on Physarum. Proceedings of the 4th European Physarum Workshop, Innsbruck-Seefeld 1979*, ed. W. Sachsenmaier, pp. 157–62. Publications of the University of Innsbruck 120. Innsbruck University Press.

Melera, P. W. (1980). Transcription in the myxomycete *Physarum polycephalum*. In *Growth and Differentiation in Physarum polycephalum*, ed. W. F. Dove & H. P. Rusch, pp. 64–97. Princeton University Press.

Mitchell, J. L. A., Campbell, H. A. & Carter, D. D. (1976). Multiple ornithine decarboxylase forms in *Physarum polycephalum;* interconversion induced by cycloheximide. *FEBS Letters*, **62**, 33–7.

Mitchell, J. L. A. & Carter, D. D. (1977). Physical and kinetic distinction of two ornithine decarboxylase forms in *Physarum. Biochimica et Biophysica Acta*, **483**, 425–34.

Mitchell, J. L. A., Carter, D. D. & Rybski, J. A. (1978). Control of ornithine decarboxylase activity in *Physarum* by polyamines. *European Journal of Biochemistry*, **92**, 325–31.

Mitchelson, K., Chambers, T., Bradbury, E. M. & Matthews, H. R. (1978). Activation of histone kinase in G2 phase of the cell cycle in *Physarum polycephalum. FEBS Letters*, **92**, 339–42.

Mitchison, J. M. (1971). *The Biology of the Cell Cycle*, pp. 159–80. Cambridge University Press.

Mittermayer, C., Braun, R. & Rusch, H. P. (1964). RNA synthesis in the mitotic cycle of *Physarum polycephalum. Biochemical and Biophysical Research Communications*, **91**, 399–405.

Mohberg, J., Dworzak, E. & Sachsenmaier, W. (1980). Thymidine kinase deficient mutants of *Physarum polycephalum. Experimental Cell Research*, **126**, 351–7.

Mohberg, J., Dworzak, E., Sachsenmaier, W. & Haugli, F. B. (1980). Thymidine kinase-deficient mutants of *Physarum polycephalum;* relationship between enzyme activity levels and ploidy. *Cell Biology International Reports*, **4**, 137–48.

Morhenn-Gruter, V. B. (1966). Versuche zum Nachweis eines Mitoseauslösenden Faktors in Plasmodien von *Physarum polycephalum*. MD Thesis, University of Heidelberg.

Muldoon, J. J., Evans, T. E., Nygaard, O. F. & Evans, H. H. (1971). Control of DNA replication by protein synthesis at defined times during the S period in *Physarum polycephalum. Biochimica et Biophysica Acta*, **247**, 310–21.

Nurse, P. & Thuriaux, P. (1977). Controls over the timing of DNA replication during the cell cycle of fission yeast. *Experimental Cell Research*, **107**, 365–75.

Nygaard, O. F., Guttes, S. & Rusch, H. P. (1960). Nucleic acid metabolism in a slime mold with synchronous mitosis. *Biochimica et Biophysica Acta*, **38**, 298–360.

Oleinick, N. L. (1972). The radiation-sensitivity of mitosis and the synthesis of thymidine kinase in *Physarum polycephalum:* a comparison of the sensitivity to actinomycin D and cycloheximide. *Radiation Research,* **51,** 638–53.

Oppenheim, A. & Katzir, N. (1971). Advancing the onset of mitosis by cell-free preparations of *Physarum polycephalum. Experimental Cell Research,* **68,** 224–6.

Rasmussen, L. & Zeuthen, E. (1962). Cell division and protein synthesis in *Tetrahymena,* as studied with *p*-fluorophenyl-alanine. *Comptes Rendus Travaux Laboratoire Carlsberg,* **32,** 333–58.

Rusch, H. P., Sachsenmaier, W., Behrens, K. & Gruter, V. (1966). Synchronization of mitosis by the fusion of plasmodia of *Physarum polycephalum. Journal of Cell Biology,* **31,** 204–9.

Sachsenmaier, W. (1964). Zur DNS- und RNS-Synthese im Teilungscyclus synchroner Plasmodien von *Physarum polycephalum. Biochemische Zeitschrift,* **340,** 541–7.

Sachsenmaier, W. (1966). Analyse des Zellcyclus durch Eingriffe in die Makromolekül-Biosynthese. In *Probleme der biologischen Reduplikation,* ed. P. Sitte, pp. 139–60. Berlin, Heidelberg & New York: Springer-Verlag.

Sachsenmaier, W. (1976). Control of nuclear mitosis in *Physarum polycephalum.* In *The Molecular Basis of Circadian Rhythms;* ed. J. W. Hastings & H.-G. Schweiger, pp. 409–20. Life Science Research Report 1, Berlin: Dahlem Konferenzen.

Sachsenmaier, W. & Becker, J. E. (1965). Wirkung von Actinomycin D auf die RNS-Synthese und die synchrone Mitosetätigkeit in *Physarum polycephalum. Monatshefte für Chemie,* **96,** 754–65.

Sachsenmaier, W., Bohnert, E., Clausnizer, B. & Nygaard, O. F. (1970*a*). Cycle dependent variation of X-ray effects on synchronous mitosis and thymidine kinase induction in *Physarum polycephalum. FEBS Letters,* **10,** 185–9.

Sachsenmaier, W., Dönges, K. H., Rupff, H. & Czihak, G. (1970*b*). Advanced initiation of synchronous mitoses in *Physarum polycephalum* following UV-irradiation. *Zeitschrift für Naturforschung,* **25 b,** 866–71.

Sachsenmaier, W. & Dworzak, E. (1976). Effects of UV and ionizing radiation on mitosis and enzyme regulation in the synchronous nuclear division cycle of *Physarum polycephalum.* In *Radiation and Cellular Processes,* ed. J. Kiefer, pp. 229–39. Berlin, Heidelberg & New York: Springer-Verlag.

Sachsenmaier, W., Finkenstedt, G., Linser, W., Madreiter, H., Woertz, G. & Wolf, H. (1973). Regulation of thymidine phosphorylating enzymes in the synchronous nuclear division cycle of *Physarum polycephalum.* In *Proceedings of the 2nd European Symposium on the Cell Cycle, Innsbruck 1973.* pp. 36–8. Publications of the University of Innsbruck 77. Innsbruck University Press.

Sachsenmaier, W., Fournier, D. v & Gürtler, K. F. (1967). Periodic thymidine kinase production in synchronous plasmodia of *Physarum polycephalum:* inhibition by actinomycin and actidion. *Biochemical and Biophysical Research Communications,* **27,** 655–60.

Sachsenmaier, W. & Hansen, K. (1973). Long- and short-period oscillations in a myxomycete with synchronous nuclear divisions. In *Biological and Biochemical Oscillators,* ed. B. Chance, A. K. Ghosh, K. E. Pye & B. Hess, pp. 429–47. New York & London: Academic Press.

Sachsenmaier, W. & Ives, D. H. (1965). Periodische Änderungen der Thymidinkinase-Aktivität im synchronen Mitosecyclus von *Physarum polycephalum. Biochemische Zeitschrift,* **343,** 399–406.

Sachsenmaier, W., Remy, H. & Plattner-Schobel, R. (1972). Initiation of synchronous mitosis in *Physarum polycephalum;* a model of the control of cell division in eukaryotes. *Experimental Cell Research,* **73,** 41–8.

Sachsenmaier, W. & Rusch, H. P. (1964). The effect of 5-fluoro-2'-deoxyuridine on synchronous mitosis in *Physarum polycephalum. Experimental Cell Research,* **36,** 124–33.

Schiebel, W., Bär, A., Hotz, E., Järvinen, P. & Murray B. (1976). DNA polymerase of *Physarum polycephalum. 3rd European Physarum Workshop, Rüttihubelbad (Bern), Sept. 14–6, 1976;* Abstr. p. 25. University of Bern.

Sedory, M. J. & Mitchell, J. L. A. (1977). Regulation of ornithine decarboxylase activity during the *Physarum* mitotic cycle. *Experimental Cell Research,* **107,** 105–10.

Solao, P. B. & Shall, S. (1971). Control of DNA replication in *Physarum polycephalum.* I. Specific activity of NAD pyrophosphorylase in isolated nuclei during the cell cycle. *Experimental Cell Research,* **69,** 295–300.

Sudbery, P. E. & Grant, W. D. (1975). The control of mitosis in *Physarum polycephalum;* the effect of lowering the DNA : mass ratio by UV irradiation. *Experimental Cell Research,* **95,** 405–15.

Thuriaux P., Nurse, P. & Carter, B. (1978). Mutants altered in control coordinating cell division with cell growth in fission yeast *Schizosaccharomyces pombe. Molecular and General Genetics,* **161,** 215–20.

Tyson, J., Garcia-Herdugo, G. & Sachsenmaier, W. (1979). Control of nuclear division in *Physarum polycephalum;* comparison of cycloheximide pulse treatment, UV irradiation and heat shock. *Experimental Cell Research,* **119,** 87–98.

Tyson, J. & Sachsenmaier, W. (1978). Is nuclear division in *Physarum* controlled by a continuous limit cycle oscillator? *Journal of Theoretical Biology,* **73,** 723–38.

Vogt, V. M. & Braun, R. (1977). The replication of ribosomal DNA in *Physarum polycephalum. European Journal of Biochemistry,* **80,** 557–66.

Wille, J. J. & Kauffman, S. A. (1975). Premature replication of late S period DNA regions in early S nuclei transferred to late S cytoplasm by fusion in *Physarum polycephalum. Biochimica et Biophysica Acta,* **407,** 158–73.

Woertz, G. & Sachsenmaier, W. (1979). Deoxyadenosine kinase in *Physarum polycephalum.* In *Current Research on Physarum, Proceedings of the 4th European Physarum Workshop, Innsbruck-Seefeld, 1979,* ed. W. Sachsenmaier, pp. 123–9. Publications of the University of Innsbruck 120. Innsbruck University Press.

Wright, M. & Tollon, Y. (1979*a*). *Physarum* thymidine kinase. A step or a peak enzyme depending upon temperature of growth. *European Journal of Biochemistry,* **96,** 177–81.

Wright, M. & Tollon, Y. (1979*b*). Regulation of thymidine kinase synthesis during the cell cycle of *Physarum polycephalum* by the heat sensitive system which triggers mitosis and S phase. *Experimental Cell Research,* **122,** 273–9.

Zeuthen, E. & Rasmussen, W. (1971). Synchronized cell division in Protozoa. In *Research in Protozoology,* ed. T. T. Chen, vol. 4, pp. 11–145. Oxford: Pergamon Press.

M.M.YEOMAN

The mitotic cycle in higher plants

Introduction

Studies on the cell cycle in higher plants have been largely restricted to four cell types: (i) the isodiametric, essentially non-vacuolate cells, so characteristic of the apical meristems of roots and shoots, (ii) cells in culture which are highly vacuolated and frequently display cytological heterogeneity, (iii) pollen mother cells, (iv) the highly vacuolated cells of storage tissues induced to divide following excision and culture (Yeoman, 1976). However, almost all of the available information on the timing and cytological characteristics of the cell cycle has come from studies on the apical meristem of the seedling root. Therefore it must be continually borne in mind that any apparent generalisations in the literature, and there are many, arise mainly from the data accumulated from studies on the cells of root meristems. It is also important to remember that there is no typical dividing cell in plants, indeed marked structural differences exist even between the constituent cells of meristems within the same plant. The cells so typical of the apical meristems bear little structural similarity to the cells of a lateral meristem such as the vascular cambium. However, they do have one important feature in common, the ability to divide, a property also shared with many other plant cells apart from meristems. To emphasise this point the range of different dividing cell types found in Angiosperms is shown in Fig. 1.

Concept of the cell cycle

The concept of a cell cycle is applied throughout the animal and plant kingdom, although it originated from studies by Howard & Pelc (1953) on root meristem cells. As the name suggests it is a cyclical process in which the parent cell passes through a series of synthetic phases and eventually divides to give two daughter cells, which in a mitotic division are identical to the original parent at the onset of the previous cycle. In meiosis the position is more complex and at the end of two cycles four cells are produced, each with half the original chromosome number of the parent and with a re-assortment of the genetic information.

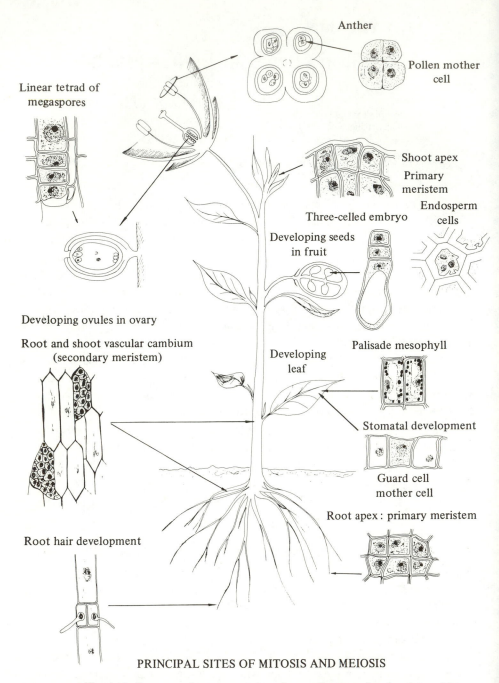

Anther

Pollen mother cell

Linear tetrad of megaspores

Shoot apex

Primary meristem

Endosperm cells

Three-celled embryo

Developing seeds in fruit

Developing ovules in ovary

Root and shoot vascular cambium (secondary meristem)

Developing leaf

Palisade mesophyll

Stomatal development

Guard cell mother cell

Root apex : primary meristem

Root hair development

PRINCIPAL SITES OF MITOSIS AND MEIOSIS

Fig. 1. A diagrammatic representation of an angiosperm showing the position and range of dividing cells. The major sites of division are within the apical (root and shoot) and lateral (cambia) meristems. (Adapted from Dyer, 1976).

General characteristics of the mitotic cell cycle

The mitotic cell cycle is divided into four phases, G1, S, G2 and M. This applies to both plant and animal cells (see Fig. 2). Within the overall cell cycle there are a number of component cycles: (*a*) the nuclear DNA cycle, (*b*) the chromosome cycle, (*c*) the nucleolar cycle, (*d*) the nuclear envelope cycle, (*e*) the spindle cycle, (*f*) the centriole cycle and (*g*) miscellaneous cycles involving plastids, mitochondria, plasma membrane and other extranuclear organelles (see Dyer, 1976). Animal and plant cells are similar with respect to most of these sub-cycles but there are some exceptions. Higher plant cells do not contain centrioles and therefore do not take part in centriole replication, and animal cells do not contain plastids. Of course details may differ as they do within plant or animal species but overall these cycles and the integrated mitotic cycle are very similar. Two further striking differences are apparent between dividing plant and

Fig. 2. A diagram showing the relationship between the cell cycle in a meristematic cell and differentiation. Cells may leave the cycle at G1, as shown above, or at G2 and pass into a G0 phase prior to subsequent differentiation. G0 may represent a biochemical condition in which metabolic processes occur that have no counterpart when the cell is cycling.

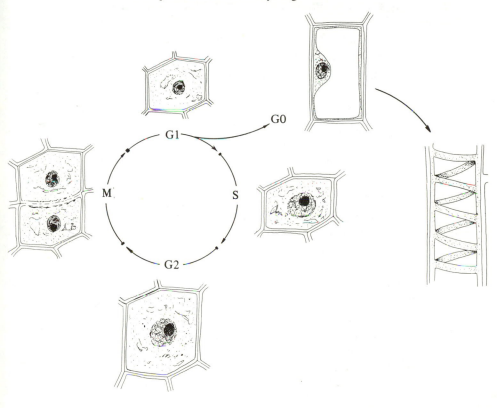

animal cells. One is the possession of asters, these are a system of radiating fibres around the centriole which appear to organise the formation of the spindle and are not observed in plant cells. The second is that no cell plate is formed in animal cell mitosis. Instead, in telophase, a furrow appears in the cell membrane forming at the equator. The furrow deepens and the two daughter cells become pinched apart. The cell plate in plants is formed from the fusion of aggregated vesicles originating from Golgi bodies and precedes the formation of a rigid primary cellulosic cell wall. The wall later thickens and a middle lamella appears which unites with that of the parent cell wall (Brown & Dyer, 1972).

It is the process of mitotis (M) which provides the most spectacular visible evidence of the division of the genetic material into two equal parts, and which has received considerable attention over the past 80 years. Mitosis is the phase of the cell cycle which is best understood (Dyer, 1976). Indeed until comparatively recently (c. 1950) it was considered that cell division consisted of two parts, interphase which was a resting period and mitosis, a period of intense cellular activity. The discovery that ended this particular misunderstanding was of the existence of an S phase in which the DNA of the cell is replicated and which is an essential pre-requisite for mitosis. The periods G1 and G2 represent temporal gaps between M, S, and M respectively. Confusions have arisen in the literature when G was taken to represent growth and not gap! It is of course true that cell growth in terms of the synthesis and accumulation of various macromolecules does take place during the gap G1 but this is coincidental. It is generally considered that a cell located in an actively growing apical meristem outside the quiescent centre cycles freely and the size of the dividing population within such a meristem stays fairly constant. This means that of the two daughters produced at each division only one remains within the meristem while the other is displaced away from the tip and differentiates into another cell type within the plant body. Immediately after leaving the cycle the cell may be said to be in the G0 phase (see Fig. 2). It has been argued that G0 represents a special biochemical condition in which processes occur that have no counterpart when the cell is cycling. Alternatively it may be that the cell is blocked at some point in G1 and that G0 is not a unique period (Roberts, 1976; Smith & Martin, 1973). Such arrested cells may proceed down a particular pathway of differentiation or begin recycling after a 'rest' period (See discussion of the G0 phase by Stiles, Cochran & Scher in this volume.)

It is now clear from the classical studies of Clowes and others (for review see Clowes, 1976) that the population of cells within an apical meristem shows a variation in cell cycle times and that these may change from one cycle to another, e.g. the quiescent promeristem. This heterogeneity has complicated the interpretation of the timing of the cell cycle and its component phases from experiments with apical meristems in which radioactive precursors of DNA, or colchicine or caffeine have been used to mark discrete populations of dividing cells.

Timing of the cell cycle and its component phases

A number of techniques have been used to time the cell cycle. The simplest possible approach is by direct observation of living cells, usually in conjunction with cinemicrography. This method relies on the direct measurement of the time between mitoses, which are visualised in living cells by phase contrast microscopy. As such it is only applicable to single cells or cell monolayers and has only been used occasionally with higher plant material. The early experiments of Jones, Hildebrandt, Riker & Wu, (1960) used this technique in conjunction with microcultures of tobacco cells, and it can be applied to isolated protoplasts in culture but has not so far been exploited.

The direct counting of cells may also be employed to give a measure of the mean cell generation time which gives an average value for the cell cycle time in the population under observation. In synchronously dividing populations, the removal of samples at intervals and subsequent counting of the total number of cells provides a means of measuring average cell cycle time. An extension of this approach using radioactive precursors or making quantitative measurements of DNA or mitotic index at intervals provides additional information about the timing of G1, S, G2 and M.

However, the best method for the determination of the timing of the cell cycle in multicellular systems such as meristems (Van't Hof, 1968) and animal or plant cell cultures (Aherne, Camplejohn & Wright, 1977) is to mark a small discrete cell population within a much larger asynchronous population and follow the course of this marked group of cells along a time axis. Three methods of labelling cells in this way have achieved popularity. The first employs colchicine which produces a small group of tetraploid cells when applied for a restricted period to the meristem of diploid seedling roots. Colchicine prevents the assembly of the spindle and this leads to the formation of a reconstitution nucleus in those cells entering, or in metaphase, at the time the colchicine is present. The proportion of dividing tetraploid cells can then be estimated at intervals after the removal of colchicine. The second approach is to use caffeine, again over a restricted time interval. This prevents cytokinesis and consequently induces a small group of binucleate cells which can be followed in much the same way as the cells treated with colchicine (see Giménez-Martin de la Torre & López-Sáez, 1977). The third technique, which has become the method of choice for cytologists (Aherne et al., 1977), is to use a DNA precursor, usually [^{3}H]thymidine ([^{3}H]dThd), to label the DNA of cells and to follow the subsequent fate of these using autoradiography. The great advantage of this last technique is that it can be used effectively to measure the component phases of the cell cycle, G1, S, G2 and M. Also these data taken together with the observed frequencies of cells in prophase, metaphase, anaphase and telophase can provide comprehensive information of the duration of the component phases of the cycle including those of mitosis. Elaborations of this 'pulse-chase' technique have been used, which include the use of two different radioactive precursors (Baserga & Nemeroff, 1962; Aherne et al., 1977), usually [³H]dThd and [¹⁴C]dThd. These procedures have proved most useful with systems, such as the shoot apex (Miller & Lyndon, 1975), where the proportion of dividing cells and the total number of individual apices available is low, and it is difficult to obtain enough labelled mitoses to give an accurate measurement of the cycle times. The double labelling technique also enables the flux rates into and out of S to be determined.

The application of these techniques to the heterogeneous cell populations of apical meristems has produced numerous elegant and informative publications. However, as has been indicated by Green (1976), great caution must be exercised in the interpretation of the results. At best the times produced represent only an average for the cell population under observation. And as pointed out many times (Van't Hof, 1968; Green, 1976) this method applies only to a meristem in an 'ideal state' where (a) all cells maintain a cycle of identical length; (b) the distribution of cells in the different phases of the cell cycle is equal to the time spent in that phase, i.e. the cells are evenly distributed throughout the cycle; (c) all cells have G1, S, G2 and M of identical length, i.e. it is in a steady state and; (d) the population is asynchronous.

It is equally clear from the studies of Clowes, Barlow, Van't Hof and others that few if any cell populations conform and there are many departures from this 'ideal state'. Cell cycles are not all of the same length, cell cycle times do vary in a particular cell from one cycle to the next and there are certainly variations in the component phases between consecutive cycles. It is also improbable that the distribution of cells in any one phase is symmetrical. A general approach to these difficulties is to look at a designated smaller population within the larger cell population within which uniformity is more likely. Studies along these lines produced the valuable concept of the quiescent pro-meristem and began to provide a better understanding of the organisation of apical meristems (Clowes, 1976).

Molecular characteristics of the cell cycle

A certain amount of information can be obtained about the timing and intensity of synthesis of DNA, RNA and protein using apical meristems. The approach using radioactive precursors of RNA ([³H]Urd, [¹⁴C]Oro), DNA ([³H]dThd, [¹⁴C]dThd) and protein ([³H]- or [¹⁴C]-Leu) is to feed the precursors individually, usually to seedling root meristems, and follow the incorporation of these precursors by autoradiography. From these studies it is known that replication of DNA in the nucleus occupies a specific part of the cell cycle, the S period, and that RNA, and protein are synthesised at particular points of the cell cycle. In addition some measure can be obtained of the rate of synthesis of these macromolecules but these techniques cannot be used to establish the overall pattern of synthesis and accumulation during the cell cycle, or to study the synthesis of particular species of RNA or proteins. Some attempt has been made to rectify this deficiency in our knowledge by the use of histochemical procedures with asynchronous systems. Here there are two basic problems, the first is the ability, or rather the inability, to recognise the point in the cell cycle reached by an individual cell in an asynchronous population, and secondly the development of reliable techniques that can be applied quantitatively to plant cells to measure individual substances. Several possibilities exist for positioning an individual in the cell cycle; however, only one method has been exploited, this is the ingenious approach adopted by Woodard, Rasch & Swift, (1961a). They showed that if the nuclear volumes of cells within the apical root meristem of *Vicia faba* were plotted against 'intermitotic time', a relationship became apparent in which the volume of the nucleus increased progressively through the cell cycle. Their assumption being that the distribution of nuclear volumes reflects the pattern of growth of the cell, an assumption itself based on the low variances of small samples of prophase and telophase volumes. Using Feulgen microdensitometry they also determined which nuclei within the population were replicating DNA and were therefore able to delimit the S period. Once this was established it was possible to investigate changes in RNA and protein using histochemical procedures and to characterise the events of G1 and G2 in the root meristem of the bean. This technique has not been widely used by other workers for a number of reasons. Lyndon (1967) working with pea root meristems has clearly shown that within this dividing cell population there is no clear relationship between nuclear volume and position of the nucleus in the cell cycle. Indeed the differences in nuclear volume between nuclei of different cells in telophase, an easily recognisable stage, with the same amount of DNA may be of the order of two or three times. Confirmation of Lyndon's observations on pea roots has been published recently by Webster (1979b). Similar results have been reported for *Vicia faba* (Bansal & Davidson, 1978; Wellwood & Davidson, 1977) and for *Tradescantia* (White & Davidson, 1976) where a considerable range of volumes exists for both prophase and telophase nuclei. A quite different result has been obtained by Doležal & Tschermak-Woess (1955) and Nagl (1968) who have shown a clear relationship between nuclear volume and position of the cell in the cycle in some species including *Allium carinatum*. It would appear that such relationships are not unusual and may depend on the particular plant under investigation (Nagl, 1977). An additional observation made by R. F. Lyndon, again with pea root meristems (personal communication), is that even in pea where there is no over-all relationship between nuclear volume and cell cycle position a relationship may exist along a file of cells.

It would appear that apart from cells in identifiable cell lineages and some exceptions (Nagl, 1977), a clear relationship does not exist between nuclear volume and percentage elapsed interphase time, i.e. the position of a cycling nucleus within the cell cycle. After a careful consideration of the available data Webster (1979a) has suggested a means of identifying early G1 and late G2 nuclei in root meristems of the pea. This identification of G1 and G2 nuclei relies on the fact that sister cells, while not dividing exactly together, have very similar cycle durations (Webster, 1979a,b) as has been noted in mammalian cells and discussed by Brooks and by Fantes & Nurse in this volume. A second approach (Nagl, 1977) relies on the fact that changes occur in the structure and dispersion of chromatin during the mitotic cycle. A feature of the interphase nuclei of the cells of many plants is the presence of chromocentres. These are segments of chromosomes that retain their condensed state following the reconstitution of new nuclei after division. There are, however, phases of the cell cycle in which the chromocentres are dispersed, the so-called 'Z' phase (Zerstäubungs-stadium, Heitz, 1928). Heitz (1929) believed that Z phase occurred at early prophase prior to the spiralisation of chromatin into distinguishable chromosomes. The possibility that the Z phase is associated with some other phases of the mitotic cycle has been denied (Tschermak-Woess, 1959; Nagl, 1970a,b) but recent work by Barlow (1976, 1977) using microdensitometry and autoradiography has shown clearly that in both *Allium flavum* and *Bryonia dioica* the Z phase is located at the end of S phase and not at early prophase as had been hitherto assumed. A valuable fact that has emerged from the work of Barlow (1977) is that in some species, e.g. *Bryonia dioica* nuclei in G1, G2 and Z phase are easily recognisable and this may subsequently be used to position nuclei within the cell cycle and allow a more meaningful autoradiographic, histochemical investigation of molecular events during the cell cycle.

Once an effective means has been found to position a cell within the cell cycle then further progress will depend on the exploitation and development of methods for the quantitative determination of specific molecules. There are already a battery of methods for the determination of the major classes of macromolecules: protein, DNA, RNA and polysaccharide. Microdensitometric measurement after Feulgen staining has been used extensively for DNA. This technique may also be linked with a fluorochrome e.g. Pararosaniline–Feulgen or Acriflavine–'Feulgen' (Ploem, 1977) and the emitted fluorescence measured using microfluorimetry; sensitivity is superior but the cost of instrumentation is high. Gallocyanin-chrome alum may be used to determine total nucleic acid content by microdensitometry (Mitchell, 1968a) and RNA by difference after RNAase treatment. Acridine orange produces green fluorescence from DNA and red fluorescence from RNA and denatured DNA (Rigler, 1966). Again a means of measuring fluorescence is required. Total protein can be easily measured by microdensitometry using a variety of techniques including dinitrofluorobenzene (Mitchell, 1968b) and the periodic acid–Schiff reaction may be used for carbohydrates. A variety of techniques have been used to detect enzymes in plant cells but even the best of these are semi-quantitative (Goldstein, 1977) because it is difficult to achieve optimal conditions for quantitative enzyme assay using tissue sections on a microscope slide. Here the techniques of immunofluorescence could provide a solution. Immunofluorescence techniques are now widely used with animal cells and micro-organisms, and there is an increasing number of publications on the application of these techniques to plants. Specific antisera can be raised to single proteins (antigen) and these can be linked to fluorochromes and a measurement made of the amount of conjugated antibody binding to the antigen within the cell (Ploem, 1977). The methodology is certainly complex and sufficient standardisation of substrate and antisera is an essential prerequisite. Instruments are available which can be used to give a quantitative evaluation of even moderately strong fluorescence, and already developments are taking place in which a laser may be used as a light source to enable the quantitation of weak fluorescence. This is certainly a technique for the future and will enable a study to be made of changes in the level of specific macromolecules during the cell cycle. However, most of the available information has been obtained with synchronously dividing pollen mother cells, high vacuolated cell populations of artichoke tissue or suspension cultures of sycamore and it is mainly this evidence that will be considered in the following section.

Patterns of synthesis and accumulation of macromolecules
Availability and suitability of systems

The timing and extent of DNA replication in higher plant cells has received a fair amount of attention over the past 20 years but little is known of the mechanism of DNA replication in plants. As has already been pointed out the majority of the research has been performed with asynchronously dividing cell populations using autoradiographic or microdensitometric methods. This is also true for studies on RNA and protein synthesis and accumulation. Apart from the approach suggested in the previous section only one other basic strategy is obvious, the use of synchronous or synchronised cell populations. Indeed the substantial advances made towards an understanding of cell division in micro-organisms and animal cells have been largely due to the free availability of relatively homogeneous populations of cells which divide synchronously (Mitchison, 1971). The exploitation of natural synchrony and the creation of artificially synchronised systems has proved of overwhelming importance in a study of the molecular events of the cell cycle. Synchrony in eukaryote systems is always imperfect (Fig. 3) but perhaps the closest approach to absolute simultaneity is to be seen within a natural process in angiosperms: the development of pollen. Here a meiotic division is followed by a mitotic one and the product is four haploid pollen grains. The synchrony observed in this pair of divisions is extremely high (see Stern, 1966). More recently other plant systems have been used (King & Street, 1973; Yeoman & Aitchison, 1973) to examine the molecular events of the cell division cycle. Other systems in which a very high degree of synchrony occurs naturally have been reported but so far these have not been exploited to study the molecular behaviour of dividing plant cells (Yeoman, Aitchison & Macleod, 1977). Attempts to enforce synchrony in root meristems have proved successful but these systems are not very suitable for biochemical studies. In contrast to the studies with meristems, tissue and cell cultures have been manipulated to produce synchronised systems (see Yeoman *et al.*, 1977) which have proved amenable to the investigation of macromolecular changes which take place as a cell traverses the cell cycle.

Fig. 3. The distribution of cells in mitosis through a single period in which the cell number doubles. (*a*) Complete synchrony, (*b*) good synchrony, (*c*) poor synchrony, (*d*) complete asynchrony. Mitosis (M) occupies 10% of the cell cycle time. Mitotic Index = percentage of cells in mitosis.

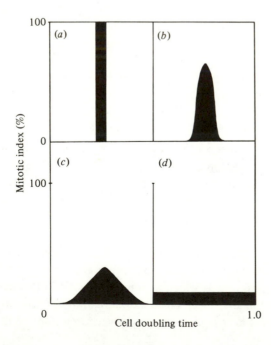

In summary, synchronous or synchronised systems have two major advantages in the study of the molecular basis of the cell cycle. (1) Large numbers of cells pass through the same metabolic sequence together. This provides an amplification system which permits sophisticated structural or chemical analyses which are impossible with a single cell or a few cells. (2) Representative samples can be withdrawn without disturbing the cyclic chain of events under study.

DNA synthesis and accumulation

Studies on the time course of DNA replication in plant cells are few (Woodard *et al.*, 1961*a*; Van't Hof, 1965; Mitchell, 1967, 1968*b*; Harland Jackson & Yeoman, 1973; Mungall, 1978). Apart from the investigations of Woodard *et al.* (1961) and Van't Hof (1965) they have all been performed with synchronous systems. It would appear from these somewhat limited results that DNA accumulates at a constant rate throughout the S phase (Mitchell, 1968*b*; Harland *et al.*, 1973). A similar linear increase in the amount of DNA per dividing cell has been demonstrated in cultures of *Paramecium* (Woodard *et al.*, 1961*b*), mouse fibroblasts (Zetterberg & Killander, 1965) and with *Euplotes* (Prescott, 1966).

Huberman & Riggs (1966, 1968) using a technique developed for *Escherichia coli* by Cairns (1962) showed that DNA replication in cultured mammalian cells occurred simultaneously at a large number of sites on the DNA template. These points are visualised as replication forks and probably represent sites of action of DNA polymerase as two daughter helices are formed from the parental template. This pattern has since been confirmed with a variety of mammalian cells (Callan, 1973). The estimated rate of replication in one direction was $0.5 - 1.2$ μm min^{-1}, less than one tenth the rate observed in *E. coli* (Cairns, 1973). The much slower rate of replication in the mammalian cells may be explained by the association of eukaryotic DNA with histones. Replication units were found to be arranged at varying intervals with a mean value of 50 μm. Replication at neighbouring origins may be initiated at staggered time intervals.

Van't Hof (1975) has used DNA fibre autoradiography to investigate DNA replication in cells of the root tip of *Pisum sativum*. From these studies he has estimated that the average distance between replication points is 54.7 μm and the rate of replication in one direction was 29 μm h^{-1} at 23°C. More recently Van't Hof (1976*a,b*) has suggested that the rate of fibre growth may not be constant throughout the S phase. The rate of fibre growth was found to increase threefold in the first two hours of the S phase in the cells of partially synchronised cultured pea root meristems. The initial rate was 4.5 μm h^{-1} but forks active after 90 min moved at nearly 18 μm h^{-1}. The faster movement was not characteristic of all replication units. Certain fibres consisted of replication units of a smaller mean size with slow-moving forks ($4.5 - 6 \mu$m h^{-1}).

It has also been reported that in some animal cells the rate of fork movement is lower in the early portion of the S phase than when the cells are in mid or late S phase. For example HeLa cells at the beginning and at the end of the S phase have rates of fork movement of $18 - 30$ and $66 \mu m \ h^{-1}$ respectively (Painter & Schaefer, 1971). Also, Chinese hamster ovary cells in early S phase have replication forks which move at $12.6 \mu m \ h^{-1}$, while those activated 7h later move at $38 \mu m \ h^{-1}$ (Houseman & Huberman, 1975). They also showed that the rate of fork movement increased 2.6 times within a 4-h period as the cells passed from early to late S phase.

It would therefore appear that the rate of DNA replication at the fibre level is not uniform throughout the S phase suggesting that a rate-limiting step exists. Linearity of net rate of synthesis throughout S suggests that there is some control over the number of sites at which DNA synthesis is occurring at any time. It is, however, worth noting that estimation of rates of synthesis based on chemical or cytochemical data can only give rather crude average rates, which may well obscure short-term variations. It is also well known from chromosome autoradiography that different regions of the genome are labelled at different times during S (Taylor, 1958; Wimber, 1961; Kuroia & Tanaka, 1970).

At the moment it has not been possible to establish the relative contribution made to the control of overall DNA synthesis by (a) the rate of DNA synthesis within a replication unit as measured by the rate of chain growth through the S phase and (b) the number of replication units active at one time. If, as suggested by Ockey & Saffhill (1976), a regulation system exists which maintains a relatively constant overall rate of chain growth over long stretches of DNA and the rate of synthesis varies through the S phase (Van't Hof, 1976a,b; Painter & Schaefer, 1971; Houseman & Huberman, 1975) it is only by the use of a system which has a high level of 'natural' synchrony that the method of controlling the overall rate of DNA synthesis may be established.

Certainly a fact that appears to determine the length of the S period is the amount of nuclear DNA to be replicated. Van't Hof (1965) and Van't Hof & Sparrow (1963) have shown this to be true for the cells of the root meristem in seven plant species. Van't Hof (1965) also showed that for this group, the total average cycle time was determined by the duration of the S Phase. These data are not necessarily in conflict with observations on large differences in cycle times in different regions of one root (Clowes, 1965), as they refer to average times for whole roots, within each of which there will be a spectrum of cycle lengths. Not all published data however, are consistent with the suggestion of Van't Hof. Troy & Wimber (1968) compared diploid and autotetraploid pairs of *Lycopersicon esculentum, Tradescantia paludosa, Ornithogalum vivens,* a *Cymbidium* culture and a series of *Chrysanthemum* species. They found that the duration of the S phase was similar between pairs and therefore independent of

DNA content. Friedberg & Davidson (1970) also found the cells within the meristems of lateral roots of *Vicia faba* to have similar mean cycle times. A possible explanation of these cases is that whatever genetic control element determines the rate of replication in the diploid will itself be present in a double dose in the tetraploid, enabling a doubling in the net rate of synthesis. Other departures from the Van't Hof rule exist but generally the rule holds. It is perhaps important to note that a highly significant relationship exists between the duration of meiosis and nuclear DNA content in higher diploid plants (Bennett, 1971, 1973).

RNA synthesis and accumulation

The protein synthesising capacity of cells, and therefore their ability to process through the cycle, rests directly on the various species of RNA that are crucial components of the machinery involved. Of these rRNA and tRNA constitute between them 90–95% of the total RNA. Therefore it might be expected that increases in the amount of protein being synthesised would be reflected in corresponding increases in the molecular machinery necessary for synthesis, and as rRNA in particular is an integral part of this machinery, it might be anticipated that levels of rRNA would correlate with established features of the metabolic pattern of the cell cycle. Early indications of the pattern of RNA accumulation in dividing plant cells came from the work of Woodard *et al.* (1961a) with *Vicia faba* root tips. They were able to correlate DNA replication with the synthesis and accumulation of RNA and protein during a cell cycle by the use of cytochemical procedures coupled with autoradiography. Jakob (1972) using a partially synchronous root meristem of *Vicia faba,* has demonstrated that the major proportion of RNA is synthesised during late G1 and early S and that at least part of the RNA synthesised is essential for normal DNA replication. A major point of interest from this and other investigations on RNA synthesis and accumulation in plants is the periodic nature of the major increases. This contrasts with the accumulation of RNA in most other organisms (Mitchison, 1971) e.g. *Chlorella* (John, Lambe, McGookin & Orr, this volume) which takes place throughout the cell cycle (Green, 1974). Rate of synthesis may change however, (Klevecz & Stubblefield, 1967), and both Woodard *et al.* (1961a), and Mitchell (1969) using synchronously dividing cells from artichoke explants, showed an approximately threefold increase during a single cycle. This increase (Mitchell, 1969) was not linear but occurred in three steps, one immediately post telophase, one during the S phase and a third very sharp increase in G2. RNA synthesis is also variable in the cell cycle of *Physarum polycephalum,* showing minima in M and early G2, and maxima during S and mid G2 (Mittermayer, Braun & Rusch, 1964). An increase of just less than threefold has been reported in *Paramecium* with a sharp rise in the rate of accumulation in pre-prophase (Woodard *et al.,* 1961b). It has not been satisfactorily explained why RNA, in these cases, shows more than the doubling in amount that would be expected in a normal cycle. Evidence for active synthesis of RNA in dividing cells has also been obtained with yeast, HeLa cells and mouse fibroblasts in culture (Mitchison & Walker, 1965; Terasima & Tolmach, 1963; Zetterberg & Killander, 1965) but in these systems the progress of RNA accumulation appears to be more or less exponential with an approximate doubling in amount during each cell cycle.

The synthesis and accumulation of RNA during the early synchronous divisions of artichoke tissue culture cells has been examined in some detail by Fraser & Loening (1974) and Macleod, Mills & Yeoman (1979). These later investigations confirmed the earlier results of Mitchell (1969) using microdensitometric methods which showed that total RNA accumulated periodically. In addition Fraser & Leoning (1974) showed that during the first two cell cycles the rate of ribosomal RNA (rRNA) synthesis increased sharply in two steps; before the onset of DNA synthesis for the first division, and early in interphase before the second division. The rate of RNA accumulation also increased sharply at these times. They also concluded that the stepwise pattern of rRNA synthesis was not caused by the replication of rRNA genes as can happen in mammalian cells. In a later investigation following on from this study of Fraser & Loening (1974), Macleod et al. (1979) were able to separate the RNA changes associated with wounding from those related to cell division and concluded that discontinuous synthesis and periodic accumulation of RNA are a feature of the cell cycle in the artichoke system. A picture that emerges from the results of Fraser & Loening (1974), Macleod et al. (1979) and A. J. Macleod (unpublished) is one of sustained synthesis of RNA, and presumably balanced degradation, with periods of increased synthesis producing transient periods of accumulation of RNA at discrete stages during the cell cycle. Within this pattern the levels of rRNA and tRNA are independently regulated. Superimposed on all of this is a pattern of periodic modification of the tRNA and a more continuous modification of the rRNA.

It has been clearly demonstrated that the nucleolus is the major site of RNA synthesis in meristematic cells and that the RNA manufactured in the nucleolus is exported to the cytoplasm (Woods, 1960; Setterfield, 1963). The nucleolus is certainly the most conspicuous component of the eukaryotic nucleus (Lafontaine, 1974). In most higher plant cells the nucleolus disappears during prophase and persistent nucleoli are observed less frequently (Braselton, 1977). Spectacular changes have been observed in both the size and structure of the nucleolus during a cell cycle. In nucleoli showing a 'dispersive' behaviour (Pickett-Heaps, 1970) the change in volume is from zero at anaphase to a maximum value at the beginning of the prophase of the next cycle. This marked increase in size is accompanied by structural changes within the nucleolus as the cell progresses through division. It becomes less compact and the fibrillar and granular regions become intermingled (Lafontaine & Chouinard, 1963; Brinkley, 1965; Birnstiel, 1967; Yeoman & Street, 1973). Frequently a very large, more electron transparent central region appears in the nucleolus which contains, and is surrounded by, granular particles similar in size to ribosomes. Fibrillar material similar to the chromatin outside the nucleolus is also found within this body, and in addition smaller electron-transparent regions are present within the fibrillar zone surrounding the central area (Braselton, 1977). All of these structural changes and rearrangements highlight the importance of the activities which are taking place in the nucleolus during the course of a cell cycle.

Protein synthesis and accumulation

Early studies by Van't Hof (1963) with asynchronously dividing cell populations in root meristems using pulse labelling techniques with [^{14}C]phenylalanine and [^{14}C]- and [^{3}H]leucine have shown that rates of incorporation of these amino acids, and therefore presumably of protein synthesis, vary during the cell cycle. The rates are highest during G2 and prophase. Woodard et al. (1961a) using cytochemical procedures with a root meristem demonstrated a doubling in protein in the nucleus coincident with the replication of DNA and showed that the accumulation of cytoplasmic protein occurred periodically during interphase. Mitchell (1968b) using the synchronously dividing artichoke explant system has demonstrated, using cytochemical techniques, that increases in total cell protein take place in three steps coincident with the three steps of RNA accumulation (Mitchell, 1969), early in G1, at the beginning of S and in G2 immediately before mitosis. This pattern may be contrasted with that obtained in cultures of mouse fibroblasts (Zetterberg & Killander, 1965; Zetterberg, 1966) and in *Chlorella* (John et al., this volume) in which protein per cell increased exponentially during interphase.

Patterns of enzyme activity

Compared with studies on micro-organisms and animal cells there are relatively few reports of enzyme patterns during the cell cycle in higher plants. This undoubtedly reflects the difficulty of obtaining synchronous or synchronised populations of plant cells. The first reports of variation in the levels of enzymes in plants which were associated with particular stages in the cell cycle exploited the natural synchrony in developing microspores of *Lillium* and *Trillium* (Stern, 1966). In a series of papers (Hotta & Stern (1961, 1963, 1965) demonstrated a distinct periodicity with regard to levels of thymidine kinase, DNAase and related enzymes. Thymidine kinase appeared transiently, before S, coincidentally with the appearance of an increased pool of deoxyribosides, and evidence was presented (Hotta & Stern, 1965) that the periodicity was due to induction by thymidine (dThd). However, it might be valid to regard the developing microspore system as an example of cell differentiation rather than displaying patterns characteristic of a normal cell cycle. This is of course a criticism which may be levelled at the only other two synchronous systems that have been used for studies of this kind, namely artichoke explants (Yeoman & Aitchison, 1973) and cell suspension cultures of sycamore (King & Street, 1973). However, unlike the developing microspore the other two systems are at least actively engaged in growth which continues beyond two divisions.

Despite the obvious limitations of the data a number of common points emerge from all three systems. Firstly, all of the enzymes studied displayed discontinuous patterns of activity and with only one exception, that of succinic dehydrogenase in sycamore cells, increased once per cycle. In other organisms, although the majority of enzymes increase in activity periodically, during one or more stages of the cell cycle, continuous patterns of activity are known (Mitchison, 1969) Perhaps this will also be shown to be true for plants as further data become available. Secondly, a close relationship exists between the rise in dThd kinase activity and the period of DNA replication. This is an almost universal feature of enzyme patterns during cell cycles in eukaryotes (e.g. Brent, Butler & Crathorn, 1965; Eker, 1965; Stubblefield & Murphree, 1967; Harland *et al.*, 1973). Additionally, in the artichoke (Yeoman & Aitchison, 1976; Yeoman *et al.*, 1977 a group of enzymes including dThd kinase, DNA polymerase and dTMP kinase show a high correlation with the amount of DNA, and increase in activity only during S. Moreover, when DNA synthesis is blocked by fluorodeoxyuridine (FdUrd), an inhibitor which blocks DNA selectively in this tissue (Harland *et al.*, 1973), with very little effect on RNA and protein synthesis, the increase in activity of DNA polymerase is also prevented. Data for dThd kinase and dTMP kinase also show that FdUrd prevents the increase in these enzymes. If it is assumed that the increased activity reflects an increased synthesis, such as has been shown by Aitchison, Aitchison & Yeoman (1976) for glucose-6-phosphate dehydrogenase, then the increased rate of synthesis of these enzymes may be dependent on the new DNA. There is therefore an interesting parallel with the requirement for DNA synthesis conversely to terminate dThd kinase synthesis in *Physarum* (Sachsenmaier, this volume).

In the artichoke system other enzymes also appear to be linked to particular phases of the cell cycle. RNAase activity displays a complicated pattern with a sixfold increase occurring just before mitosis. In contrast ATP glucokinase and glucose-6-phosphate (Glc-6-*P*) dehydrogenase reach a maximum value at the end of G1 and then decrease reaching a minimum value at about the time of mitosis (Aitchison & Yeoman, 1973; Yeoman & Aitchison, 1973). As far as dThd kinase, dTMP kinase, Glc-6-*P* dehydrogenase and ATP glucokinase are concerned there is reasonable evidence that the changes are not due to changes in the presence of activators or inhibitors (Harland *et al.*, 1973; Aitchison & Yeoman, 1974) and that in the case of Glc-6-*P* dehydrogenase at least, the increase can be wholly accounted for by *de novo* synthesis (Aitchison *et al.*, 1976). It has also been shown (Aitchison & Yeoman, 1973) that RNA synthesis is necessary for the regulation of levels of Glc-6-*P* dehydrogenase during G1 although no conclusions could be drawn as to the class of RNA involved.

Cell cycle regulation

The terms control and regulation are used repeatedly in attempts to explain how the cell cycle works. However, we are still largely ignorant of the mechanisms which control or regulate the cell cycle in plants, since much of the basic information essential to form a platform on which hypotheses of regulation can be built is still not available. In contrast, work with prokaryotes, lower eukaryotes and animal tissue cultures has provided a good factual framework from which it is possible to appreciate how the cell cycle is regulated. These ideas may be usefully applied to higher plant cells. It is against this background of insufficient data and general uncertainty that we must view current theories of how the cell cycle is controlled in plants.

The principal control point hypothesis (the Van't Hof theory). This theory was based on evidence from cytokinetic and spectrophotometric analyses of the primary seedling meristem and was published in 1972 (Van't Hof & Kovacs, 1972). It is basically a very simple theory and postulates that the cell cycle is controlled at two positions, one located in G1 and another in G2. These controls determine whether cells will proceed from G1 to S and from G2 to M. They are called principal controls because subsequent events such as DNA synthesis and mitosis, which in themselves are highly regulated, do not take place until the requirements of the principal controls are satisfied. To put it another way, a cell arrested in G1 need not synthesise molecules required for S unless it is metabolically committed to progress to S; likewise, a cell in G2 need not prepare for mitosis unless it is going to divide. A survey of the literature soon reveals that cells tend to 'park' in G1 and G2 and not in S or M (Van't Hof, 1974): this non-random distribution is characteristic of almost all cells including the potentially proliferative cells of the radicle.

Van't Hof in a series of elegant cytokinetic investigations with *Pisum* (see Van't Hof, 1974) has shown that cells previously arrested in G1 may also be manipulated to stop in G2 and that a substance exists which originates in the cotyledons and promotes cell arrest in G2. This does no more than begin to suggest a molecular basis for control and that proteins are involved in such a mechanism. The Brown hypothesis however goes further and attempts to come to grips with possible molecular mechanisms.

The Brown hypothesis. The basis of the Brown hypothesis (Brown, 1976) is that the genes governing the events of the cell cycle are arranged in groups. These groups correspond with the four phases of the cell cycle G1, S, G2 and M (Fig. 4). Only the genes in any one group are transcribed at any one time. Each group is controlled by a master gene (G) and within the group a set of operator genes (g) is activated in a temporal sequence. The master gene itself requires activation and when this has been effected the group of genes which it dominates is sensitised (s) and the first operator gene (g_1) in the series is activated. Activation of the master gene or the operator gene generates a depressor (d) which in due course deactivates the unit (G or g) by which it is produced. Activation of g also leads to the release of an enzyme complex (e). The train of sensitisation, activation and repression proceeds until the last operator gene in the group (g_n). Subsequently it is envisaged that the activator (a_n) acts on the master gene (G) for the next stage in succession. Thus the master for S phase is activated by the activator (a) from the final operator gene (g_n) in the G1 series and so on until mitosis (M). Since the process is cyclical the situation requires that the master gene (G) for G1 is activated by the appropriate agent (a_n) from the final operator in the mitotic series (M).

Fig. 4. The Brown Model. (A) An overall view of the sequential activation of genes through a cell cycle, and (B) details of the activation, sensitisation and depression of genes within one phase of the cell cycle (G1). G, master gene; g, operator gene; a, activator; s, sensitiser; d, depressor; e, enzyme complex. For further explanation see text. (Adapted from Brown, 1976.)

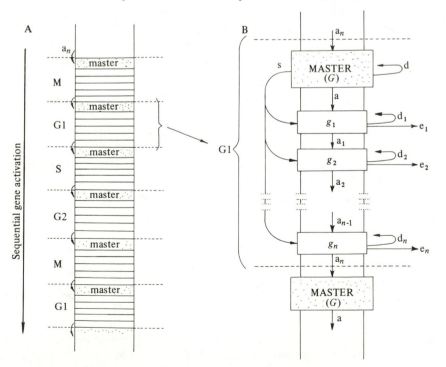

In summary, there are several pieces of evidence which are consistent with this hypothesis.

(1) Studies with inhibitors show that events of the cell cycle are mainly controlled at the level of transcription.

(2) Cells tend to 'park' in G1 or in G2, which suggests that special control points exist at the G1/S, S/G2, G2/M and M/G1 boundaries; in other words a device exists which can induce a transition from one transcription pattern to the next in a series.

(3) Structural and molecular patterns, especially of proteins are characteristic of each phase of the cycle, i.e. a particular group of enzymes will be associated with G1, S, G2 or M.

(4) Very tentative evidence that enzymes are released in succession within a phase.

(5) The cell cycle cannot be reversed, i.e. $G1 \leftarrow S \leftarrow G2 \leftarrow M$.

(6) The existence of phase specific inhibitors, e.g. FdUrd, colchicine, and the possibility that specific inhibitors may be present for each enzyme.

The model, as stated by Brown (1976), 'is clearly inadequate and requires elaboration' and the evidence is fragmentary but consistent with the model. However, it provides us with a good working hypothesis on which to plan future investigations.

I wish to thank Mr Keith Lindsey for his most excellent drawings and Miss Moira McCagie for typing and processing the manuscript so efficiently. I am also extremely grateful to Professor Robert Brown, Dr Adrian Dyer and the Academic Press for permitting me to include adaptations of two previously published diagrams.

References

Aherne, W. A., Camplejohn, R. S. & Wright, N. A. (1977). *An Introduction to Cell Population Kinetics*. London: Edward Arnold.

Aitchison, P. A., Aitchison, J. M. & Yeoman, M. M. (1976). Synthesis and loss of activity of glucose-6-phosphate dehydrogenase in dividing plant cells. *Biochimica et Biophysica Acta*, **451**, 393–407.

Aitchison, P. A. & Yeoman M. M. (1973). The use of 6-methylpurine to investigate the control of glucose-6-phosphate dehydrogenase levels in cultured artichoke tissue. *Journal of Experimental Botany*, **24**, 1069–83.

Aitchison, P. A. & Yeoman, M. M. (1974). Control of periodic enzyme synthesis in dividing plant cells. In *Cell Cycle Controls*, ed. G. M. Padilla, I. L. Cameron & A. Zimmerman, pp. 251–63. New York: Academic Press.

Bansal, J. & Davidson, D. (1978). Analysis of growth of tetraploid nuclei in roots of *Vicia faba*. *Cell and Tissue Kinetics* **11**, 193–200.

Barlow, P. W. (1976). The relationship of the dispersion phase of chromocentric nuclei in the mitotic cycle to DNA synthesis. *Protoplasma, ***90**, 381–92.

Barlow, P. W. (1977). Changes in chromatin structure during the mitotic cycle. *Protoplasma,* **91,** 207–11.

Baserga, R. & Nemeroff, K. (1962). Two-emulsion autoradiography. *Journal of Histochemistry Cytochemistry,* **10,** 628–35.

Bennett, M. D. (1971). The duration of meiosis *Proceedings of the Royal Society of London, Series B,* **178,** 277–99.

Bennett, M. D. (1973). The duration of meiosis. In *The Cell Cycle in Development and Differentiation,* ed. M. Balls & F. S. Billett, pp. 111–31. Symposia of the British Society for Developmental Biology 1. Cambridge University Press.

Birnstiel, M. (1967). The nucleolus in cell metabolism. *Annual Review of Plant Physiology,* **18,** 25–58.

Braselton J. P. (1977). The nucleolus during mitosis in plants: ultrastructure of persistent nucleoli in mung beans. In *Mechanisms and Control of Cell Division,* ed. T. L. Rost & E. M. Gifford, Jr., pp. 194–211. Stroudsburg: Dowden, Hutchinson & Ross.

Brent, T. P., Butler, J. A. V. & Crathorn, A. R. (1965). Variations in phosphokinase activities during the cell cycle in synchronous populations of HeLa cells. *Nature, London,* **207,** 176–7.

Brinkley, B. R. (1965). The fine structure of the nucleolus in mitotic division of Chinese Hamster cells *in vitro. Journal of Cell Biology,* **27,** 411–22.

Brown, R. (1976). Significance of division in the higher plant. In *Cell Division in Higher Plants,* ed. M. M. Yeoman, pp. 3–46. London & New York: Academic Press.

Brown, R. & Dyer, A. F. (1972). Cell division in higher plants. In *Plant Physiology,* ed. F. C. Steward, vol. 6c, pp. 49–90. New York & London: Academic Press.

Cairns, J. (1962). A minimum estimate for the length of the DNA of *Escherichia coli* obtained by autoradiography. *Journal of Molecular Biology,* **4,** 407–9.

Cairns, J. (1973). DNA synthesis. *British Medical Bulletin,* **29,** 188–91.

Callan, H. G. (1973). Replication of DNA in eukaryotic chromosomes. *British Medical Bulletin,* **29,** 192–5.

Clowes, F. A. L. (1965). The duration of the G1 phase of the mitotic cycle and its relation to radiosensitivity. *New Phytologist,* **64,** 355–9.

Clowes, F. A. L. (1976). The Root Apex. In *Cell Division in Higher Plants,* ed. M. M. Yeoman, pp. 253–84. London & New York: Academic Press.

Doležǎl, R. & Tschermak-Woess, E. (1955). Verhalten von Eu-und Heterochromatin, und Interphasisches Kernwachstum bei *Rhoeo discolor* vergleich von Mitose und Endomitose. *Österreichische Botanische Zeitschrift,* **102,** 158–85.

Dyer, A. F. (1976). The visible events of mitotic cell division. In *Cell Division in Higher Plants,* ed. M. M. Yeoman, pp. 49–110. London & New York: Academic Press.

Eker, P. (1965). Properties and assay of thymidine deoxyribonucleotide phosphatase of mammalian cells in tissue culture. *Journal of Biological Chemistry,* **240,** 419–22.

Fraser, R. S. S. & Loening U. E. (1974). RNA synthesis during synchronous cell divisions in cultured explants of Jerusalem artichoke tuber. *Journal of Experimental Botany,* **25,** 847–59.

Friedberg, S. H. & Davidson, D. (1970). Duration of S-phase and cell cycles in diploid and tetraploid cells of mixoploid meristems. *Experimental Cell Research,* **61,** 216–8.

Giménez-Martin, G., de la Torre, C. & López-Sáez, J. F. (1977). Cell division in higher plants. In *Mechanisms and Control of Cell Division*, ed. T. L. Rost & E. M. Gifford, Jr., pp. 261–307. Stroudsburg: Dowden, Hutchinson & Ross.

Goldstein, D. J. (1977). Integrating microdensitometry. In *Analytical and Quantitative Methods in Microscopy*, ed. G. A. Meek & H. Y. Elder, pp. 117–36. Society for Experimental Biology Seminar Series 3. Cambridge University Press.

Green, H. (1974). Ribosome synthesis during preparation for division in the fibroblast. In *Control of Proliferation in Animal Cells*, pp. 743–55. New York: Cold Spring Harbor Laboratory.

Green, P. B. (1976). Growth and cell pattern formation on an axis: critique of concepts, terminology, and modes of study. *Botanical Gazette*, **137**, 187–202.

Harland, J., Jackson, J. F. & Yeoman, M. M. (1973). Changes in some enzymes involved in DNA biosynthesis following induction of division in cultured plant cells. *Journal of Cell Science*, **13**, 121–38.

Heitz, E. (1928). Das Heterochromatin der Moose. I. *Jahrbuch für wissenschaftliche Botanik, Berlin*, **69**, 762–818.

Heitz, E. (1929). Heterochromatin, Chromocentren, Chromomeren. *Berichte Deutsche Botanische Gesellschaft*, **47**, 276–84.

Hotta, Y. & Stern, H. (1961). Transient phosphorylation of deoxyribosides and regulation of deoxyribonucleic acid synthesis. *Journal of Biophysical and Biochemical Cytology*, **11**, 311–19.

Hotta, Y. & Stern, H. (1963). Molecular facets of mitotic regulation. I. Synthesis of thymidine kinase. *Proceedings of the National Academy of Sciences, USA.*, **49**, 861–5.

Hotta, Y. & Stern, H. (1965). Inducibility of thymidine kinase by thymidine as a function of interphase time. *Journal of Cell Biology*, **25**, 99–108.

Houseman, D. & Huberman, J. A. (1975). Changes in the rate of DNA replication fork movement during 'S' phase in mammalian cells. *Journal of Molecular Biology*, **94**, 173–81.

Howard, A. & Pelc, S. (1953). Synthesis of deoxyribonucleic acid in normal and irradiated cells and its relation to chromosome breakage. *Heredity*, **6** (*Suppl*), 261–73.

Huberman, J. A. & Riggs, A. D. (1966). Autoradiography of chromosomal DNA fibers from Chinese Hamster cells. *Proceedings of the National Academy of Sciences, USA*, **55**, 599–606.

Huberman, J. A. & Riggs, A. D. (1968). On the mechanism of DNA replication in mammalian chromosomes. *Journal of Molecular Biology*, **32**, 327–41.

Jakob, K. M. (1972). RNA synthesis during the DNA synthesis period of the first cell cycle in the root meristem of germinating *Vicia faba*. *Experimental Cell Research*, **72**, 370–6.

Jones, L. E., Hildebrandt, A. C., Riker, A. J. & Wu, J. H. (1960). Growth of somatic tobacco cells in microculture. *American Journal of Botany*, **47**, 468–75.

King, P. J. & Street, H. E. (1973). Growth patterns in cell cultures. In *Plant Tissue and Cell Culture*, ed. H. E. Street, pp. 269–337. Botanical Monographs 11. Oxford: Blackwell Scientific Publications.

Klevecz, R. R. & Stubblefield, E. (1967). RNA synthesis in relation to DNA replication in synchronized Chinese Hamster cell cultures. *Journal of Experimental Zoology*, **165**, 259–68.

Kuroia, T. & Tanaka, N. (1970). DNA replication patterns in somatic chromosomes of *Crepis capillaris*. *Cytologia*, **35**, 271–9.

Lafontaine, J. G. (1974). The nucleus. In *Dynamic Aspects of Plant Ultrastructure*, ed. A. W. Robards, pp. 1–51. London: McGraw-Hill.

Lafontaine, J. G. & Chouinard, L. A. (1963). A correlated light and electron microscope study of the nucleolar material during mitosis in *Vicia faba*. *Journal of Cell Biology*, **17**, 167–201.

Lyndon, R. F. (1967). The growth of the nucleus in dividing and non-dividing cells of the pea root. *Annals of Botany*, **31**, 133–46.

Macleod, A. J., Mills, E. D. & Yeoman, M. M. (1979). Seasonal variations in the pattern of RNA metabolism of tuber tissue in response to excision and culture. *Protoplasma*, **98**, 343–54.

Miller, M. B. & Lyndon, R. F. (1975). The cell cycle in vegetative and floral shoot meristems measured by a double labelling technique. *Planta, Berlin*, **126**, 37–43.

Mitchell, J. P. (1967). DNA synthesis during the early division cycles of Jerusalem artichoke callus cultures. *Annals of Botany*, **31**, 427–35.

Mitchell, J. P. (1968a). Quantitative microspectrophotometry of RNA in plant tissue. *Histochemical Journal*, **1**, 106–123.

Mitchell, J. P. (1968b). The pattern of protein accumulation in relation to DNA replication in Jerusalem artichoke callus cultures. *Annals of Botany*, **32**, 315–26.

Mitchell, J. P. (1969). RNA accumulation in relation to DNA and protein accumulation in Jerusalem artichoke callus cultures. *Annals of Botany*, **33**, 25–34.

Mitchison, J. M. (1969). Enzyme synthesis in synchronous cultures. *Science*, **165**, 657–63.

Mitchison, J. M. (1971). *The Biology of the Cell Cycle*. Cambridge University Press.

Mitchison, J. M. & Walker, P. M. B. (1965). RNA synthesis during the cell life cycle of a fission yeast, *Schizosaccharomyces pombe*. *Experimental Cell Research*, **16**, 49–58.

Mittermayer, C., Braun, R. & Rusch, H. P. (1964). RNA synthesis in the mitotic cycle of *Physarum polycephalum*. *Biochimica et Biophysica Acta*, **91**, 399–405.

Mungall, J. T. K. (1978). A study of DNA synthesis in cultured plant tissues. M.Phil. Thesis, University of Edinburgh.

Nagl, W. (1968). Der mitotische und endomitotische Kernzyklus bei *Allium carinatum*. I. Struktur, Volumen und DNS gehalt der Kerne. *Österreichische Botanische Zeitschrift*, **115**, 322–53.

Nagl, W. (1970a). Correlation of chromatin structure and interphase stage in nuclei of *Allium flavum*. *Cytobiologie*, **1**, 395–8.

Nagl, W. (1970b). The mitotic and endomitotic nuclear cycle in *Allium carinatum*. II. Relations between DNA replication and chromatin structure. *Caryologia*, **23**, 71–8.

Nagl, W. (1977). Nuclear structures during cell cycles. In *Mechanisms and Control of Cell Division*, ed. T. L. Rost & E. M. Gifford, Jr., pp. 147–93. Stroudsburg: Dowden, Hutchinson & Ross.

Ockey, C. H. & Saffhill, R. (1976). The comparative effects of short-term DNA inhibition on replicon synthesis in mammalian cells. *Experimental Cell Research*, **103**, 361–73.

Painter, R. B. & Schaefer, A. W. (1971). Variation in the rate of DNA chain growth through the 'S' phase in HeLa cells. *Journal of Molecular Biology*, **58**, 289–95.

Pickett-Heaps, J. D. (1970). The behaviour of the nucleolus during mitosis in plants. *Cytobios*, **6**, 69–78.

Ploem, J. S. (1977). Quantitative fluorescence microscopy. In *Analytical and Quantitative Methods in Microscopy*, ed. G. A. Meek & H. Y. Elder, pp. 55–89. Society for Experimental Biology Seminar Series 3. Cambridge University Press.

Prescott, D. M. (1966). The synthesis of total macronuclear protein, histone and DNA during the cell cycle in *Euplotes eurystomus*. *Journal of Cell Biology*, **31**, 1–9.

Rigler, R. (1966). Microfluorometric characterization of intracellular nucleic acids and nucleoproteins by Acridine orange. *Acta Physiologica Scandinavica (Suppl.)*, **267**, 1–122.

Roberts, L. W. (1976). *Cytodifferentiation in plants*, pp. 38–40. Developmental and Cell Biology Series 2. Cambridge University Press.

Setterfield, G. (1963). Growth regulation in excised slices of Jerusalem artichoke tuber tissue. In *Symposia of the Society for Experimental Biology*, **17**, 98–126.

Smith, J. A. & Martin, L. (1973). Do cells cycle? *Proceedings of the National Academy of Sciences, USA*, **70**, 1263–7.

Stern, H. (1966). The regulation of cell division. *Annual Review of Plant Physiology*, **17**, 345–78.

Stubblefield, E. & Murphree, S. (1967). Synchronized mammalian cell cultures. II. Thymidine kinase activity in colcemid synchronized fibroblasts. *Experimental Cell Research*, **48**, 652–6.

Taylor, J. H. (1958). The mode of chromosome duplication in *Crepis capillaris*. *Experimental Cell Research*, **15**, 350–7.

Terasima, T. & Tolmach, L. J. (1963). Growth and nucleic acid synthesis in synchronously dividing populations of HeLa cells. *Experimental Cell Research*, **30**, 344–362.

Troy, M. R. & Wimber, D. E. (1968). Evidence for a constancy of the DNA synthetic period between diploid-polyploid groups in plants. *Experimental Cell Research*, **53**, 145–54.

Tschermak-Woess, E. (1959). Die DNS-Reproduktion in ihrer Beziehung zum endomitotischen Strukturwechsel. *Chromosoma, Berlin*, **10**, 497–503.

Van't Hof, J. (1963). DNA, RNA and protein synthesis in the mitotic cycle of pea root meristem cells. *Cytologia*, **28**, 30–5.

Van't Hof, J. (1965). Relationships between mitotic cycle duration, S period duration and the average rate of DNA synthesis in the root meristem cells of several plants. *Experimental Cell Research*, **39**, 48–58.

Van't Hof, J. (1968). Experimental procedures for measuring cell population kinetic parameters in plant root meristems. In *Methods in Cell Physiology*, vol. **3**, ed. D. M. Prescott, pp. 95–117. New York & London: Academic Press.

Van't Hof, J. (1974). Control of the cell cycle in higher plants. In *Cell Cycle Controls*, ed. G. M. Padilla, I. L. Cameron & A. Zimmerman, pp. 77–85. New York & London: Academic Press.

Van't Hof, J. (1975). DNA fiber replication in chromosomes of a higher plant (*Pisum sativum*). *Experimental Cell Research*, **93**, 95–104.

Van't Hof, J. (1976a). DNA fiber replication of chromosomes of pea root cells terminating S. *Experimental Cell Research*, **99**, 47–56.

Van't Hof, J. (1976b). Replicon size and rate of fork movement in early S of higher plant cells (*Pisum sativum*). *Experimental Cell Research* **103**, 395–403.

Van't Hof, J. & Kovacs, C. J. (1972). Mitotic cycle regulation in the meristem of cultured roots: the principal control point hypothesis. In *The Dynamics of Meristem Cell Populations,* ed. M. W. Miller & C. C. Kuehnert, pp. 15–32. Advances in Experimental Medicine and Biology 18. New York & London: Plenum Press.

Van't Hof, J. & Sparrow, A. A. (1963). A relationship between DNA content, nuclear volume and minimum mitotic cycle time. *Proceedings of the National Academy of Sciences, USA,* **49,** 897–902.

Webster, P. L. (1979*a*). Variation in sister-cell cycle durations and loss of synchrony in cell lineages in root apical meristems. *Plant Science Letters,* **14,** 13–22.

Webster, P. L. (1979*b*). Heterogeneity of nuclear volumes and interphase nuclear growth in cells of root apical meristems. *Plant Science Letters,* **14,** 23–9.

Wellwood, C. A. & Davidson, D. (1977). Growth of nuclei in methylxanthine induced binucleate cells of *Vicia faba. Canadian Journal of Genetics and Cytology,* **19,** 461–9.

White, R. L. & Davidson, D. (1976). Growth of pollen grain nuclei of *Tradescantia paludosa. Canadian Journal of Genetics and Cytology,* **18,** 385–93.

Wimber, D. E. (1961). Asynchronous replication of DNA in root tip chromosomes of *Tradescantia paludosa. Experimental Cell Research,* **23,** 402–7.

Woodard, J., Rasch, E. & Swift, H. (1961*a*). Nucleic acid and protein metabolism during the mitotic cycle of *Vicia faba. Journal of Biophysical and Biochemical Cytology,* **9,** 445–462.

Woodard, J., Gelber, B. & Swift, H. (1961*b*). Nucleoprotein changes during the mitotic cycle in *Paramecium aurelia. Experimental Cell Research,* **23,** 258–64.

Woods, P. S. (1960). *The Cell Nucleus.* London: Butterworths.

Yeoman, M. M. (1976). *Cell Division in Higher Plants,* ed. M. M. Yeoman. London & New York: Academic Press.

Yeoman, M. M. & Aitchison, P. A. (1973). Changes in enzyme activities during the division cycle of cultured plant cells. In *The Cell Cycle in Development and Differentiation,* ed. M. Balls & F. S. Billett, pp. 185–201. Symposia of the British Society for Developmental Biology 1. Cambridge University Press.

Yeoman, M. M. & Aitchison, P. A. (1976). Molecular events of the cell cycle: a preparation for division. In *Cell Division in Higher Plants,* ed. M. M. Yeoman, pp. 111–33. London & New York: Academic Press.

Yeoman, M. M., Aitchison, P. A. & Macleod, A. J. (1977). Regulation of enzyme levels during the cell cycle. In *Regulation of Enzyme Synthesis and Activity in Higher Plants,* ed. H. Smith, pp. 63–81. Phytochemical Society Symposium 14. London & New York: Academic Press.

Yeoman, M. M. & Street, H. E. (1973). General cytology of cultured cells. In *Plant Tissue and Cell Culture,* ed. H. E. Street, pp. 121–60. Botanical Monographs 11. Oxford: Blackwell Scientific Publications.

Zetterberg, A. (1966). Synthesis and accumulation of nuclear and cytoplasmic proteins during interphase in mouse fibroblasts *in vitro. Experimental Cell Research,* **42,** 500–11.

Zetterberg, A. & Killander, D. (1965). Quantitative cytochemical studies on interphase growth. II. Derivation of synthesis curves from the distribution of DNA, RNA and mass value of individual mouse fibroblasts *in vitro. Experimental Cell Research,* **39,** 22–32.

P.C.L.JOHN, C.A.LAMBE, R.McGOOKIN & B.ORR

Accumulation of protein and messenger RNA molecules in the cell cycle

The two topics of messenger RNA (mRNA) accumulation and protein accumulation are considered together in this chapter because their study encounters similar possible artifacts from methods of cell culture and also because a comparison of messenger populations with proteins which are concurrently being synthesised, allows a test of whether mRNA level or post-transcriptional control regulates protein synthesis.

Objectives

The cell cycle involves biosynthetic events such as DNA synthesis, and ultrastructural events such as mitosis and cytokinesis. These events must be initiated and terminated in the correct sequence for successful division. Since proteins are both catalysts and structural components it is an evident possibility that changes in the level of some individual proteins act as molecular switches controlling cycle events. This possibility is strengthened if the cell cycle is viewed as a developmental process like sporulation which clearly does involve changes in enzyme composition, for example in *Dictyostelium* (Sussman & Brackenbury, 1976).

One objective of the study of proteins in the cell cycle is therefore to determine whether any proteins that are involved in periodic cycle events are themselves accumulated periodically. These could be controllers of the cell cycle. However, any protein which controls a cycle event need not necessarily be synthesised periodically, and we discuss the possible role of periodic activation or protein turnover in changing activity level.

Another objective is to test the converse possibility; that progress through the cell cycle may influence the ability of a cell to synthesise individual proteins that are not directly involved in cycle events. In an extreme case the synthesis of

185

each protein might be restricted within the cell cycle and many observations of enzyme accumulation in synchronous cultures of bacteria, yeasts and algae can be taken to indicate that this is the case.

Our first major topic in this chapter will be to consider whether indeed the majority of proteins, which are not directly involved in cycle events, do normally accumulate periodically as cells progress through the cell cycle. If they do then any control of cell cycle events by changes in level of a protein can be seen as an exploitation of a commonly-occurring phenomenon. Conversely, if the majority of proteins accumulate continuously, they can be maintained at optimum proportions for efficient metabolism or formation of structures, but any which do control cell cycle events by periodic changes in their levels will require the operation of special periodic controls not employed with other proteins.

Periodic accumulation

A predominantly periodic accumulation of enzyme activities has been seen in synchronously dividing cultures of bacterial, yeast and algal cells. The data are tabulated in reviews by Mitchison (1971) and by Halvorson, Carter & Tauro (1971). We shall not present again here the individual observations, but will discuss the possible significance of periodic enzyme accumulation in cell cycle control and some possible mechanisms by which periodic accumulation could be achieved.

Patterns

Protein accumulation can most easily control changes in metabolism or structure if the proportion of a functional protein changes abruptly in relation to other proteins. Mitchison (1969) has described two patterns, termed 'step' and 'peak', which both fulfill this criterion. In both cases the functional protein accumulates during only part of the cycle, but is then increasing more rapidly than total protein with a consequent increase in the proportion of the periodic protein. In the case of a step pattern the individual protein then remains at a constant level, but its proportion declines slowly as other proteins accumulate, while in the case of a peak pattern, decline after the period of accumulation is accelerated because the protein is inactivated and falls to low levels between periods of accumulation. This extreme periodic fluctuation seems most suitable for catalysing periodic cycle events.

Activation and turnover

A periodic accumulation of enzyme activity need not indicate that synthesis of the enzyme is restricted to the period in which its activity accumulates. A phenomenon which can entirely dissociate the appearance of enzyme activity from enzyme synthesis is the activation, or inactivation, of catalytic function.

Although only a minority of enzymes are subject to this sort of control, the phenomenon is now well established and mechanisms are known to include the binding of epi-proteins, and enzyme-catalysed chemical modifications. Control by activation is found among enzymes regulating key metabolic branch points, for example, pyruvate dehydrogenase in respiration (Wieland *et al.*, 1973), pyruvate carboxylase in anaplerosis (Cazzulo, Sundaram & Kornberg, 1970) and ornithine transcarbamoylase in biosynthesis (Wiame, 1971). The regulated enzymes catalyse reactions whose contribution to metabolism can be changed by physiological circumstance, between being essential, to wasteful or even, as in the case of glutamine synthetase, a threat to viability (Schutt & Holzer, 1972).

There is no evidence that the cell cycle requires such changes in activation among enzymes involved in general metabolism. Indeed an absence of change in activity state of glutamine synthetase in *Candida,* when care was taken to minimise metabolic disturbance, provided Folkes *et al.* (cited by Mitchison, 1977) with an early indication of the potential distortion of activity patterns in synchronous culture. The accompanying data in this chapter, which show that rates of photosynthesis and respiration in *Chlorella* are in balance with the levels of key enzymes involved in these processes unless the cells are disturbed, also suggest that progress through the cycle does not involve fluctuations, either in the levels or state of activation, of enzymes involved in general metabolism. Therefore, while further studies of enzymes capable of activation or inactivation would be welcome, we suspect that enzymes involved in general metabolism will show that the cell cycle requires few changes in activation state.

However, enzymes which are directly involved in cycle events are more likely candidates for activation during the cell cycle and instances are already known. Periodic changes in DNA polymerase activity have been observed to occur in *Saccharomyces cerevisae* without periodic changes in rate of enzyme synthesis (Golombek, Wolf & Wintersberger, 1974) and Matthews describes in the present volume the appearance of a 15-fold increase in histone kinase activity without proportionate increase in enzyme synthesis.

Protein turnover is another phenomenon which could produce a pattern of enzyme activity different from the pattern of synthesis. The occurrence of turnover would not remove a requirement for enzyme synthesis at the time of accumulation but it could impose a more periodic accumulation than would be produced by change in rate of synthesis alone. Donachie & Masters (1969) point out that changes in rate of turnover are a potential cause of periodicity, for example a decrease in the rate of breakdown would result in a more rapid accumulation. Even if the proportion of enzyme lost by breakdown is held constant, turnover will accentuate periodicity. A protein whose level would otherwise form a step will tend towards a peak pattern, because of its decline when not synthesised. Further a protein which would show a linear pattern of accumulation with a rate change will, if subject to turnover, tend towards a step pattern provided the half is brief compared with the duration of the cell cycle (equations derived by Schimke (1973) apply). Schmidt (1974) has suggested that step patterns could be accounted for economically without requiring further specialised periodic controls, if proteins do turn over rapidly and if their synthesis does increase in response to replication of the structural genes. However there are indications that these two conditions are rarely met. Study of a wide range of microorganisms shows that most proteins do not turn over rapidly when cells are simply growing and dividing, although rapid turnover can occur if cells are starved (Goldberg & St John, 1976) or are differentiating (Hames & Ashworth, 1974). Neither is gene replication the cause of faster synthesis of many enzymes, since in eukaryote cells many changes in rate of enzyme accumulation lie outside the S phase (see for example Fig. 3*a,c*) and although some are coincident with S phase (Schmidt, 1974) the effects of inhibiting DNA synthesis do not reveal that the preceding replication is necessary; either for the periodic increase in inducible synthesis of glutamate dehydrogenase in the thermophilic *Chlorella* (Israel, Gronostajski, Yeung & Schmidt, 1977), or for the periodic constitutive synthesis of tricarboxylic acid cycle enzymes in light–dark synchronised *Chlorella* 211–8p (B. G. Forde, E. Cole & P. C. L. John, unpublished). A similar independence of enzyme synthesis from DNA replication has been revealed in *Schizosaccharomyces pombe* by mutants that are temperature-conditional for DNA synthesis (Benitez, Nurse & Mitchison, 1980).

For these reasons we think rate change of synthesis together with rapid turnover are not a common cause of step patterns, however there are some indications that protein breakdown does occur in some cell cycles. An individual protein subject to breakdown is tubulin, which has been measured by immune precipitation through the cell cycle in light–dark-synchronised *Chlamydomonas* (Piperno & Luck, 1977) and is seen to decline after the cells are darkened at 14 h. Tubulin is involved in cycle events as the major constituent of microtubules. These comprise the mitotic spindle (Burns, this volume) and the phycoplast (which orientates cytokinesis in some algae, Pickett-Heaps, 1972), but there is no evidence that change in tubulin level initiates any of these events since its decline begins before its role in division is complete. The decline is exactly coincident with darkening of the culture (which immediately halts photosynthesis, Bassham, 1973) and may therefore be caused by carbon starvation rather than the cell cycle. We suspect that if a rapid general turnover of proteins occurs in a cell cycle it will be of brief duration for reasons of metabolic economy. Such a phase of turnover, occupying about a tenth of the cycle, occurs during cytokinesis in *Chlorella* as will be illustrated later in this chapter.

Therefore the general observation that total protein does not turn over rapidly in growing asynchronous populations still allows the possibility that a minority of proteins is unstable throughout the cycle or that the majority may be transiently unstable, as at cytokinesis in *Chlorella*.

We have acknowledged that both activation and turnover can produce greater periodicity of protein function than might otherwise result from changes in rate of synthesis. However, these types of control are metabolically more expensive since they require the synthesis of ancillary activating proteins or, in the case of turnover, require additional synthesis to counterbalance breakdown, so there is an economic advantage if periodic changes in protein level result from periodic synthesis of the proteins.

Enzymes and synchronous culture

The study of individual protein levels and of synthesis through the cell cycle presents technical difficulties which experimenters have been aware of for a long time but may only recently have begun to solve. The key problem is how to get enough material for biochemical assay at each phase of the cell cycle. Some measurements can be made on individual cells as mentioned in the Introduction by Mitchison; for example enzyme level has been followed by microscopic assay of β-galactosidase in single cells of *Saccharomyces lactis*, (Yashphe & Halvorson, 1976). However, for the majority of measurements, including the estimation of individual proteins, material from millions rather than single cells is required.

The commonest solution has been to set up synchronously-dividing populations which yield sufficient material for assay in successive samples taken through the division cycle. With synchronous cultures there is the problem of achieving sufficiently exact synchrony to minimise the blurring of periodic changes, while ensuring that any changes seen are truly involved in the cell cycle and are not a response to the procedures used to obtain synchronous division. This reservation has directed attention away from use of inhibitors which can induce synchrony by causing cells in an initially asynchronous population to accumulate at a blocked event, such as DNA synthesis and then to divide synchronously when the inhibitor is removed. Such cells are likely to have suffered side effects of the inhibitor and will have reached abnormal sizes while inhibited (Mitchison, 1971). More favourable systems have been devised but at the cost of reducing the choice of cell available for study. Mammalian cells can be selected by mitotic detachment but yields are low and it is difficult to devise controls for the effects of detachment (Klevecz & Kapp, 1973), therefore attention has focussed on more robust microbial cells.

Many unicellular algae can be synchronised by periodic darkening at intervals which are frequently close to the diurnal cycle (Tamiya *et al.*, 1953; Lorenzen & Hess, 1974). Darkness stops photosynthesis and so halts cells which are in the growth phase of the cycle, but cells in division continue through the process at an undiminished rate, by drawing on reserves, and their daughters and any halted cells then synchronously commence new cycles when re-illuminated. Repeated light–dark cycles give highly synchronised populations, of *Chlamydomonas* and *Chlorella,* which have played a part in establishing the early view that most proteins accumulate periodically in the cell cycle. These algae usually grow more than twofold in volume and mass in a cycle and then produce more than two daughters. The corespondingly-great increase in enzyme activity per cycle has made periodic enzyme accumulation an unmistakable feature in light–dark synchronised *Chlamydomonas* (Kates & Jones, 1967) and *Chlorella* (John *et al.,* 1973; Schmidt, 1974). However, it has been much harder to establish whether the evident periodicity is caused by the light–dark alternation, or by the cell cycle. We shall return to this problem after considering evidence for the periodic accumulation of enzyme activity in other organisms.

Most studies of enzyme accumulation have been performed on synchronous populations of bacterial or yeast cells which have been prepared by sucrose density-gradient selection of small cells. In this procedure cells from an exponentially-growing asynchronous culture are pelleted, suspended in a small volume of growth medium and layered onto a sucrose gradient which retards smaller cells during centrifugation. The small cells are collected from the upper layers and, when grown on in fresh medium, show a synchrony which diminishes in successive cycles.

Such synchronous cultures have been prepared from a number of bacteria and among the eukaryotes largely from the budding and fission yeasts. A majority of enzymes in these cultures show a periodic pattern approximating to a step and there has been a tendency to allow for the incomplete synchrony, and assume that the periodicity may be sharper in individual cells. The extensive literature, involving more than two dozen bacterial enzymes and more than a hundred eukaryote enzymes, has been thoroughly described in earlier reviews (Mitchison, 1969, 1971; Halvorson, Carter & Tauro, 1971), which also deal with possible mechanisms. Two suggested mechanisms will be described briefly here and are named according to Mitchison (1969).

Oscillatory repression

This hypothesis suggests that mechanisms which control genes in response to changes in environment, also operate periodically in the cycle. The idea originates from studies of synchronous bacterial cells (Kuempel, Masters & Pardee, 1965; Donachie & Masters, 1969). It is suggested that metabolites which act as inducers or repressors of enzyme synthesis, change in level as the cycle progresses and produce oscillations in rate of enzyme synthesis. For example an excess concentration of an end-product which acts as gene repressor could repress enzymes involved in its synthesis until growth of the cell increased metabolic demand and the consequent fall in concentration of end product allowed a renewed enzyme synthesis. A weakness of the theory is that it is not clear why periods of enzyme synthesis should not damp-out but persistently result in excess of enzyme, or should occur with the periodicity of the cell cycle. The hypothesis is difficult to test because measurements of changes in repressor level will be obscured by the presence of storage or organelle pools and any changes which are seen may be a consequence rather than a cause of periodic enzyme synthesis. It is certainly possible to demonstrate that the timing of periodic synthesis seen in synchronous culture can be altered or obliterated by deliberate experimental manipulation of metabolic levels. A clear example is the step in the level of aspartate transcarbamoylase in *B. subtilis* which can be delayed 0.5 of a cycle by addition of the end-product uracil (Masters & Donachie, 1966) and an important early indication that eukaryote cells can also change their pattern of enzyme accumulation in response to change in metabolism was provided by Molloy & Schmidt (1970) who showed that change in rate of photosynthesis could alter the pattern of ribulose-1,5-bisphosphate carboxylase accumulation in synchronous *Chlorella*. The capacity for altered enzyme accumulation in response to changes in metabolic state does argue against a programmed sequence of gene expression, such as proposed by the next hypothesis, but leaves open the possibility that metabolic disturbance, caused by the preparation of the synchronous cultures, may be the true cause of oscillations in rate of synthesis.

Linear reading

A bold and stimulating hypothesis, proposed to account for periodic enzyme synthesis in *Saccharomyces cerevisiae,* suggests that each gene on an individual chromosome is transcribed once per cell cycle in a linear sequence beginning from one end of the chromosome. The key observation in support of the hypothesis is a correlation between the sequence of four genes on the fifth chromosome and the sequence of stepwise accumulation of the enzymes that they code for (Tauro, Halvorson & Epstein, 1968). There is an appreciable chance that this correlation is coincidental, since there are only $4 \times 3 \times 2 \times 1$, i.e. 24, different possible sequences of four events. Since chromosome reading could begin at either end, there are effectively only twelve different sequences and so the probability that the correlation is a coincidence, is above the conventional limit of one in twenty. However the hypothesis is indirectly supported by evidence consistent with the view that gene expression is programmed in some way. For example, different genes coding for β-glucosidase or α-glucosidase (Tauro & Halvorson, 1966), when present in the same diploid cell, retain separate times of expression, but the different genes could be responding to different concentrations of metabolite reached at different times. Further evidence suggesting the importance of gene position has come from a comparison of two strains of *S*.

cerevisiae, in which a greater separation of two genes in one strain correlates with a greater separation of the time of enzyme synthesis (Cox & Gilbert, 1970), but again the results could indicate that the two strains differ in the times at which metabolite levels change.

Other enzyme patterns are not easily reconciled with linear reading. The ability to synthesise adaptive enzymes throughout the cycle (illustrated in Fig. 9) is common in eukaryotes (Schmidt, 1974; Aasberg, Lien & Knutsen, 1974) so the hypothesis must be modified to admit access to genes for adaptive enzymes throughout the cycle (Sebastian, Carter & Halvorson, 1973). Similarly the expression of genes for constitutive enzymes can be modulated through the cycle by changes in environment. An early example is provided by Molloy & Schmidt (1970) and we shall argue, in the section beginning below that a majority of proteins fall into this category.

The attractively-simple model of RNA polymerase molecules traversing each chromosome within a cycle cannot be sustained since nucleotides are incorporated into RNA at 30 s^{-1} in *S. cerevisiae* (Lacroute, 1973) and careful preparation has revealed that the chromosomes contain single DNA molecules of around 10^6 base pairs (Newlon, Petes, Hereford & Fangman, 1974) therefore transcription from one end would take 500 min. Comparison with cycle times in *S. cerevisiae* (Carter, this volume) shows that internal initiations will be necessary to allow complete reading of the chromosome in a cycle. There is also evidence against the assumption implicit in the linear-reading hypothesis, that the population of mRNA molecules changes with progress through the cell cycle, since there is evidence, presented in Fig. 7c, that no such changes occur in the predominant messengers of *Chlorella*. However, the main evidence against a major role for linear reading, is an indication that the majority of proteins accumulate continuously in the cell cycle of *S. cerevisiae* (Elliott & McLaughlin, 1978) when no opportunity is allowed for metabolic disturbance to produce periodicity. An implication is therefore that the periodicity, which the linear reading hypothesis seeks to explain, is caused by disturbance. This theme will be emphasised in the next section.

Continuous synthesis

An early indication that enzyme synthesis might be continuous through the bacterial cell cycle was obtained by Bellino (1973) who exploited the unique property of *E. coli* strain B/r to bind to millipore filters and release newborn daughter cells into a flow of growth medium, so providing synchronous cultures with minimum metabolic disturbance. In such cultures synthesis of aspartate transcarbamoylase is continuous through the cell cycle, unless the intracellular concentration of uridine is deliberately perturbed. In earlier studies which have employed the re-inoculation of stationary cultures (Masters & Donachie, 1966) or sucrose-gradient selection (Kuempel *et al.* 1965) to obtain synchrony, the accumulation of this enzyme is periodic, but the observation of continuous synthesis with the gentler method of membrane synchronisation provided an early indication that periodicity may have been an artefact of the synchronising procedure.

Artefacts from synchronising have been avoided in a study of pulse-labelled proteins by Lutkenhaus, Moore, Masters & Donachie (1979). An asynchronous population of *E. coli* was fractionated into phases of the cycle by sucrose gradient centrifugation, the cells killed and the proteins resolved by two-dimensional electrophoresis. All of the approximately 750 individual proteins seen in autoradiograms were seen to be synthesised throughout the cell cycle and for 30 proteins, which were well resolved from their neighbours, quantitative estimates of incorporation showed no significant variation in the rate of synthesis through the cycle. Furthermore the periodic incorporation of a transport protein into membrane, previously seen in carefully prepared synchronous cultures (Gudas, James & Pardee, 1976) was shown to be absent and is therefore an artefact of cell manipulation (Boyd & Holland, 1977).

Many of the periodic patterns of enzyme accumulation which were seen in earlier synchronous cultures may also have been artefacts, especially since synchrony was commonly obtained by transferring stationary cells into fresh medium (Masters & Donachie, 1966). The possibility of artificial enzyme periodicity has been directly investigated in the eukaryote *Schizosaccharomyces pombe* (Mitchison, 1977). Effects of the standard method of selecting small cells for synchronous culture, by sucrose density gradient centrifugation, were studied by remixing gradients and following enzyme accumulation in the resulting asynchronous control cultures. Perturbations of accumulation in these controls were most marked in the case of the small number of enzymes which had shown stepwise accumulation in synchronous culture, therefore periodicity in these enzymes was probably a response to the sucrose selection. Mitchison concludes that of 19 enzymes studied in this organism, 18 are now believed to be exponential or linear with rate change and it may be significant that, of the enzymes studied, the only enzyme which shows a periodicity not attributable to other causes than the cell cycle is thymidine monophosphate kinase, which is directly involved in the cell division process.

There has been no report of a similar study in *S. cerevisiae,* to test the effect of sucrose gradient selection on enzyme activity patterns. However, in a study which avoided the development of such artefacts by use of pulse-labelling in asynchronous exponential phase culture, and separation of cells according to phase of the cycle with an elutriator rotor before two-dimensional electrophoresis of soluble proteins, Elliott & McLaughlin (1978) observed continuous synthesis through the cycle in each of 550 proteins resolved. This study pre-dates that of

Lutkenhaus *et al.* (1979) and is perhaps more remarkable because, in spite of the greater structural complexity of the cell cycle in eukaryote cells, no periodic proteins were detected.

Bacteria, budding yeasts and algae have provided many examples of periodic increase in enzyme activity, yet now the first two groups show by contrast a continuous accumulation of proteins and in consequence both Elliott & McLaughlin (1978) and Lutkenhaus *et al.*, (1979) have raised the possibility that periodic activation of proteins is common in the cell cycle. We have attempted to resolve the apparent paradox of different patterns of accumulation for activity and protein, by studying the effects of synchronising procedures on both processes in *Chlorella*.

Accumulation of enzyme activity and protein

In a previous study with batch cultures, we found that periodic accumulation of enzyme activity persists into a cycle that has had no preceding dark period (Forde & John, 1973) but we were then unable to prevent changes in culture density from sustaining enzyme periodicity. In the present studies we have transferred cells to conditions of constant illumination and constant density, under control of a turbidostat, and compared them with sister cells held under the usual synchronising conditions of batch growth and periodic illumination. Constant optical density was maintained in the turbidostat culture by an inflow of medium, which is monitored and used to calculate growth rate, while in the batch culture a single manual dilution was performed at the end of the cell cycle to return the cells to their original density (Fig. 1*a*). Illumination was continuous in the turbidostat culture but the batch culture was exposed to the standard synchronising regime of 9 h darkness from 15 h to the end of each 24 h period. Darkening prevents net growth as seen by the immediate cessation of total protein accumulation from 15 h until 24 h when illumination was resumed (Fig. 1*b*). Two differences conferred by constant illumination of the turbidostat culture are therefore the accumulation of protein beyond 15 h up to the onset of cytokinesis at 18 h and the immediate commencement of growth in daughter cells as cytokinesis is completed at 20 h (Atkinson, Gunning & John, 1972; Atkinson, John & Gunning, 1974). Daughters formed during a dark period are also autonomous by 20 h, but darkness delays their growth until 24 h. This 4 h delay is reflected in the later release of daughters, with a mid point (R) at 26 h compared with 22 h (Fig. 1*c*), and in the later initiation of DNA synthesis; mid S phase (S) being at 36 h compared with 32 h in cells not delayed by a dark period (Fig. 1*d*). A further effect of the different illumination received in the two treatments is revealed by comparing the rate of protein accumulation when both cultures are receiving light. In the first 15 h of culture there is a progressive slowing of protein accumulation in the increasingly-dense batch culture, which contrasts with continued exponential accumulation over the same period in the turbidostat culture. Cells in both cultures have come from the same population, which was previously in batch culture synchronised by dark periods, and several strands of evidence show that the first cycle under continuous illumination at constant turbidity is a cycle of recovery from the relatively starved synchronising conditions.

Fig. 1. (*a*) Turbidity, (*b*) protein, (*c*) cell number and (*d*) DNA, in cells which have been synchronised in batch culture by ten previous cycles of 15 h light and 9 h dark (McCullough & John, 1972*a*) and during the sampling period divided and exposed to either (△) continuation of the synchronising regime in batch culture, with dilution in fresh medium at 0 h and 24 h, or (○) held at constant density by inflow of fresh sterile medium, at the growth temperature of 25 °C, under the control of a turbidostat. Cells in the synchronising regime were darkened for 9 h, as in their previous cycles, between 15 h and 24 h and between 39 h and 48 h, but cells in turbidostat culture received continuous illumination. Both cultures were incubated and aerated in parallel with 0.5% (v/v) CO_2 in air. The mid times of cell release (R) and DNA synthesis (S) are marked by broken vertical lines for the batch light–dark cells and by solid vertical lines for the turbidostat cells. Turbidity was measured at 680 nm, with a 10 mm path cuvette and cell number was estimated with a Coulter Counter model ZB. Protein was estimated in a precipitate, formed by 5% (w/v) trichloroacetic acid (TCA), from a total cell extract obtained by breaking the cells in a French pressure cell, and DNA was estimated by the diphenylamine reaction; both procedures as described by McCullough & John (1972*a*).

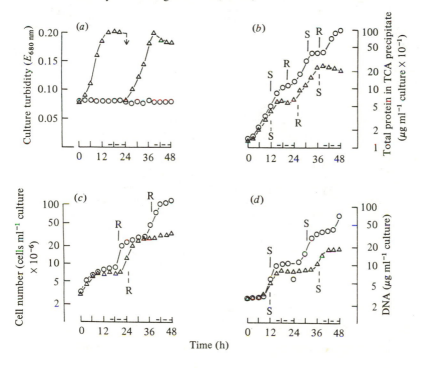

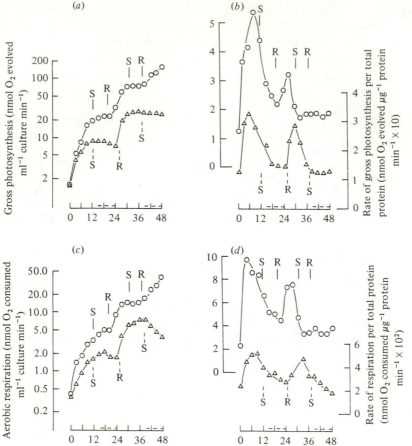

Fig. 2. (a) Rate of gross photosynthetic oxygen evolution, (b) rate of gross photosynthesis per total protein, (c) rate of aerobic respiration and (d) rate of respiration per total protein, in the cells analysed in Fig. 1, which were either in batch culture with periodic illumination (△), or in turbidostat culture with continuous illumination (○). Mid times of cell release (R) and of DNA synthesis (S), which were determined in Fig. 1, are marked for the batch light–dark culture by broken vertical lines and for the turbidostat culture by solid vertical lines. For estimation of respiration and photosynthesis, cells in 40 ml of culture were pelleted by centrifugation at 2000g for 1 min, resuspended in 4 ml of aerated medium at 25 °C and transferred to an oxygen electrode where NaHCO₃ was added to a final concentration of 1 mM and glucose to a final concentration of 5 mM. The cells were darkened and a linear rate of oxygen uptake due to respiration was measured for 5 min, then illumination at 930 nEinsteins of quanta in the 400–700 nm range was provided and a linear rate of oxygen evolution due to net photosynthesis was measured for 5 min. The illumination was adequate to obtain light saturation in spite of the culture density in the oxygen electrode, but not sufficiently intense to cause damage. Estimates of net photosynthesis were increased by the amount of respiration, to calculate gross photosynthesis. An identical rate of gross photosynthesis was indicated

Part of the recovery is the faster accumulation of protein in the first turbidostat cycle. This is an invariable observation and results, as do all succeeding turbidostat cycles, in the formation of daughter cells with a mean of 6 pg protein, in contrast with a mean of 4 pg protein in daughters formed in light–dark synchronised batch cultures. Note that the pause in protein accumulation at the time of cytokinesis (18–21 h) which just precedes cell release, is also evident in the second turbidostat cycle when metabolic and enzymic evidence now to be presented suggests that there is no disturbance of the cycle. The pause is also present in selection-synchronised cells (Fig. 6) but not in control asynchronous cells (Fig. 5) and is therefore a normal part of the cell cycle.

Evidence that metabolic balance is disturbed by the synchronising conditions is provided by study of photosynthesis and respiration. Higher rates of photosynthesis are attained under turbidostat conditions than under batch conditions when both cultures are illuminated, as seen between 0 and 15 h (Fig. 2a), and a particularly large increase in rate of photosynthesis is seen in the first cycle in turbidostat, again supporting the view that this is a cycle in which there is recovery of protein levels previously depressed by the synchronising regime. The dark periods imposed in the synchronising regime between 15 and 24 h and between 39 and 48 h halt photosynthetic carbon assimilation within a few seconds (Bassham, 1973). Although cells transferred to light can resume photosynthesis, their latent capacity for photosynthesis does not increase while they are held in darkness (Fig. 2a). The carbon starvation imposed by these 9 h periods of photosynthetic deprivation is revealed by the declining levels of starch reserves (Atkinson, et al. 1974) and by the declining rates of respiration which the cells are able to sustain in darkness (Fig. 2c). Evidence for a contrasting development of metabolic balance under turbidostat conditions comes from the stabilisation of rates of photosynthesis and respiration per unit of cell protein (Fig. 2b,d). The proportion of resources devoted to photosynthetic and respiratory apparatus, and the demand for their products in relation to cell mass, stabilises through the cell cycle under turbidostat conditions, but under synchronising conditions daughters are formed with initially high rates of photosynthesis and respiration in relation to their protein content, reaching maxima at 6–9 h which cannot be sustained as culture density increases. Under synchronising conditions these fluctuations are

Caption to Fig. 2. (cont.)
throughout the experiment by parallel samples in which photosynthesis was immediately measured without prior centrifugation by fixation of $^{14}CO_2$ in cells illuminated beside the parent culture vessel. No photosynthesis occurred in the darkened batch cultures. Rates shown at these times refer only to cells that were removed and provided with standard illumination conditions to test their latent capacity for photosynthesis. Rates of photosynthesis and respiration per total protein are calculated using the protein levels shown in Fig. 1(b). In (b) and (d), left hand ordinates refer to the turbidostat culture.

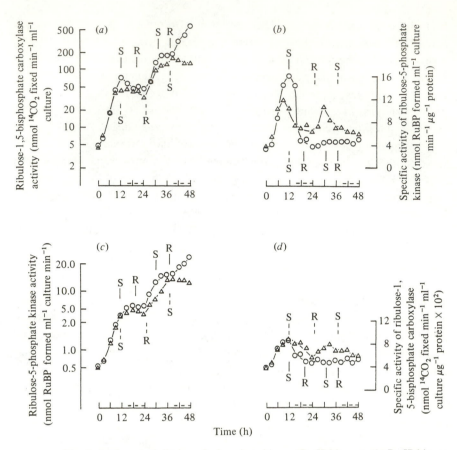

Fig. 3. (*a*) Level of ribulose-5-phosphate kinase (Ru5P kinase), (*b*) Ru5P kinase specific activity, (*c*) level of ribulose-1,5-biphosphate carboxylase (RuBP carboxylase), (*d*) RuBP carboxylase specific activity, in the cells analysed in Fig. 1, which were either in batch culture with periodic illumination (△) or in turbidostat culture with continuous illumination (○). Mid times of cell release (R) and of DNA synthesis (S), which were determined in Fig. 1, are marked for the batch light–dark culture by broken vertical lines and for the turbidostat culture by solid vertical lines. For estimation of enzyme activity, cells were washed and suspended at 0 °C in 50 mM Tris–HCl, 5 mM $MgCl_2$, 1 mM EGTA and 5 mM dithiothreitol pH 7.7 and smashed at 0 °C with a French pressure cell. For assay of Ru5P kinase, volumes of extract containing proteins from 10, 20 and 50 μl of culture were then incubated, in assays of 0.5 ml final volume which contained, 50 μmol Tris–HCl giving a final pH 8.0; 2.5 μmol $MgCl_2$; 1 μmol dithiothreitol; 0.5 μmol ATP; 5 μmol (2 μCi) of $NaH^{14}CO_3$ and an excess of RuBP carboxylase capable of converting at least 20 times the amount of RuBP formed during the assay of Ru5P kinase, by reaction with $^{14}CO_2$, to form radioactive 3-phosphoglyceric acid. The carboxylase was prepared as an assay reagent for estimation of Ru5P kinase, from preparative extracts of *Chlorella* by ammonium sulphate fractionation and G-200 Sephadex chromatography. Assays of Ru5P kinase were performed at 30 °C in gas-tight ampoules and the reaction was started by addition of 0.1 μmol Ru5P and terminated after 10 min, by addition

repeated in successive cycles (note the maxima at 30–33 h in the second cycle) and they reveal the continual changes in metabolism through the cell cycle under synchronisation.

Periodic increase in enzyme activity is a common feature of the cell cycle in algae synchronised by periodic illumination (Kates & Jones, 1967; reviewed by Lorenzen & Hess, 1974). In the strain of *Chlorella* 211-8p used here, although we have detected continuous patterns of increase in enzyme activity for gluta-mate oxaloacetate transminase, glycollate dehydrogenase and nitrate reductase (F. Haldane, B. Taylor & P. C. L. John, unpublished) a majority of enzymes have periodic patterns in periodic light. As representatives of the periodic en-zymes we have followed tricarboxylic acid cycle (Cole, 1978) and also photo-synthetic enzymes into the second cycle of turbidostat growth. Both groups show similar patterns, which we illustrate here by describing the behaviour of two photosynthetic enzymes that were measured in the same cells as analysed in Figs. 1–3. Ribulose-1, 5-bisphosphate carboxylase (RuBP carboxylase) fixes in-organic carbon dioxide and its substrate is generated by ribulose-5-phosphate kinase (Ru5P kinase). Both enzymes are unique to the reductive pentose phos-phate cycle of photosynthesis (Bassham, 1973). Both enzyme activities accu-mulate periodically under synchronising conditions; kinase activity increases up to 9 h and up to 33 h and, since each increase stops 6 h before total protein accumulation ceases at the dark period, the pattern could be considered a step imposed by the cell cycle (Fig. 3a). Similarly RuBP carboxylase, with a mid-point of increase at 7 h and 31 h (Fig. 3c) can be considered a later step enzyme. In consequence the specific activities of the two enzymes regularly increase at the beginning of the cycle and later decline (Fig. 3b,d).

These periodic changes are not, however, truly part of the cell cycle since there is evidence that increasing density of the batch culture was responsible for the early halt of kinase accumulation, because its accumulation continues for a further 3 h under turbidostat conditions (Fig. 3a). More evidence for the perturb-

Caption to Fig. 3. (*cont.*)
of acetic alcohol for determination of radioactive 3-phosphoglyceric acid, as described by Forde & John (1973). For assay of RuBP carboxylase volumes of cell extract containing material from 0.1, 0.2 and 0.5 ml culture were incubated in assays of 0.5 ml final volume which contained, 50 μmol Tris–HCl giving a final pH of 8.0; 2.5 mM $MgCl_2$; 1 μmol dithiothrietol; 5 μmol (2 μCi) NaH $^{14}CO_3$. Assays were performed in gas-tight ampoules and enzyme was pre-incubated at 30 °C for 10 min so that the effect of in-vitro conditions on the activity state of the enzyme (Andrews, Badger & Lorimer, 1973) would be constant in all samples, before starting the reaction by addition of 0.1 μmol RuBP. The reaction was stopped after 10 min by addition of acetic alcohol for determination of radioactive 3-phosphoglyceric acid as described by Forde & John (1973). Specific enzyme activities are calculated using protein levels shown in Fig. 1(b).

ing effect of the synchronising regime comes from comparison of enzyme specific activities in the first and second cycles under constant turbidostat conditions. Periodic changes in specific activity which are seen in the first cycle (Fig. 3b,d) are adaptive and accompany the recovery of the cells from the less favourable synchronising conditions but in the second cycle under constant conditions a balance has been struck between rates of synthesis of individual proteins and so specific activities have stabilised. Thus daughter cells formed in the second cell cycle are released (R) and synchronously commence growth at 40 h (Fig. 1b) without an increase in specific activity of kinase (Fig. 3b) or carboxylase (Fig. 3d). The stable specific activities of these two enzymes indicate that a constant proportion of cell protein synthesis is probably devoted to both enzymes through the cell cycle and that the ratio of their two activities has also been stabilised. This stable ratio appears to be more efficient for these two sequentially-acting enzymes, since their specific activities stabilise at lower values than are transiently induced during (or in recovery from) synchronising conditions (Fig. 3b,d).

A general observation may now be appropriate. Synchronous cultures may remain essential for the study of rates of metabolic processes, such as respiration and photosynthesis, and for the unambiguous study of rates of adaptive enzyme synthesis in the cell cycle, but the setting up of synchronous cultures can cause perturbations which do not quickly subside. In photosynthetic algae the changes of shading which occur in batch culture and the periods of darkness which are normally used to impose synchrony disturb both metabolic processes and enzyme accumulation. Even study of events in the first cycle of turbidostat culture is also subject to error since this first cycle involves adaptive changes which recover a stable balance that can then be maintained in subsequent cycles.

The cell cycle of *Chlorella* can therefore proceed with two contrasting patterns of enzyme activity. Under the synchronising conditions of batch growth and periodic darkness most enzyme activities are periodic, but in the second cycle of constant light and turbidity all the activities which we have studied increase more steadily in close parallel total cell protein (Cole, 1978; Fig. 3). To test whether the two different activity patterns are caused by a greater prevalence of periodic enzyme activation in synchronised cells or by a change in pattern of synthesis from predominantly periodic to predominantly continuous, we have pulse-labelled cells in both cycle conditions with $^{35}SO_4$ and [^{3}H]leucine and then resolved polypeptides by electrophoresis.

Under the synchronising conditions of batch culture and periodic darkening, identical to those employed in Figs. 1–3, the inorganic tracers (and organic tracers, not shown) reveal changes in the rate of synthesis of the majority of soluble proteins through the cell cycle (Fig. 4a). There is therefore a correlation with the periodic accumulation of enzyme activity prevalent under these condi-

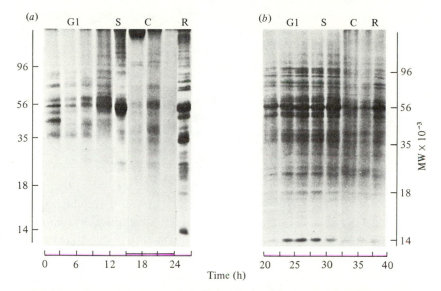

Fig. 4. Soluble proteins synthesised in successive 3 h periods of pulse label, (*a*) through the standard synchronising cycle, of batch culture with 15 h illumination followed by 9 h darkness and (*b*) through the second cycle under continuous illumination in a turbidostat. Identical cells are described in Figs. 1, 2 and 3 where cells in batch culture between 0 and 24 h or between 24 and 48 h correspond with (*a*) in this Figure, and cells in turbidostat culture between 20 and 40 h correspond with (*b*). Phases indicated here were G1, S, cytokinesis (C) and cell release (R). Cells in both cultures were grown from a small inoculum in medium with low, but excess, SO_4^{2-} (Lambe & John, 1979) and this medium was employed throughout the experiment. For pulse labelling, cells were pelleted by centrifugation at 2000g for 2 min and resuspended in aerated medium at 25 °C containing $^{35}SO_4^{2-}$ at a final concentration of 0.1 μmol and 0.2 μCi ml^{-1}, and during labelling were illuminated and aerated in parallel with the parent cultures, in vessels also of light path 68 mm. Pulse labelling of parallel subcultures with [^{3}H]leucine at 4.2 μmol and 0.2 μCi ml^{-1} confirmed the differences revealed by ^{35}S labelling. Cells undergoing the longer cell cycle in batch culture were labelled for 3 h periods, while cells in turbidostat culture were labelled for 2.5 h periods. After labelling, cells were pelleted and smashed in cold Tris–HCl, pH 7.7, 5mm MgCl$_2$ in a French pressure cell. Part of the resulting extract was subject to centrifugation at 100 000g for 1 h and the supernatant was retained as the soluble protein fraction. Electrophoresis was performed using 15% (w/v) acrylamide gel under the dissociating conditions of Laemmli (1970) and isotope was detected by autoradiography. Separations of the soluble protein fraction from equal volumes of culture are illustrated here. Analysis of the total protein fraction including membranes confirmed the prevalence of periodic synthesis in batch culture and of continuous synthesis in turbidostat. Note in (*a*) the quiescent phase, between 21 h and 24 h, when cells have completed a cycle but must wait for re-illumination to commence their next cycle. In (*a*) the tracks carrying proteins labelled between 15–18 h and 24–27 h, have been printed at reduced photographic exposure to improve visibility of bands, because increasing density in the batch culture resulted in high radioactive incorporation at these times.

tions. Similar changes in rates of individual protein synthesis through the cell cycle in batch cultured synchronised *Chlamydomonas* have been observed (Howell, Posakony & Hill, 1977). In *Chlorella* such periodicity is revealed to be an artefact of the culture conditions, because in the second cell cycle under constant conditions, identical with that seen between 20 h and 40 h in Figs. 1–3, there are no extensive changes in rates of synthesis of the predominant individual proteins detected by either tracers (Fig. 4*b*). The larger number of minor proteins are not resolved in the gel shown, but two-dimensional separations reveal none that are periodic. Again proteins and enzymes show the same general pattern and, in the undisturbed cycle, are continuously accumulated.

As in the studies of *S. cerevisiae* (Elliot & McLaughlin, 1978) and *E. coli* (Lutkenhaus *et al.*, 1979) we cannot in general identify enzyme activities with individual proteins. However, RuBP carboxylase comprises a tenth of the soluble proteins and comparison with enzyme purified from our organism allows us to identify its usual two subunits, of 56 000 and 14 000 daltons. The synchronising conditions, which cause an increase in specific activity of this enzyme early in the cycle (Fig. 3*d*), also cause a preferential synthesis of the enzyme early in the cycle (Fig. 4*a*) which is particularly clear in the larger, and so more heavily labelled, subunit. This specific instance and the general prevalence of periodic protein synthesis when periodic accumulation of activities is common, indicates that there is no need to invoke periodic activation for most proteins in synchronised *Chlorella*.

A similar correlation between enzyme synthesis and the accumulation of activity is seen after perturbations have subsided, in the second cycle under turbidostat conditions. Synthesis of the majority of proteins, including the two subunits of RuBP carboxylase, is then continuous through the cell cycle (Fig. 4*b*) and this continuous synthesis correlates with the continuous increase in activity of the carboxylase (Fig. 3*d*) and kinase Fig. 3*b*) and of fumarase and citrate synthase (not shown). Further evidence for continuous accumulation of activity in unperturbed cells is provided by cells selected from asynchronous culture by continuous flow centrifugation during turbidostat growth (as employed in Fig. 6) which show continuous accumulation of fumarase, citrate synthase and succinate dehydrogenase (John, Cole, Keenan & Rollins, 1980).

A brief deviation from continuous exponential increase in enzyme activities (Fig. 3*a,c*) and protein (Fig. 1*b*) is caused by an acceleration of protein breakdown at cytokinesis, which will be discussed in the next section, but the most significant feature of the present data is that both individual proteins and enzyme activities accumulate continuously through most of the cycle. It is likely that the same correlation exists in the cell cycle of other organisms. Mitchison (1977) suggests that when cells are selected with sucrose gradients, the high cell density results in starvation. In *S. cerevisiae* three enzymes do show the same pattern after synchronisation by deliberate starvation or by gradient selection, (Tauro & Halvorson, 1966). Therefore just as starvation by the synchronising dark period is shown here to cause periodicity of both enzyme activity and synthesis of proteins in *Chlorella,* a similar periodicity of both activity and proteins was probably inadvertently caused by sucrose gradient selection of bacteria and yeasts in the earlier studies. By contrast, the true undisturbed pattern in the cell cycle is one of continuous synthesis of most proteins, as seen in Fig. 4*b* and by Elliott & McLaughlin (1978) and Lutkenhaus *et al.,* (1979), with a corresponding continuous accumulation of most enzyme activities as reported here in *Chlorella* and in *S. pombe* (Mitchison, 1977).

There is probably a similar situation in mammalian cells but there are difficulties in assessing data from cultured mammalian cells, because if synchrony is obtained by washing cells in mitosis from agar into liquid medium metabolic disturbance may cause false periodicity. Conversely use of inhibitors can synchronise division but may fail to synchronise events involved in growth which should be genuinely part of the cell cycle. With these reservations it is interesting to note that Milcarek & Zahn (1978), by use of inhibitor-induced synchrony, also observed continuous synthesis of the majority of individual proteins in HeLa cells.

The cell cycle does not therefore commonly restrict the synthesis of individual proteins relative to others but there is evidence that the cell cycle does have effects on general rates of protein accumulation.

Phases of the cycle and accumulation of mRNA and protein

It is common for cycle phases to affect the rate of increase in cell volume or the rate of RNA or protein synthesis but it is not certain what the underlying mechanisms are. If protein increases linearly, with a change in rate in successive cycles, this pattern would be consistent with a limitation of protein synthesis by a component which increases at the time of rate change. Whereas if protein accumulates exponentially, this suggests that its synthesis is limited by components which are increasing through the cycle. These two patterns are difficult to discriminate. Incorporation of radioactive precursor into protein at intervals through the cycle, provides a sensitive measure of rate of synthesis but incorporation is very easily distorted by changes in rate of precursor uptake and by changes in level of endogenous non-radioactive precursor within the cell. Direct chemical assay of protein is difficult to perform with the necessary accuracy since cells which only double their protein content in a cycle can have a maximum difference in slope between exponential and linear patterns of only 3%. These difficulties are reviewed by Mitchison (1971).

It is likely that examples of both patterns will be established since a linear accumulation with rate change has been established by statistical analysis of several enzymes in *Schizosaccharomyces pombe* (Mitchison & Creanor, 1969) where total protein accumulation may also be linear (Wain & Staatz, 1973). However, an exponential pattern is consistent with present evidence from *S. cerevisiae* (Elliott & McLaughlin, 1978) and an exponential increase is clearly detectable by protein assay during the higher fold of growth per cycle in *Chlorella* (Schmidt, 1969; Fig. 1*b* this chapter).

To clarify the pattern of protein accumulation and to determine whether the level of poly(A)$^+$-containing messenger RNA (mRNA) influences the rate of protein accumulation in *Chlorella,* synchrony was obtained in cells never subjected to the light–dark regime. Small cells were selected by continuous-flow centrifugation from an asynchronous population growing under constant conditions and, to avoid effects of change in growth rate, the cells were returned to the same constant conditions for synchronous growth. An important feature of this procedure is that a control experiment can be performed to test whether the method of synchronisation causes perturbations in the accumulation of RNA or protein. For the control experiment, cells were subjected to centrifugation, as in selection, but an asynchronous population was retained. Because of their importance in assessing results from selected cells, data for the control cells are shown in Fig. 5 and it is evident that the selection procedure does not cause periodicity in the accumulation of RNA or protein. In the synchronous cells (Fig. 6), there is a clear periodicity in the accumulation of poly(A)$^+$ RNA. DNA replication occurs between 5 and 10 h but the increase in poly(A)$^+$ RNA is delayed until G1 of the next cycle, commencing at 12 h when the increase in cell number reveals that daughters have begun to grow and rupture the mother cell wall. There is similarly a stepwise accumulation of poly(A)$^+$ RNA in the second cycle of turbidostat growth (as employed in Figs. 1-3) when the sharper synchrony reveals a greater time separation between phases of DNA synthesis and mRNA accumulation. We do not know what causes the stepwise increase in mRNA, but it is unlikely that DNA replication is a direct cause since the increase is out of phase. There is no evidence that the sudden increase in mRNA level increases protein accumulation since rates of accumulation in the semi-log plot (Fig. 6) are parallel in the first cycle ending at 12 h, and in the second cycle from 15 h; both showing a smooth exponential increase. A similar exponential increase in protein, in spite of a sharp step in messenger level, is also seen in the second cycle of turbidostat growth.

In *Chlorella* therefore, mRNA is present in excess and the rate of protein synthesis is limited by other factors which are increasing steadily, such as ribosomes (see total RNA in Fig. 6), enzymes (see typical proteins in Fig. 4*b*) or products of the chloroplast which maintains close to 40% of cell volume through the cell cycle (Atkinson, *et al.,* 1974). There are indications of a parallel in *S. cerevisiae,* where mRNA appears to accumulate periodically (Fraser & Carter, 1976) and available evidence is consistent with exponential increase in protein

Fig. 5. Effect of harvesting and mock continuous-flow selection centrifugation on the accumulation of (○) cell number, (△) DNA, (□) total protein, (●) total RNA and (▲) poly(A)⁺RNA. An asynchronous 20 l culture was grown in continuous light under turbidostat control at an extinction of 0.1 measured at 680 nm, in a culture vessel of 68 mm light path. At 0 h the cells were harvested by passage through a Sharples continuous-flow centrifuge and resuspended in growth medium aerated at 25 °C. This suspension was passed through a continuous-flow rotor which can be used for the purification of small cells by the selective sedimentation of larger cells. The principle has been described by Lloyd, John, Edwards & Chagla (1975) and for the rotor a nylon disc was turned to form a shallow cup of 80 mm diameter, with a constricted opening and 11 pockets were machined in its inner face. The speed of rotation was governed by a rheostat and in this control experiment, speed was set so that cells were not pelleted with a through-flow rate of 120 ml min⁻¹. After the mock selection, cells were returned to a 2 l culture vessel, again of 68 mm light path at a turbidity of 0.1, for continued turbidostat growth and were sampled for cell number, DNA and protein as in Fig. 1. Cells taken for RNA analysis were immediately processed by suspension in 2% sodium tri-*iso*-propyl-napthalenesulphate; 10 mM NaCl; 50 mM Tris–HCl, pH 7.8 at 0 °C and broken at 0 °C with a French pressure cell, the effluent from the press being passed directly through a 400 mm length of 1 mm diameter stainless steel tubing in an ice water bath to re-cool the suspension. RNA was immediately purified from the extract as described by Lambe & John (1979) and stored as a precipitate under ethanol at −70 °C. Total RNA was estimated by absorption at 260 nm and poly(A)⁺RNA was estimated in 20 μg samples of total RNA, by hybridisation with excess [³H]poly(U) as described by Fraser & Carter (1976).

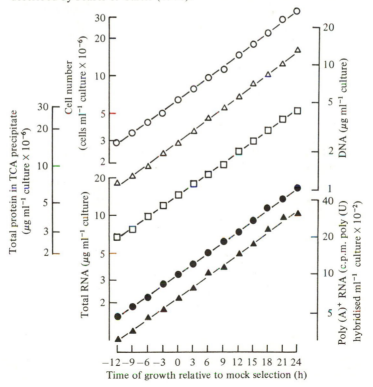

(Elliott & McLaughlin, 1978), but it is too early to speculate whether periodic excess of mRNA is common. In *Physarum* mRNA synthesis is concentrated in S phase (Sachsenmaier; Matthews, this volume) but in other cell types accumulation may be continuous, or if it is periodic the increase may be directly reflected in a rate change of protein synthesis, as could be the case in *Schizosaccharomyces pombe* (Fraser & Nurse, 1978).

Fig. 6. Accumulation of cell number (○), DNA (△), total protein (□), total RNA (●) and poly(A)⁺RNA (▲) in cells selected by continuous-flow centrifugation from a 20-l asynchronous population. Cells were grown under turbidostat culture and sampled as described in Fig. 5, except that rotor speed in the selection centrifuge was adjusted so that a little over 40% of cells were pelleted. The pelleted cells were discarded and the selection was repeated three times, leaving one tenth of the original cell number. The selection procedure occupied the same period of time as in the control experiment (Fig. 5) and in both cases the cells which were subsequently cultured were those not pelleted during passage through the selection rotor. Cells in their second division cycles under constant turbidostat conditions, equivalent to those in Figs. 1, 2 and 3, confirm that total RNA accumulation is linear with a rate change at the beginning of G1 phase and that accumulation of poly(A)⁺RNA forms a step in early G1, whether measured by hybridisation, or by purification with oligo(dT)cellulose (Lambe & John, 1979) for estimation at 260 nm.

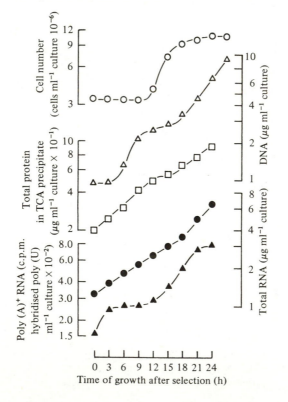

A further feature of many cell cycles, which is illustrated by the data presented here, is a pause in growth at division. The incidence of such changes in growth in many cell types is reviewed by Mitchison (1971) and it is now emerging that different underlying mechanisms are involved.

Among higher eukaryotes a depression of growth at division is common and results from a depression of protein synthesis (Scharff & Robbins, 1966). The cause of this depression is complex. A concurrent halt in total RNA synthesis at division is a common feature (Schiff, 1965; Van't Hof, 1963; Sachsenmaier, this volume) but this does not directly cause the slowing of protein synthesis because some mammalian cells continue protein synthesis through mitosis (Konrad, 1963) and there is evidence that, where protein synthesis does slow down, there is an inhibition of translation from mRNAs which are still present in the cytoplasm (Fan & Penman, 1970).

In *S. pombe* growth in volume ceases at division, but in this case there is no cessation of protein accumulation and instead, as the region of wall growth shifts to the developing septum, there is an increase in cell density (Mitchison, 1963).

Chlorella shows another means of achieving a pause in growth at division. Protein ceases to accumulate at cytokinesis and this as a normal part of the cell cycle since it is seen in the second cycle of turbidostat culture (Fig. 1 *a*) and in selected synchronous cells at 13 h (Fig. 6) but not in control cells. During the pause protein synthesis continues, as can be judged qualitatively from the labelling of protein bands in Fig. 4 (*b*), but protein fails to accumulate because there is an acceleration of protein breakdown. This correlates with a maximum loss of protein at this time if resynthesis is blocked with cycloheximide (M. Rollins & P. C. L. John, unpublished). Turnover involves a majority of proteins at cytokinesis, causing a pause in the accumulation of Ru5P kinase (Fig. 3*a*), in the accumulation of several tricarboxylic acid cycle enzymes (not shown) and in the accumulation of RuBP carboxylase which is being synthesised, as revealed by pulse labelling of the 56 000 and 14 000 dalton peptides at 36 h (Fig. 4*b*), although breakdown prevents net accumulation of activity (Fig. 3*c*).

We wish to emphasise the similarity between accumulation of proteins in *Chlorella* and *S. cerevisiae* because similarity in these different eukaryotes probably indicates a common situation. In both organisms the G1 phase is one in which the majority of individual proteins accumulate in constant proportions with one another and so entry to S phase and commitment to divide occur without the background of successive metabolic changes which are seen in other developmental situations. The few differences in proteins that accumulate at cytokinesis compared with other phases of the cycle in *Chlorella* may result from differences in susceptibility of the proteins to the rapid turnover at this time, rather than changes in the population being synthesised, since there is no evidence of changes in the population of quantitatively predominant mRNAs at cytokinesis; as we show in the next section.

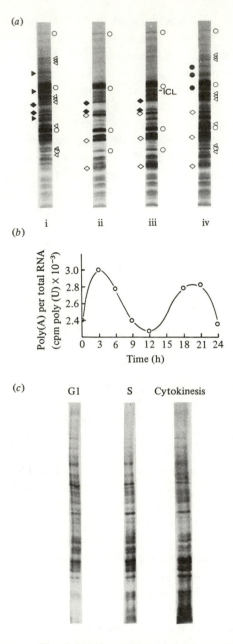

Fig. 7. Messenger RNA populations studied in relation to mode of carbon nutrition and progress through the cell cycle. In (a) and (c) levels of individual mRNA molecules are estimated indirectly, by translation in the wheat germ cell-free protein synthesising system, as described in Fig. 9, in the presence of [³⁵S]methionine. Proteins coded by the RNA molecules were separated by electrophoresis (Laemli, 1970) and incorporation was detected by autoradiography.

The mRNA population

We have investigated the population of mRNA molecules through the cell cycle by first translating them *in vitro* and then separating the products by electrophoresis. To test whether *Chlorella* cells can change their population of mRNAs and whether our methods would be able to detect any such changes, we compared in-vitro translation products of total poly(A)$^+$ RNA taken from cells either starved, or growing photoautotrophically, or growing heterotrophically on either glucose or acetate. Clear differences in mRNA population are seen in Fig. 7 (a). The levels of mRNAs coding for some proteins are influenced by whether growth is autotrophic or heterotrophic; other mRNAs are influenced by growth rate, which is highest in autotrophic cultures and lowest with these grown on acetate, while others are specific to the particular form of heterotrophic metabolism: glycolysis from glucose or glyconeogenesis from acetate.

Although the cell can change its mRNA population there is no evidence that it needs to do so to progress through the cell cycle. The same cells which were described in Fig. 6, were also analysed for their mRNA population in G1, S and cytokinesis, by translation and electrophoresis. No differences were detected (Fig. 7c). The mRNA population seen in late G1, when the periodic increase in total mRNA is complete (Fig. 6) and the proportion of messenger in total RNA

Caption to Fig. 7. (*cont.*)

(a) Messenger RNA populations revealed by translation of the poly(A)$^+$ RNA fraction taken from batch-grown autotrophic cells which were aerated for 6 h before harvest in (i) continued autotrophic growth (ii) darkness without added carbon source (iii) darkness with 10 mM sodium acetate and (iv) darkness with 10 mM glucose. Mass doubling times under these conditions are: (i) 8 h; (ii) infinity, but some growth by use of reserves; (iii) 60 h; (iv) 12 h. Some mRNAs appear to be present in translatable form under all of those four conditions (○) (although the identity of the proteins has not been tested by peptide mapping), but the majority change in level under different physiological conditions, many being present predominently at higher growth rates regardless of carbon source; i.e. present in (i) and (iv) (Δ); others being specific for autotrophic growth (▶); or for growth on glucose (●); while others are specific for heterotrophic growth, whether on acetate or glucose (◇); and others are specifically repressed by glucose (◆). ICL, mRNA for isocitrate lyase. (b) Levels of poly(A)$^+$ RNA relative to total RNA, in cells analysed in Fig. 6 which were synchronised by selection of small cells without exposure to periodic illumination. Messenger RNAs are at a maximum relative level by 3 h, which is at the end of the period of accumulation in G1, and are at a minimum relative level by 12 h, which is in cytokinesis and prior to the next round of accumulation in G1 of the subsequent cycle. Messenger RNA populations, at these extremes of proportion, in G1 and cytokinesis and also during S phase at 6 h, are analysed below. (c) Messenger RNA species revealed by translation of the poly(A)$^+$ RNA fraction, from cells analysed in Fig. 6, taken at 3 h in G1, at 6 h in S and at 12 h in cytokinesis. Preparation of poly(A)$^+$ RNA and cell-free translation were as described in Fig. 9.

is at a maximum (Fig. 7b), is the same as at other times, indicating that all mRNAs share in the periodic accumulation.

There is therefore a consistent correlation between constant populations of mRNAs, of proteins being synthesised, and of accumulating enzyme activities through the cell cycle. We conclude that the cell cycle proceeds with minimum disturbance of events involved in growth of the cell.

Potential for adaptive enzyme synthesis

One situation in which the phase of the cycle may influence the rate at which a cell can synthesise an individual protein is during adaptation to changed physiological circumstances. If a cell must make a new enzyme to resume its growth, it is more likely then to fully induce or derepress the necessary genetic information and in this case gene replication could double the rate of transcription. This simple relationship does hold in the bacterial cell cycle and there is a correlation between time of replication of the genes for β-galactosidase, tryptophanase and D-serine deaminase and the doubling time of their rates of adaptive synthesis (Donachie & Masters, 1969; Lutkenhaus et al., 1979). In view of the essentially transcriptional nature of control over protein synthesis in bacteria it is also clear that in this case an increase in mRNA production underlies the increase in adaptive synthesis.

In eukaryote cells the underlying mechanisms which limit adaptive enzyme synthesis are far from clear. One problem is the possibility of post-transcriptional control, since there are established instances in which mRNA level is not reflected in rate of enzyme synthesis in animal cells (reviewed by Revel & Groner, 1978) and in plant cells (Scragg, John & Thurston, 1975). We have therefore studied the level of isocitrate lyase mRNA and the amount of isocitrate lyase protein produced, when *Chlorella* cells were transferred at intervals through the cell cycle to growth in darkness with acetate as their sole carbon source. Isocitrate lyase mRNA was estimated by in-vitro translation of the total poly(A)$^+$ RNA fraction and isocitrate lyase was purified from the products by specific antiserum precipitation and electrophoresis. The method gives a recovery, free of contaminating proteins, which is linearly proportional to amount of lyase present after in-vitro synthesis (Fig. 8)

In the growing cells the amount of isocitrate lyase protein formed can be estimated from the appearance of enzyme activity, because in this organism the enzyme is stable during growth on acetate (John, Thurston & Syrett, 1970), the increase in activity correlates with an incorporation of amino acids into the immune-precipitable enzyme protein (Dunham & Thurston, 1978) and there is a parallel formation of enzyme activity and enzyme protein during adaptation through the cell cycle (Syrett, 1966; McCullough & John, 1972a). In cells which have recovered from the synchronising regime and are progressing through their second cell cycle under constant conditions, adaptive synthesis of the enzyme

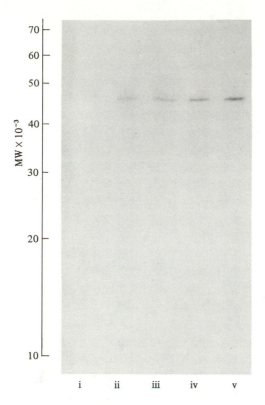

Fig. 8. Quantitative immune recovery of isocitrate lyase synthesised *in vitro*.
Poly(A)$^+$ RNA was purified as described in Fig. 9 from cells growing auto-
trophically by rapid photosynthesis and from cells which had been transferred for
6 h to darkness with 10 mM sodium acetate. Aliquots of each preparation containing 0.5 μg RNA were translated by the wheat germ cell-free system, as
described in Fig. 9. For immune recovery of isocitrate lyase, the in-vitro products were diluted in 10 mM phosphate buffer pH 7.5, 0.15 M NaCl and rabbit
IgG, raised against purified isocitrate lyase (John & Syrett, 1967), was allowed
to complex for 30 min at 37 °C and then recovered by precipitation overnight at
4 °C with goat anti-rabbit IgG serum (Miles). The complex recovered by centrifugation was subject to electrophoresis (Laemli, 1970) and the isocitrate lyase
band, at 47 000 molecular weight (MW) was located by autoradiography and
excised for estimation by scintillation counting. Background activity was determined by measurement of a segment of equal weight taken immediately
below the lyase band. This procedure was applied in track (i) to the in-vitro
products of autotrophic cell mRNAs and other tracks show recovery from mixtures with increasing proportions of products coded by mRNAs of cells adapting
to acetate, which comprised in (ii) 25%, (iii) 50%, (iv) 75% and (v) 100% of the
proteins subject to immune fractionation. Counts in isocitrate lyase were; (i) 33,
(ii) 1422 (iii) 2229, (iv) 3710 and (v) 4751 per minute. Analysis of this in-
vitro-synthesised enzyme, after excision from the gel, by partial proteolysis
with chymotrypsin, papain or *Streptomyces griseus* protease, shows peptides
which correlate in electrophoretic mobility with those obtained, by the corresponding protease, from authentic in-vivo-synthesised isocitrate lyase.

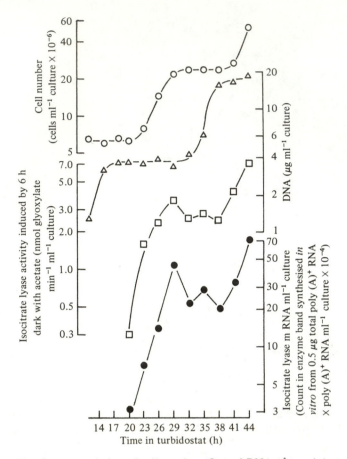

Fig. 9. Accumulation of cell number (○) and DNA (△) and the capacity for adaptive accumulation of isocitrate lyase activity (□) and isocitrate lyase messenger RNA (●), in cells which were previously synchronised by periodic illumination, but during the experimental period were transferred to constant conditions of turbidostat growth. The first cycle in turbidostat culture is one of recovery and was not sampled, but cells were sampled in the second cycle for cell number and DNA, as described in Fig. 1. At intervals during the sampling period, 1 l volumes of culture were removed, darkened and supplemented with sodium acetate to 10 mM final concentration. Aeration was continued and after 6 h cells in the subculture were sampled for estimation of isocitrate lyase activity by cell-free assay, as described by McCullough & John (1972a) and for RNA, which was immediately extracted, purified and precipitated as described in Fig. 5. Isocitrate lyase activity and isocitrate lyase mRNA level are plotted at the time of subculture because there is relatively little growth during the first 6 h of adaptation to acetate and, during G1 at least, progress through the cycle halts during adaptation. For estimation of isocitrate lyase mRNA the amount of total RNA present in each sample was measured and relative levels of the total poly(A)+RNA population in each sample were estimated by hybridisation with [³H]poly(U), as described in Fig. 5. The poly(A)+RNA fraction was then prepared from 1 mg aliquots of total RNA by binding to oligo(dT)cellulose

can be obtained at all stages through the cycle (Fig. 9). Activity detected after 6 h of growth with acetate does indicate adaptive synthesis, because basal levels of the enzyme are very low in rapidly-photosynthesising cells (Thurston, 1977); less than 0.002% of levels reached after 6 h adaptation.

This experiment reveals that rates of adaptive synthesis are another property that can be distorted by perturbations arising from synchronisation. Cells being synchronised by light–dark in batch culture show an increase of adaptive synthesis in early G1 phase similar to that shown in Fig. 9, but then enter a non-inducible phase from late G1 through cytokinesis (McCullough & John, 1972 a). A possible explanation for the period of non-inducibility of batch-grown cells is that their slower growth rate in late G1 as culture density increases, results in a slower consumption of the sugar pools (which repress synthesis of the enzyme; Syrett, 1966; McCullough & John, 1972 b) when cells are shifted to darkness with acetate.

The amount of isocitrate lyase which can be induced increases stepwise during early G1 phase and there follows a slight, but reproducible, decline in inducibility, to a level which is maintained through S phase and cytokinesis (Fig. 9). This pattern, therefore, contrasts with the steady increase in capacity for general protein accumulation throughout G1 and S phases. The specific limitation of isocitrate lyase synthesis is probably the capacity for accumulation of its mRNA, since the amount of lyase mRNA, measured at the end of each 6 h period of adaptation to acetate, correlates closely with the amount of enzyme synthesised; both increase in early G1 phase between 20 and 21 h and then in the next cycle from 38 h, and both decline slightly during late G1 and S phases (Fig. 9).

In this particular instance, control in a eukaryote cell resembles the simple situation in prokaryote cells, where level of mRNA determines the pattern of adaptive enzyme synthesis, but in other situations post-transcriptional control may be found to produce a rate of enzyme synthesis different from mRNA level.

Caption to Fig. 9. (*cont.*)
(Lambe & John, 1979). Each sample of RNA was offered to two samples of oligo(dT)cellulose to obtain complete recovery of poly(A)$^+$RNA. This fraction was then translated into protein *in vitro,* by the wheat germ system (Roberts & Paterson, 1973) in which endogenous in RNAs were eliminated by the method of Pelham & Jackson (1976). The wheat germ preparation was linear in response up to 1 μg of added mRNA and in this experiment 0.5 μg amounts of mRNA were added for optimum linearity and recovery of product. In-vitro-synthesised isocitrate lyase was purified by double immune precipitation and SDS electrophoresis, as described in Fig. 8. The amount of isocitrate lyase coded in 0.5 μg samples of poly(A)$^+$RNA provided an estimate of the level of this mRNA relative to others in an individual sample and, for comparison with other culture samples, was multiplied by the amount of total poly(A)$^+$RNA in each.

One difference from prokaryote cells concerns the effect of gene replication on mRNA accumulation. In *Chlorella* endogenous levels of isocitrate lyase mRNA are low in rapidly-growing autotrophic cells and are below the limits accurately detectable by our assay (Fig. 8), therefore levels measured after 6 h adaptation to acetate (Fig. 9) show the amount of mRNA which has accumulated in response to acetate. There is no indication that new gene copies result in a greater capacity to synthesise mRNA since cells transferred to growth on acetate at 38 h have four times as many gene copies as cells that are transferred at 32 h but they accumulate only the same amount of isocitrate lyase mRNA and synthesise the same amount of enzyme.

There are a number of instances of coincidence between time of S phase and time of increase in capacity for adaptive enzyme synthesis in the thermophilic *Chlorella* (Schmidt, 1974; Israel *et al.*, 1977) but it remains to be established whether an increase in mRNA synthesis is the cause, and instances of poor correlation with S phase such as that presented here and reported in *S. pombe* (Mitchison, 1977) in *Chlorella* 8b (Aasberg, *et al.* 1974), and in hepatoma cells (Martin, Tomkins & Granner, 1969) make it clear that the control of adaptive synthesis is more complex in eukaryote cells, where it may be less common for genetic information to be fully unmasked for transcription during adaptation and where the requirement for packaging of DNA into chromatin, or for maturation of transcript into functional message (Crick, 1979), may delay the expression of new DNA copies.

Enzymes which allow continuation of growth by adjusting metabolism to use available nutrients can be synthesised adaptively in all phases of the cell cycle. There are obvious survival advantages in this, but there are indications that where new enzymes will initiate a change in phase of life cycle, (as in bacterial sporulation, Young & Mandelstam, 1979) or tissue differentiation (as in globin synthesis by erythrocytes, Harrison, Conkie, Rutherford & Yeoh, 1978), there may then be a necessity for passage through S phase before the new enzymes can be produced. It is tempting to speculate that genes for some developmental proteins may be securely masked unless the developmental signal is present as they replicate.

Are any proteins periodic?

Only a small proportion of proteins at most, are synthesised periodically in the cell cycle, whereas in other developmental processes these are common, for example in the sporulation of *Dictyostelium* (Tuchman, Alton & Lodish, 1974), with consequent changes in numerous enzyme levels (Franke & Sussman, 1973). This difference suggests that the cell cycle should be viewed more as a means for dividing existing resources between daughters, than as a process requiring the successive synthesis of numerous new chemical constituents.

Studies could now usefully be concentrated on those enzymes which are directly involved in cycle events such as DNA synthesis. It is relevant that the one enzyme so far shown to be periodic as a normal part of the cell cycle in *S. pombe* is deoxythymidine monophosphate kinase (Mitchison, 1977). This enzyme is involved in synthesis of DNA precursors and such enzymes also show peak patterns in the cell cycle of *Physarum* (see Sachsenmaier; Matthews, this volume) in animal (Gelbard, Kim & Perez, 1969) in plant (Yeoman, this volume) and in algal cells (Shen & Schmidt, 1966) but the influence of cell manipulation in these and other instances remains to be tested. An analysis of control signals for the thymidine kinase peak in the *Physarum* cycle, is illustrated in Sachsenmaier's chapter.

Among the questions which remain are: whether the periodic synthesis of proteins, such as the enzymes which synthesise DNA precursors, initiates the overall process of division; and whether later periodic syntheses initiate the successive later dependent events within division (events illustrated for example by Carter, this volume). It is already clear that some proteins, which are essential to division, cannot trigger events by their synthesis because they are present in excess. If their synthesis is blocked by higher temperature, the level already present will allow more divisions, until the successively smaller amounts inherited by daughter cells do become limiting (Hartwell, 1978). Such excess proteins could nonetheless function periodically if they were subject to periodic activation or were denied access to substrate or to site of function for part of the cycle. To detect a situation in which synthesis of a protein must occur, for the cycle in progress to continue, is more difficult. The commonest cause of arrest within one cycle of temperature shift, is loss of function by all the molecules of mutant protein, regardless of whether synthesis continues. The situation is clarified by use of temperature-sensitive nonsense suppressors (as described in this volume by Donachie). Such mutants do make clear that, at the non-permissive temperature, it is further synthesis that is lost, and for some proteins – termed 'cell cycle control proteins' – in *E. coli,* lack of their synthesis does block the cycle; but this form of analysis does not directly reveal whether the essential synthesis is continuous or periodic.

Lutkenhaus *et al.,* (1979) make the stimulating suggestion that synthesis of cell cycle control proteins will be continuous in *E. coli* because, from the available genetic information, they argue that the 750 proteins they studied were more than half of all the proteins, and all were synthesised continuously. If there is indeed no periodic synthesis in *E. coli* then greater attention must be given to how the cell cycle may be controlled by attainment of threshold protein levels (Fantes & Nurse discuss general mechanisms in this volume) or by periodic activation, as discussed earlier. However, of the hundreds of proteins that have not yet been studied some might be synthesised periodically, and there is an indirect indication that the cell cycle control protein *ftsA* (Donachie, this volume) might be synthesised preferentially at its time of function in septation because its mRNA level appears to be maximal then (Lutkenhaus & Donachie, unpublished).

The eukaryote cell cycle however, does involve the periodic synthesis of at least the histone proteins. This periodicity is not detected in analysis of poly(A)$^+$ RNA (Fig. 7c), because histone mRNAs are poly(A)$^-$; nor is it detected in analysis of the majority of proteins, because the histones are basic and are not resolved. But the histones can be labelled and readily resolved by appropriate methods, and it is clear that they are predominantly synthesised in S phase (Matthews, this volume).

Enzymes which are influenced by DNA synthesis are also probably genuinely periodic. Perhaps the clearest example is thymidine kinase, since completion of S phase is necessary to terminate its synthesis in *Physarum* (Sachsenmaier, this volume) and is necessary to stimulate its synthesis in artichoke (Yeoman, this volume). This particular enzyme is not essential to the cell cycle, but other enzymes associated with DNA metabolism are, and their possible control by initiation or termination of S phase is an area of study for the future.

Another objective will be the identification of proteins, which in mutant form can affect progress through the cycle, either as spots on gels, or as enzymes, so that they can be measured through the cycle. The more interesting proteins could be particularly difficult to measure because the cell cycle may be controlled by interacting proteins (Sachsenmaier; Fantes & Nurse, this volume) and the variability of the cell cycle (Brooks, this volume) may indicate that there is a large random element in their collision and so only a very small number may be present in each cell.

We thank the Science Research Council for support.

References

Aasberg, K. E., Lien, T. & Knutsen, G. (1974). Induction of isocitrate lyase in synchronous cultures of *Chlorella fusca*. *Physiologia Plantarum,* **31,** 245–51.

Andrews, T. J., Badger, M. R. & Lorimer, G. H. (1973). Factors affecting the interconversion between kinetic forms of ribulose diphosphate carboxylase-oxygenase from spinach. *Archives of Biochemistry and Biophysics,* **171,** 93–103.

Atkinson, A. W., Jr., Gunning, B. E. S. & John, P. C. L. (1972). Sporopollenin in the cell wall of *Chlorella* and other algae: ultrastructure, chemistry and incorporation of ^{14}C-acetate studied in synchronous culture. *Planta, Berlin,* **107,** 1–32.

Atkinson, A. W., Jr., John, P. C. L. & Gunning, B. E. S. (1974). The growth and division of the single mitochondrion and other organelles during the cell cycle of *Chlorella*, studied by quantitative stereology and three-dimensional reconstruction. *Protoplasma*, **81,** 77–109.

Bassham, J. A. (1973). Control of photosynthetic carbon metabolism. In *Rate Control of Biological Processes*, ed. D. D. Davies, pp. 461–83. Cambridge University Press.

Bellino, F. L. (1973). Continuous synthesis of partially derepressed asparate transcarbamylase during the division cycle of *Escherichia coli* B/r. *Journal of Molecular Biology*, **74,** 223–38.

Benitez, T., Nurse, P. & Mitchison, J. M. (1980). Arginase and sucrase potential in the fission yeast *Schizosaccharomyces pombe. Journal of Cell Science*, **46,** 399–431.

Boyd, A. & Holland, I. B. (1977). Protein d, an iron-transport portein induced by filtration of cultures of *E. Coli. FEBS Letters*, **76,** 20–4.

Cazzulo, J. J., Sundaram, T. K. & Kornberg, H. L. (1970). Mechanism of pyruvate carboxylase formation from the apo-enzyme and biotin in a thermophilic *Bacillus. Nature, London*, **233,** 1103–5.

Cole, E. M. A. (1978). Regulation of tricarboxylic acid cycle enzyme accumulation during the cell cycle of *Chlorella*. PhD thesis, The Queen's University, Belfast.

Cox, C. G. & Gilbert, J. B. (1970). Non-identical times of gene expression in two strains of *Saccharomyces cerevisiae* with mapping differences. *Biochemical and Biophysical Research Communications*, **38,** 750–7.

Crick, F. H. C. (1979). Split genes and RNA splicing. *Science*, **204,** 264–71.

Donachie, W. D. & Masters, M. (1969). Temporal control of gene expression in bacteria. In *The Cell Cycle: Gene–Enzyme Interactions*, ed. G. M. Padilla, G. L. Whitson & I. L. Cameron, pp. 37–76. London & New York: Academic Press.

Dunham, S. M. & Thurston, C. F. (1978). Control of isocitrate lyase synthesis in *Chlorella fusca* var. *vacuolata*. Rate of enzyme synthesis in the presence and absence of acetate, measured by ^{35}S methionine labelling and immune precipitation. *Biochemical Journal*, **176,** 179–85.

Elliott, S. G. & McLaughlin, C. S. (1978). Rate of macromolecular synthesis through the cell cycle of the yeast *Saccharomyces cerevisiae. Proceedings of the National Academy of Sciences, USA*, **75,** 4384–8.

Fan, H. & Penman, S. (1970). Regulation of protein synthesis in mammalian cells II. Inhibition of protein synthesis at the level of initiation during mitosis. *Journal of Molecular Biology*, **50,** 655–70.

Forde, B. G. & John, P. C. L. (1973). Stepwise accumulation of autoregulated enzyme activities during the cell cycle of the eucaryote *Chlorella. Experimental Cell Research*, **79,** 127–35.

Franke, J. & Sussman, M. (1973). Accumulation of uridine diphosphoglucose pyrophosphorylase in *Dictyostelium discoideum* via preferential synthesis. *Journal of Molecular Biology*, **81,** 173–85.

Fraser, R. S. S. & Carter, B. L. A. (1976). Synthesis of polyadenylated messenger RNA and ribosomal RNA during the cell cycle of *Saccharomyces cerevisiae. Journal of Molecular Biology*, **104,** 223–42.

Fraser, R. S. S. & Nurse, P. (1978). Novel cell cycle control of RNA synthesis in yeast. *Nature, London*, **271,** 726–30.

Gelbard, A. S., Kim, J. H. & Perez, A. G. (1969). Fluctuation in deoxycytidine-monophosphate deaminase activity during the cell cycle in synchronous populations of HeLa cells. *Biochimica et Biophysica Acta*, **182,** 564–6.

Goldberg, A. L. & St John, A. C. (1976). Intracellular protein degradation in mammalian and bacterial cells: part 2. *Annual Review of Biochemistry*, **45,** 747–803.

Golombek, J., Wolf, W. & Wintersberger, E. (1974). DNA synthesis and DNA-polymerase activity in synchronised yeast cells. *Molecular and General Genetics,* **132,** 137–45.

Gudas, L. J., James, R. & Pardee, A. B. (1976). Evidence for the involvement of an outer membrane protein in DNA initiation. *Journal of Biological Chemistry,* **251,** 3470–9.

Halvorson, H. O., Carter, B. L. A. & Tauro, P. (1971). Synthesis of enzymes during the cell cycle. *Advances in Microbial Physiology,* **6,** 47–106.

Hames, B. D. & Ashworth, J. M. (1974). The metabolism of macromolecules during the differentiation of myxamoebae of the cellular slime mould *Dictyostelium discoideum* containing different amounts of glycogen. *Biochemical Journal,* **142,** 301–15.

Harrison, P. R., Conkie, D., Rutherford, T. & Yeoh, G. (1978). Molecular aspects of erythroid cell regulation. In *Stem Cells and Tissue Homeostasis,* ed. B. I. Lord, C. S. Potten & R. J. Cole. pp. 241–57. British Society for Cell Biology Symposium 2. Cambridge University Press.

Hartwell, L. H. (1978). Cell division from a genetic perspective. *Journal of Cell Biology,* **77,** 627–37.

Howell, S. H., Posakony, J. W. & Hill, K. R. (1977). The cell cycle program of polypeptide labelling in *Chalmydomonas reinhardtii. Journal of Cell Biology,* **72,** 223–41.

Israel, D. W., Gronostajski, R. M., Yeung, A. T. & Schmidt, R. R. (1977). Regulation of accumulation and turnover of an inducible glutamate dehydrogenase in synchronous cultures of *Chlorella. Journal of Bacteriology,* **130,** 793–804.

John, P. C. L., Cole, E. M. A., Keenan, P. & Rollins, M. J. (1980). Mitochondrial development in the cell cycle of *Chlorella. The Society for General Microbiology Quarterly,* **8,** 26–27.

John, P. C. L., McCullough, W., Atkinson, A. W. Jr., Forde, B. G. & Gunning, B. E. S. (1973). The cell cycle in *Chlorella.* In *The Cell Cycle in Development and Differentiation,* ed. M. Balls & F. S. Billett, pp. 61–76. Cambridge University Press.

John, P. C. L. & Syrett, P. J. (1967). The purification and properties of isocitrate lyase from *Chlorella. Biochemical Journal,* **105,** 409–18.

John, P. C. L., Thurston, C. F. & Syrett, P. J. (1970). Disappearance of isocitrate lyase enzyme from cells of *Chlorella pyrenoidosa. Biochemical Journal,* **119,** 913–9.

Kates, J. R. & Jones, R. F. (1967). Periodic increases in enzyme activity in synchronised cultures of *Chlamydonomas reinhardtii. Biochimica et Biophysica Acta,* **145,** 153–8.

Klevecz, R. R. & Kapp, L. N. (1973). Intermittent DNA synthesis and periodic expression of enzyme activity in the cell cycle of WI-38. *Journal of Cell Biology,* **58,** 564–73.

Konrad, C. G. (1963). Protein synthesis and RNA synthesis during mitosis in animal cells. *Journal of Cell Biology,* **19,** 267–77.

Kuempel, P. L., Masters, M. & Pardee, A. B. (1965). Bursts of enzyme synthesis in the bacterial duplication cycle. *Biochemical and Biophysical Research Communications,* **18,** 858–67.

Lacroute, F. (1973). RNA and protein elongation rates in *Saccharomyces cerevisiae. Molecular and General Genetics,* **125,** 319–27.

Laemmli, U. K. (1970). Cleavage of structural proteins during the assembly of the head of bacteriophage T_4. *Nature, London,* **227,** 680–5.

Lambe, C. A. & John, P. C. L. (1979). Action of 6-methylpurine on RNA synthesis: incorporation of the adenine analogue into coding sequences and polyadenylate tracts of messenger RNA in *Chlorella. New Phytologist*, **83**, 321–41.

Lloyd, D., John, L., Edwards, C. & Chagla, A. H. (1975). Synchronous cultures of micro-organisms: large scale preparation by continuous-flow size selection. *Journal of General Microbiology*, **88**, 153–8.

Lorenzen, H. & Hess, M. (1974). Synchronous cultures. In *Algal Physiology and Biochemistry*, ed. W. D. P. Stewart, pp. 894–908. Oxford: Blackwell Scientific Publications.

Lutkenhaus, J. F., Moore, B. A., Masters, M. & Donachie, W. D. (1979). Individual proteins are synthesised continuously throughout the *Escherichia coli* cell cycle. *Journal of Bacteriology*, **138**, 352–60.

Martin, D., Tomkins, G. M., & Granner, D. (1969). Synthesis and induction of tyrosine aminotransferase in synchronised hepatoma cells in culture. *Proceedings of the National Academy of Sciences, USA*, **62**, 248–55.

Masters, M. & Donachie, W. D. (1966). Repression and the control of cyclic enzyme synthesis in *Bacillus subtilis. Nature, London*, **209**, 476–9.

McCullough, W. & John, P. C. L. (1972*a*). A temporal control of de-novo synthesis of isocitrate lyase during the cell cycle of the eukaryote *Chlorella pyrenoidosa. Biochimica et Biophysica Acta*, **269**, 287–96.

McCullough, W. & John, P. C. L. (1972*b*). Control of de-novo isocitrate lyase synthesis in *Chlorella. Nature, London*, **239**, 402–5.

Milcarek, C. & Zahn, K. (1978). The synthesis of ninety proteins including actin throughout the HeLa cell cycle. *Journal of Cell Biology*, **79**, 833–8.

Mitchison, J. M. (1963). Pattern of synthesis of RNA and other cell components during the cell cycle of *Schizosaccharomyces pombe. Journal of Cellular and Comparative Physiology*, **62** Suppl. 1, 1–13.

Mitchison, J. M. (1969). Enzyme synthesis in synchronous cultures. *Science*, **165**, 657–63.

Mitchison, J. M. (1971). *The Biology of the Cell Cycle*, Cambridge University Press.

Mitchison, J. M. (1977). Enzyme synthesis during the cell cycle. In *Cell Differentiation in Microorganisms Plants and Animals*, ed. L. Nover & K. Mothes, pp. 377–401. Jena: VEB Gustav Fischer Verlag.

Mitchison, J. M. & Creanor, J. (1969). Linear synthesis of sucrase and phosphatases during the cell cycle of *Schizosaccharomyces pombe. Journal of Cell Science*, **5**, 373–91.

Molloy, G. R. & Schmidt, R. R. (1970). Studies on the regulation of ribulose-1,5-diphosphate carboxylase synthesis during the cell cycle of the eucaryote *Chlorella. Biochemical and Biophysical Research Communications*, **40**, 1125–33.

Newlon, C. W., Petes, T. D., Hereford, L. M., & Fangman, W. L. (1974). Replication of yeast chromosomal DNA. *Nature, London*, **247**, 32–5.

Pelham, H. R. B. & Jackson, R. J. (1976). An efficient mRNA-dependent translation system from reticulate lysates. *European Journal of Biochemistry*, **67**, 247–56.

Pickett-Heaps, J. D. (1972). Variation in mitosis and cytokinesis in plant cells: its significance in the phylogeny and evolution of ultrastructural systems. *Cytobios*, **5**, 59–77.

Piperno, G. & Luck, D. J. L. (1977). Microtubular proteins of *Chlamydomonas reinhardtii. Journal of Biological Chemistry*, **252**, 383–91.

Revel, M. & Groner, Y. (1978). Post-transcriptional and translational controls of gene expression in eukaryotes. *Annual Review of Biochemistry*, **47**, 1079–126.

Roberts, B. E. & Paterson, B. M. (1973). Efficient translation of Tobacco

Mosaic Virus RNA and rabbit globin 9S RNA in a cell-free system from commercial wheat germ. *Proceedings of the National Academy of Sciences, USA,* **70,** 2330–4.

Scharff, M. D. & Robbins, E. (1969). Polyribosome disaggregation during metaphase. *Science,* **151,** 992–5.

Schiff, S. O. (1965). Ribonucleic acid synthesis in neuroblasts of *Chortophage viridifasciata* (de Geer) as determined by observation of individual cells in the mitotic cycle. *Experimental Cell Research,* **40,** 264–76.

Schimke, R. T. (1973). Control of enzyme levels in mammalian tissues. *Advances in Enzymology,* **37,** 135–87.

Schmidt, R. R. (1969). Control of enzyme synthesis during the cell cycle of *Chlorella.* In *The Cell Cycle: Gene–Enzyme Interactions,* ed. G. M. Padilla, G. L. Whitson & I. L. Cameron, pp. 159–77, London & New York: Academic Press.

Schmidt, R. R. (1974). Transcriptional and post-transcriptional control of enzyme levels in eucaryotic microorganisms. In *Cell Cycle Controls,* ed. G. M. Padilla, I. L. Cameron & A. H. Zimmerman, pp. 201–33. London & New York: Academic Press.

Schutt, H. & Holzer, H. (1972). Biological function of the ammonia-induced inactivation of glutamine synthetase in *Escherichia coli. European Journal of Biochemistry,* **26,** 68–72.

Scragg, A. H., John, P. C. L. & Thurston, C. F. (1975). Post-transcriptional control of isocitrate lyase induction in the eukaryote alga *Chlorella fusca. Nature, London,* **257,** 498–501.

Sebastian, J., Carter, B. L. A. & Halvorson, H. O. (1973). Induction capacity of enzyme synthesis during the cell cycle of *Saccharomyces cerevisiae. European Journal of Biochemistry,* **37,** 516–22.

Shen, S. R-C. & Schmidt, R. R. (1966). Enzymic control of nucleic acid synthesis during synchronous growth of *Chlorella pyrenoidosa.* II. Deoxycytidine monophosphate deaminase. *Archives of Biochemistry and Biophysics,* **115,** 13–20.

Sussman, M. & Brackenbury, R. (1976). Biochemical and molecular-genetic aspects of cellular slime mould development. *Annual Review of Plant Physiology,* **27,** 229–65.

Syrett, P. J. (1966). The kinetics of isocitrate lyase formation in *Chlorella:* Evidence for the promotion of enzyme synthesis by photophyosphorylation. *Journal of Experimental Botany,* **17,** 641–54.

Tamiya, H., Iwamura, T., Shibata, K., Hase, E. & Nihei, T. (1953). Correlation between photosynthesis and light-independent metabolism in growth of *Chlorella. Biochimica et Biophysica Acta,* **12,** 12–40.

Tauro, P. & Halvorson, H. O. (1966). Effect of gene position on the timing of enzyme synthesis in synchronous cultures of yeast. *Journal of Bacteriology,* **92,** 652–61.

Tauro, P., Halvorson, H. O. & Epstein, R. L. (1968). Time of gene expression in relation of centromere distance during the cell cycle of *Saccharomyces cerevisiae. Proceedings of the National Academy of Sciences, USA,* **59,** 277–84.

Thurston, C. F. (1977). Control of isocitrate lyase synthesis in *Chlorella fusca* var. *vacuolata.* The basal activity of the enzyme and the kinetics of induction.*Biochemical Journal,* **164,** 147–51.

Tuchman, J., Alton, T. & Lodish, H. F. (1974). Preferential synthesis of actin during early development of the slime mould *Dictyostelium discoideum. Developmental Biology,* **40,** 116–28.

Van't Hof, J. (1963). DNA, RNA and protein synthesis in the mitotic cycle of pea root meristem cells. *Cytologia,* **28,** 30–5.

Wain, W. H. & Staatz, W. D. (1973). Rates of synthesis of ribosomal protein and total ribonucleic acid through the cell cycle of the fission yeast *Schizosaccharomyces pombe*. *Experimental Cell Research,* **81,** 269–78.

Wiame, J. M. (1971). The regulation of arginine metabolism in *Saccharomyces cervisiae*. In *Current Topics in Cellular Regulation,* **4,** 1–38.

Wieland, O. H., Siess, E. A., Weiss, L., Löffler, G., Patzelt, C., Portenhauser, R., Hartmann, U. & Schirman, A. (1973). Regulation of mammalian pyruvate dehydrogenase complex by covalent modification. In *Rate Control of Biological Processes,* ed. D. D. Davies, pp. 371–400. Cambridge University Press.

Yashphe, J. & Halvorson, H. O. (1976). β-D-galactosidase activity in single yeast cells during the cell cycle of *Saccharomyces lactis*. *Science,* **191,** 1283–4.

Young, M. & Mandelstam, J. (1979). Early events during bacterial endospore formation. *Advances in Microbial Physiology,* **210,** 104–62.

H.R.MATTHEWS

Chromatin proteins and progress through the cell cycle

Introduction

The replication and ordered division of the genetic material of a cell is fundamental to cell proliferation. Replication of DNA, and histone synthesis, occur during S phase while the chromatin is in a dispersed state. This is followed by G2 phase and then mitosis during which stage the chromosomes condense into characteristic metaphase chromosomes which divide into two equal sets, one for each new cell. Sachsenmaier (this volume) has discussed the mitotic clock in *Physarum*. Chromosome condensation may be a primary part of its timer and is certainly an essential feature when mitosis is initiated. This chapter concentrates on (i) the initiation of chromosome condensation and mitosis and (ii) the control of transcription in the cell cycle. Factor (i) concerns the events of G2 phase and the control of this part of the cell cycle at the molecular level and factor (ii) leads towards the final stages of the solution of the cell cycle problem, namely the control exerted by the nucleotide sequence of the chromatin DNA.

Chromatin composition and subunit structure

Chromatin is organised into subunits, called nucleosomes, each of which contains a continuous 200 base pair length of DNA and nine histone molecules, the major basic proteins of chromatin. The histone complement is two each of H2A, H2B, H3 and H4 plus one H1 molecule. The group $(H3 \cdot H4)_2 \cdot H2A_2 \cdot H2B_2$ is bound together by hydrophobic interactions to form a core which is closely associated with about 140 base pairs of DNA. This histone DNA complex is a well defined, universal, structure which has been studied widely (reviewed in Nicolini, 1979). X-ray diffraction studies of crystals are expected to reveal the detailed structure but the outline is already known from neutron scattering and other studies. In solution the core is a flat disc of protein 5.5 nm thick and 7.0 nm in diameter with DNA wound round the edge to give an 11 nm diameter disc of DNA/protein that is called the core particle (Hjelm *et al.*, 1977).

223

A similar structure is found in the crystals (Finch *et al.*, 1977). The positions of the remaining 60 base pairs of DNA (called linker DNA) are not known and may depend on the higher order packing of the nucleosomes. Histone H1 is probably bound directly to the linker DNA near the core particle since it protects some linker DNA from digestion by nuclease (Noll, 1976). The linker DNA may also be bound to core particle histones but this is not yet clear. The length of the linker DNA varies substantially between species, between cell types and within a cell type, although the length of DNA in the core particle is constant (Compton, Bellard & Chambon, 1976; Johnson *et al.*, 1978).

A single metaphase chromosome may contain of the order of 0.1 m of DNA with its associated proteins in the form of nucleosomes (Vogt & Braun, 1976). The nucleosomes are packed together in a compact structure that can form the metaphase chromosome. It has been proposed that this structure is organised round a protein matrix or scaffold which can be seen by electron microscopy of histone-depleted chromosomes. The protein scaffold retains the shape of the metaphase chromosome and the DNA is partly released to form loops of length 30 000 to 90 000 base pairs (Adolph, Cheng & Laemmli, 1977; Paulson & Laemmli, 1977). Domains of comparable lengths have also been deduced from studies of restriction nuclease digestion of interphase chromatin (Igo-Kemenes, Greil & Zachau, 1977; Igo-Kemenes & Zachau, 1977). It is assumed that in the presence of histones many core particles are formed in each loop and these interact with one another and with H1 histone and linker DNA to form the metaphase chromosome. These interactions can be modified either to cause the metaphase chromosome to disperse or to re-form by condensation of dispersed chromatin.

Short lengths of chromatin (about 200 nucleosomes) will pack *in vitro* to form a higher order structure, the 30 nm coil, which has also been described in electron micrographs of native chromatin. Neutron scattering and electron microscopic studies of the 30 nm coil suggest a flat helix of nucleosomes of pitch 10 nm with about six nucleosomes per turn (Carpenter, Baldwin, Bradbury & Ibel, 1976; Finch & Klug, 1976). There is also evidence for a higher order structure based on groups of about eight nucleosomes (called 'super-beads') (Renz, Nehls & Hozier, 1977). The relationship between these two structures is not clear. The formation of the 30 nm coil depends on the presence of divalent cations. It will form in the absence of H1 histone, at high chromatin concentration, but requires H1 histone at low concentration, suggesting that H1 histone stabilises or initiates the formation of the 30 nm coil.

The suggestion that H1 histone is involved in chromosome packing *in vivo* is supported by earlier data (e.g. Littau, Burdick, Allfrey & Mirsky, 1965). *In vitro,* sodium chloride induces a substantial contraction of chromatin gels which is completely absent in H1-depleted chromatin and a similar effect is seen in dilute solutions of chromatin where sodium chloride induces an H1-dependent turbidity in the solution (Bradbury, Carpenter & Rattle, 1973; Corbett, 1979). Since H1 interacts directly with DNA in chromatin, similar effects are obtained with H1·DNA complexes (Matthews & Bradbury, 1978). The turbidity and gel contraction changes correlate with changes in the nuclear magnetic resonance spectrum of the H1 in the complex (Bradbury *et al.*, 1973).

Much less is known about the structural roles of the other chromatin proteins, the non-histone chromosomal proteins. The most detailed characterisation has been carried out for a group of highly charged proteins called HMG proteins (for high mobility group) by Johns and coworkers (Johns, Goodwin, Walker & Sanders, 1975; Goodwin, Walker & Johns, 1978). HMG proteins are thought to have a structural role related to transcriptionally active chromatin (Levy-W, Wong & Dixon, 1977; Vidali, Boffa & Allfrey, 1977; Weisbrod & Weintraub, 1979). The other non-histone proteins constitute a heterogeneous group, a few of which have known enzymatic functions such as polymerase complexes and histone modifying enzymes, but the majority remain unstudied in any depth.

Synthesis of chromatin components in the cell cycle

DNA synthesis occurs in S phase as has been discussed in previous chapters (Mitchison; Sachsenmaier) although extra-chromosomal DNA may be replicated throughout the cell cycle. Changes in the patterns of RNA synthesis have been measured during the cell cycle but the measurements do not normally distinguish between changes in transcription and changes in RNA processing or sometimes between these and specific changes in translation efficiency. Total RNA synthesis appears to stop during mitosis and dormancy but RNA is synthesised throughout the remainder of the cycle (Hall & Turnock, 1976; Fink & Turnock, 1977). Transcription of histone genes occurs throughout the cell cycle in HeLa cells although histone mRNA only accumulates in S phase, implying control at the level of processing (Melli, Spinelli & Arnold, 1977; Wilkes, Bernie & Old, 1978). In *Physarum polycephalum* there is evidence for a different pattern of transcription between S and G2 phases that is consistent with predominant rRNA synthesis in G2 phase and predominant mRNA synthesis in S phase (Mittermayer, Braun & Rusch, 1964; Braun, Mittermayer & Rusch, 1966; Cummins, Weisfeld & Rusch, 1966; Mittermayer, Braun & Rusch, 1966; Cummins, Rusch & Evans, 1967; Zellweger & Braun, 1971; Grant, 1972; Fouquet & Braun, 1974; Fouquet *et al.*, 1974; Fouquet & Sauer, 1975; Fouquet *et al.*, 1975). Net tRNA synthesis appears to be uniform through the cycle except for a cessation during mitosis (Fink & Turnock, 1977).

Histone synthesis occurs in S phase and is loosely coupled to DNA synthesis (Marks, Paik & Borun, 1973). Nucleosome assembly may require acetylation of H4 before assembly followed by de-acetylation after assembly. It seems likely, in spite of some contradictory evidence, that histones segregate semiconservatively in that old histones are found on the old DNA strand and new histones on the new DNA strand during chromatin replication (Tsanev & Russev, 1974; Elgin & Weintraub, 1975; Jackson, Granner & Chalkley, 1975; Seale, 1976; Jackson, Granner & Chalkley, 1976a; Cremise, Chestier & Yaniv, 1977; Freelender, Taichman & Smithies, 1977; Leffak, Grainger & Weintraub, 1977; Tan, 1977; Russev & Tsanev, 1979).

Activity of chromatin enzymes through the cell cycle

Histone-modifying enzymes are clearly important for the control of chromosome structure and function and the cell cycle dependence of one of these enzymes, the mitosis-associated histone kinase, has been determined (Bradbury, Inglis & Matthews, 1974). Fig. 1 shows the activity of crude nuclear kinase with H1 as substrate through the naturally synchronous cell cycle in *Physarum*. There is a 15-fold increase in kinase activity between S phase and G2 phase. Two components of this kinase activity were separated and partly purified by ion-exchange chromatography. Both components showed a large increase in activity in G2 phase compared with S phase although one had a maximum activity 1 h earlier in G2 phase than the other. The two enzymes had rather different substrate specificities; for example protamine was an excellent substrate for one enzyme and an extremely poor substrate for the other enzyme. It was suggested that the two enzymes might phosphorylate different sites on H1 *in vivo* so providing a sequential phosphorylation of specific sites on H1 during G2 phase (Hardie, Mat-

Fig. 1. Cell cycle variation of nuclear histone kinase activity in *Physarum*. The units of kinase activity are (mol ^{32}P transferred per nucleus in 15 min) $\times$ 10^{18}. The maxima occur about 2 h before metaphase, in late G2 phase.

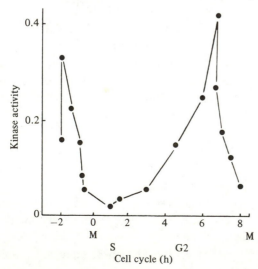

thews & Bradbury, 1976). This suggestion has recently received support from
the finding that the two enzymes do phosphorylate different sites on calf H1 *in
vitro* (T. Chambers & H. R. Matthews, unpublished).

Measurements of enzyme activity do not distinguish between changes in the
amount of enzyme present and changes in the activity of pre-existing enzyme.
This distinction was made for histone kinase by the procedure of ^{2}H labelling as
follows. Synchronous plasmodia of *Physarum* were grown on normal medium
up to the second mitosis after fusion. They were then transferred either to fresh
normal medium or to medium in which the normal amino acid mixture had been
replaced with a mixture of ^{2}H-labelled amino acids. The cultures were harvested
7 h later, in G2 phase, when the nuclear kinase activity had increased 15-fold.
Extracts containing the nuclear kinase activity were prepared and analysed by
isopycnic centrifugation on preformed Metrizamide gradients in ^{2}HOH. The kinase from plasmodia grown on normal medium gave three bands, probably corresponding to different levels of hydration, and the kinase from plasmodia grown
on ^{2}H-labelled medium gave six bands, three corresponding to normal kinase
and three shifted 5 mg ml^{-1} to higher density corresponding to newly synthesised
^{2}H-labelled kinase. These bands are shown in Fig. 2 and it can be seen that the

Fig. 2. Isopycnic centrifugation of nuclear histone kinase activity on Metrizamide–^{2}HOH gradients. (●), Kinase from plasmodia grown on normal
medium; (○), kinase from plasmodia grown on ^{2}H-labelled medium from
metaphase to late G2 phase.

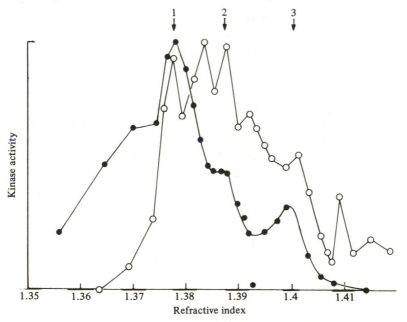

[2]H-labelled kinase bands are approximately the same size as the normal kinase bands, showing that the 15-fold increase in kinase activity after transfer to [2]H-labelled medium was not primarily due to synthesis of new enzyme and must therefore have been due to activation or transport of pre-existing enzyme (Mitchelson, Chambers, Bradbury & Matthews, 1978). The mechanism of activation is unknown; the kinase is not affected by cyclic AMP (Hardie *et al.*, 1976; Matthews, 1979).

There are also substantial cell cycle fluctuations in nuclear histone phosphatase activity, and histone phosphate content changes accordingly as described below. These changes appear to be major events in the progress of the nucleus through G2 phase and mitosis (Matthews, 1977).

RNA polymerase activity has been studied as a function of the cell cycle. Eukaryotic nuclei contain three types of polymerase: polymerase A or polymerase I which is insensitive to α-amanitin and transcribes the DNA coding for the large ribosomal RNA sequences; polymerase B or polymerase II which is sensitive to α-amanitin and transcribes the DNA coding for mRNA and associated sequences; and polymerase C or polymerase III which is sensitive to high concentrations of α-amanitin and transcribes DNA coding for tRNA and 5S rRNA (Chambon, 1975). In *Physarum* plasmodia or nuclei, polymerase A shows a peak of activity in G2 phase and polymerase B shows a peak of activity in S phase. The enzymes were distinguished by their sensitivity to α-amanitin (Grant, 1972) or by their RNA products (Mittermayer *et al.*, 1966). In contrast, the activity of the polymerases when isolated and separated by chromatography showed no changes through the cell cycle (Hildebrandt & Sauer, 1976). In the case of polymerase B this may be due to the presence of a stimulation factor which occurs primarily in S phase (Ernst & Sauer, 1977).

Poly (adenosinediphosophoribose) polymerase (poly(ADPR)polymerase) is a chromatin-associated enzyme which has received a substantial amount of attention. Its activity *in vivo* is inversely correlated with DNA replication when assayed in minimally perturbed cells (Berger *et al.*, 1978). The activity can be stimulated by damaging the DNA with DNase 1; by the procedures used for isolating nuclei; or by some cell synchronising procedures such as hydroxyurea treatment. It has been suggested, on the basis of these correlations, that poly(ADPR) is involved in DNA repair. Poly(ADPR) has been found in nuclei both free and bound to protein but its function remains unclear.

Histone modification

Histones are subject to a number of reversible modifications which are likely to change their interactions with DNA within and between nucleosomes and to be involved in cell cycle dependent changes in chromosome structure.

The best studied modifications are phosphorylation of H1 and acetylation of H3 and H4.

Phosphorylation of H1 during the cell cycle needs to be clearly distinguished from phosphorylation of H1 in response to hormones mediated by cyclic AMP (Malette, Neblett, Exton & Langan, 1973). The latter was the first clear demonstration of phosphorylation of a specific site (namely serine-37) on H1 (Langan, 1969). This site is phosphorylated by the cyclic AMP dependent protein kinase which is also found in the cytoplasm and may be associated with changes in patterns of transcription (Langan, 1978a; Watson & Langan, 1973).

Phosphorylation of H1 at a group of other specific sites has been clearly associated with mitosis (Bradbury, Inglis, Matthews & Sarner, 1973) and this will be referred to as 'mitosis-associated phosphorylation'. The sites involved include threonine-16, -136 and -153 and serine-180 (Langan, 1978a). Threonine-16 is in the N-terminal region of H1 which has an extended structure that does not bind tightly to DNA. The other three identified sites are in the C-terminal region of H1 which contains mostly lysine, alanine and proline and forms an extended structure that does bind tightly to DNA (Hartman, Chapman, Moss & Bradbury, 1977). Phosphorylation makes it bind less tightly (Matthews & Bradbury, 1978). None of the phosphorylation occurs in the globular region of H1, residues 41 to 118 approximately, where a rigid three-dimensional structure forms (Chapman et al., 1978).

Phosphorylation of serine-106 can be achieved in vitro with a nuclear histone kinase (Langan, 1978b) and this greatly reduces the stability of the H1 globular region (Rattle, Langan, Danby & Bradbury, 1977) but such phosphorylation has not been observed in vivo. Phosphorylation specific to S phase may occur and reversible phosphorylation of an S-phase specific site has been reported (Dolby, Ajiro & Borun, 1979). Phosphorylation of other histones also occurs but no detailed studies have been carried out (Gurley et al., 1978). Phosphorylation of non-histone proteins is a major process that has been linked with changing patterns of transcription (Kleinsmith, Stein & Stein, 1976; Thomson, Stein, Kleinsmith & Stein, 1977) but cell cycle specific phosphorylations may also occur.

Acetylation of all four core histones has been reported and the sites of acetylation determined (Ogawa et al., 1969; Sung & Dixon, 1970; Candido & Dixon, 1971; Candido & Dixon, 1972a,b; Marzluff & McCarty, 1972; Hooper, Smith, Summer & Chalkley, 1973; Sautière et al., 1974; Dixon et al., 1975): H2A (lysine-5); H2B (lysine-5, -10, -13 and -18); H3 (lysine-9, -14, -18 and -23) and H4 (lysine-5, -8, -12 and -16). All sites of acetylation are in the N-terminal regions of these four histones as are other modifications such as phosphorylation of serines and non-reversible methylations of lysines. The N-terminal regions of these four histones are basic and probably not directly involved in the hydrophobic interactions that form the nucleosome core (Rattle *et al.*, 1979). These regions probably bind to DNA and affect its binding to the protein core. They are probably important for the packing of nucleosomes into the 30 nm coil. The acetate groups on H3 and H4 turnover rapidly during S phase and the deacetylase activity, responsible for removing the acetate, can be inhibited with sodium butyrate thus allowing the acetylated forms to accumulate (Ruiz-Carillo, Waugh & Allfrey, 1975; Jackson, Shires, Tanphaichitr & Chalkley, 1976*b*; Riggs, Whittaker, Neumann & Ingram, 1977; Candido, Reeves & Davie, 1978; Sealy & Chalkley, 1978; Vidali, Boffa, Bradbury & Allfrey, 1978). The effect is shown in Fig. 3 for H4 from HeLa cells. It has been proposed that the acetylation is associated with chromosome replication (Sung & Dixon, 1970). Acetylation of H4 may also be associated with transcription (Allfrey, Faulkener & Mirsky, 1964) and this is discussed below together with data on the cell cycle dependence of H4 acetylation.

The larger HMG proteins, HMG-1 and HMG-2, are acetylated but the effect of the acetylation is not yet known (Sterner, Vidali, Henrikson & Allfrey, 1978).

Fig. 3. Polyacrylamide gel electrophoresis of histone H4 from HeLa cells grown in normal medium or in medium to which 7 mM sodium butyrate was added for 22 h after the cell density reached 2×10^5 cells ml^{-1}.

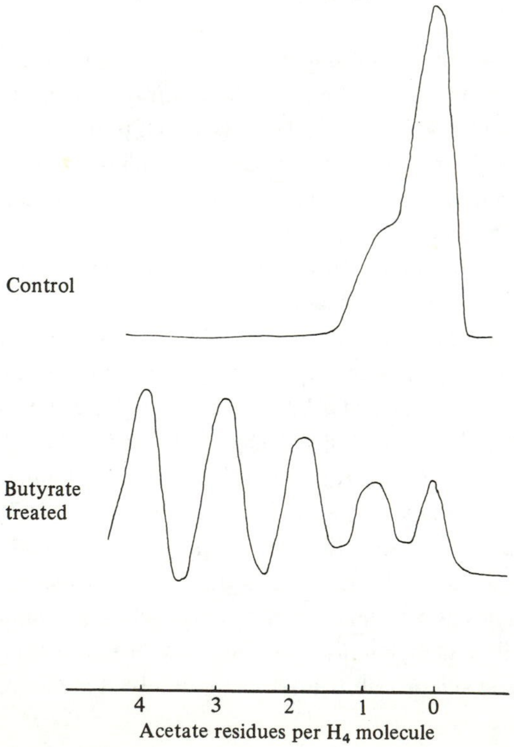

Another type of modification that has received some attention is the addition of poly(ADPR) to proteins, particularly histone H1. Dixon, Wong & Poirier, (1976) reported the binding of poly (ADPR) to a specific glutamic acid residue in H1 and Stone, Lorimer & Kidwell, (1977) reported a dimer of two molecules crosslinked by poly (ADPR). The chemical nature of the bonds remains unclear and both these reports refer to compounds formed *in vitro* by isolated nuclei. Adamietz, Bredehorst & Hilz, (1978) reported that neither of the above compounds could be detected in extracts from HeLa cell cultures, although a small amount of poly (ADPR) was associated with H1. The very small amounts of poly (ADPR) found *in vivo* and the difficulty of labelling it specifically have so far prevented any definitive experiments on its role in the cell cycle or elsewhere (Hilz & Stone, 1976; Hayaishi & Ueda, 1977).

Phosphorylation of histone H1 correlates with chromosome condensation

Fig. 4(*b*) shows the variation in H1 phosphate content through the cell cycle in *Physarum*. The large peak in prophase correlates with the beginning of chromosome condensation as seen by phase contrast microscopy. The data from *Physarum* are fully in agreement with the observation in mammalian cells that H1 phosphate content is high for cells growing rapidly and low for cells growing slowly, since cells growing rapidly have a short G1 phase and so spend proportionately more time in G2 phase where the H1 phosphate content is high (Balhorn, Balhorn, Morris & Chalkley, 1972). The precise time of the maximum is hard to check in mammalian cells because of the difficulty in obtaining synchrony in G2 phase. However, detailed studies have been carried out by Gurley *et al.*, (1978) using CHO cell cultures that were blocked in S phase by hydroxyurea and isoleucine deprivation and then allowed to proceed through S phase and into G2 phase where colcemid was used to block the cells near metaphase (Gurley *et al.*, 1978). Under these conditions complex mixtures of cells at different stages were obtained, their phosphorylation states estimated by gel electrophoresis and partly corrected for microheterogeneity of the H1. Clearly a direct comparison of the *Physarum* and CHO cell data is difficult but it is possible to make the necessary transformations and the results are shown in Fig. 4(*a*). They suggest that the maximum phosphate content in CHO cells occurs in mitosis, as it does in *Physarum*. In CHO cells the phosphate content appears to remain very high at metaphase whereas in *Physarum* it is falling at metaphase. This may be a consequence of the use of colcemid in the CHO cell cultures or of the lack of a G1 phase in *Physarum*. Colcemid is known to produce excessive condensation of chromosomes, which may be related to the high phosphate content in colcemid arrested metaphase.

In *Physarum*, mitosis-associated H1 phosphorylation is correlated with the initiation of chromosome condensation (Bradbury *et al.*, 1973). This correlation is supported by the pattern of histone kinase activity in the cell cycle (Fig. 1) which shows that H1 phosphorylation is due to increase of kinase activity and is not merely a consequence of chromosome condensation (Bradbury *et al.*, 1974). The correlation with initiation is also supported by the findings (i) that chromosomes

Fig. 4. Phosphate content of histone H1 through the cell cycle. (*a*) Data from CHO cells (Gurley *et al.*, 1978) (–––) obtained by synchronisation procedures as indicated together with predicted results for *Physarum* based on the same mixtures of cell cycle stages as were observed with CHO cells. (*b*) Actual data for *Physarum* in which there are no problems of synchronisation procedures or mixtures of cell cycle stages.

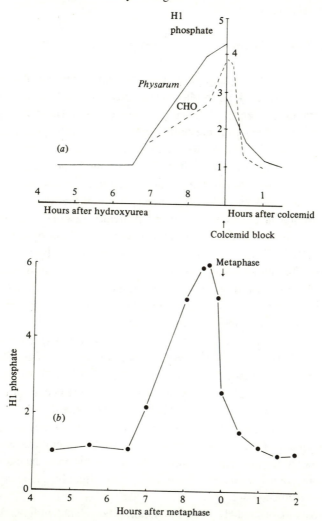

can disperse at least partially into G1 phase even if dephosphorylation is inhibited by zinc chloride (Tanphaichitr, Moore, Granner & Chalkley, 1976); (ii) that *Tetrahymena* macronuclei have phosphorylated H1 although they divide amitotically without fully condensed chromosomes; and (iii) that *Tetrahymena* micronuclei have permanently condensed chromosomes but no H1 (Nilsson, 1970; Zeuthen & Rasmussen, 1972; Gorovsky, Keevert & Plager, 1974; Gorovsky & Keevert, 1975).

Prior & Cantor (unpublished, see Matthews, 1979) have shown that *Physarum* will take up H3 histone from the growth medium and use it in its chromatin. If this also applies to histone kinase then the correlation between phosphorylation and mitosis can be tested by adding histone kinase to growing plasmodia. Fig. 5 shows that added histone kinase brought forward the time of metaphase by up to 40 min depending on the concentration of histone kinase used (Inglis *et al.*, 1976). Unfortunately the kinase preparations available were not pure so, in spite of the fact that many control substances had no effect on the timing of mitosis, the experiments do not *prove* that histone kinase itself brought forward mitosis. A further indication that the mitotic stimulator was actually histone kinase was obtained by studying the sensitivity of the plasmodia to added kinase preparations as a function of the mitotic cycle (Bradbury, Inglis, Matthews & Langan, 1974). Fig. 6 shows that the period of sensitivity corresponded to the period of G2 phase when the endogenous histone kinase activity was increasing. This correlation greatly strengthens the interpretation that histone kinase was entering the nucleus and acting on H1 histone there to raise its phosphate content and initiate chromosome condensation. S phase immediately follows mitosis in *Physarum* plasmodia and the importance of chromosome condensation in controlling the cell cycle is indicated by the fact that histone kinase brought forward the time of S phase as well as the time of mitosis as shown in Table 1.

Fig. 5. Advance of mitosis by addition of histone kinase to *Physarum* plasmodia.

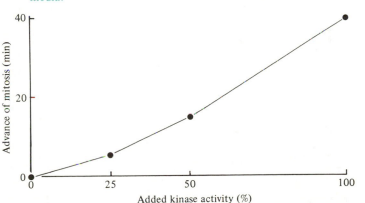

The in-vivo correlation of H1 phosphorylation with chromosome condensation is also supported by in-vitro experiments. Isolated chromatin undergoes a reversible contraction or aggregation as a function of salt concentration. The aggregation correlates with changes in the nuclear magnetic resonance spectrum of H1 in chromatin and no aggregation is seen in chromatin depleted of H1 (Bradbury et al., 1973). Technical difficulties with reconstitution methods (Steinmetz, Streeck & Zachau, 1978) have prevented the use of phosphorylated H1 with chromatin depleted of normal H1 but the aggregation process appears to be very similar when only H1 and DNA are present and this provides an excellent model

Fig. 6. Cell cycle dependence of histone H1 phosphate content (———); nuclear histone kinase activity (– – –); and ability of added histone kinase to advance mitosis (———).

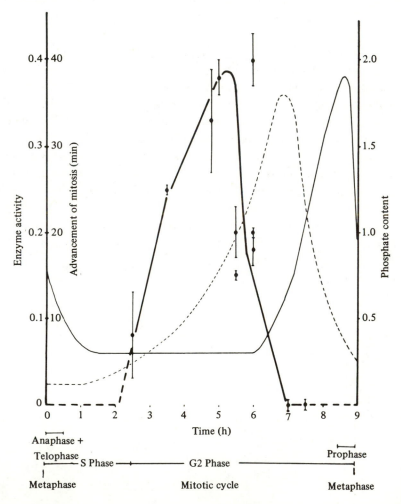

Table 1. *Advanced initiation of S phase*

Physarum plasmodia were labelled continuously with [¹⁴C]thymidine. Six hours after the second mitosis after fusion the plasmodia were treated with buffer or an extract containing histone kinase. Separate plasmodia were pulse labelled with [³H]thymidine as indicated and harvested at the end of the pulse. Their DNA was purified and synthesis estimated from the $^3H/^{14}C$ ratio. Synthesis, like metaphase, began 20 min earlier in the plasmodia treated with histone kinase. Metaphase occurred 2h 40 min after treatment for kinase treated plasmodia and 3h 0min after treatment for buffer-treated plasmodia.

	Radioactivity in DNA			Time of pulse (h and min after treatment)	
	^{14}C d.p.m (continuous)	^3H d.p.m (pulse)	$^3H/^{14}C$	Begin	End
Treated	1 426	2 561	1.8	2:20	2:40
Control	1 019	1 396	1.4	2:40	3:00
Treated	3 468	42 984	12	2:50	3:10
Control	2 374	34 294	14	3:10	3:30
Treated	3 030	52 559	17	3:40	4:00
Control	3 125	49 282	16	4:00	4:20

Fig. 7. Turbidity of H1·DNA mixtures. The maximum turbidity obtained with various phosphorylated H1 molecules is shown as a percentage of the control turbidity in each case. Kinase M is the growth-associated histone kinase prepared by T. A. Langan from Ehrlich ascites cells; it phosphorylates H1 *in vitro* at a selection of the mitosis associated sites.

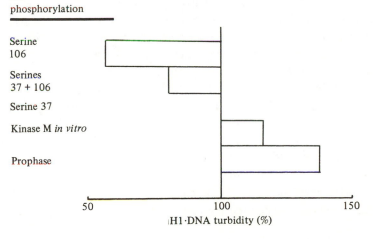

H1 phosphorylation

Serine 106

Serines 37 + 106

Serine 37

Kinase M *in vitro*

Prophase

50   100   150

H1·DNA turbidity (%)

system for studying the effect of phosphorylation on H1·DNA complexes. Phosphorylation weakens the interaction between H1 and DNA but the effect is specific for different sites of phosphorylation so that the aggregation behaviour of the complex is either diminished or enhanced according to the sites of phosphorylation (Matthews & Bradbury, 1978). Fig. 7 shows that phosphorylation of serine-106 greatly reduces the ability of H1 to aggregate DNA whereas phosphorylation at the mitosis-associated sites greatly enhances the ability of H1 to aggregate DNA. Recently, it has become possible to prepare purified H1 from specific stages of the cell cycle in *Physarum*, and in Fig. 8 H1 from mid S phase and from prophase are compared. Up to 0.4 M NaCl the prophase (phosphorylated) H1 shows a greater ability to aggregate DNA. Above 0.4 M NaCl the H1 is being released from the DNA and the mid S phase H1 is seen to be bound more tightly (Corbett, 1979). These experiments need to be repeated with the new techniques now available for gentle removal of H1 from chromatin but it is already clear that phosphorylation of H1 at the mitosis-associated sites is correlated with an increased ability to aggregate DNA.

Acetylation of histone H4 correlates with RNA synthesis in the cell cycle

Acetylation studies are not so far advanced as phosphorylation studies of H1 but the discovery that butyrate allows the accumulation of highly acetylated H3 and H4 has stimulated interest in the role of H3 and H4 acetylation. It may be that acetylation weakens the interactions between nucleosomes that form the 30 nm coil, so making the DNA more accessible for replication and transcription. Conversely, de-acetylation could encourage the formation of the 30 nm coil structure as a preliminary to chromosome condensation in mitosis (Matthews, 1980).

Rapid turnover of acetate groups has been observed in S phase (Jackson *et al.*, 1976*b*) but more detailed cell cycle studies have not been published. In *Physarum*, acetyl content of H4 is being studied using naturally synchronous plasmodia. Histone H4 is isolated and purified and then analysed by polyacrylamide gel electrophoresis in acetic acid–urea (Corbett *et al.*, 1977). Fig. 9 shows scans of the H4 regions of Sample gels from specific cell cycle stages (S. Chahal, unpublished; see Matthews, 1979). It is clear that most H4 is in the monoacetylated stage (52–63%) at all stages with substantial amounts being non-acetylated (10–20%) or di-acetylated (15–24%). The amounts of tri-acetylated (4–7%) and tetra-acetylated (0.7–3.8%) forms are smaller. It is not known if all states of acetylation have specific functions or if some states are intermediates between the states with specific functions. Tetra-acetylated H4 seems likely to have specific function(s) since it is not an obvious intermediate form. The pattern of tetra-acetylated H4 through the cell cycle is shown in Fig. 10 (S. Chahal, unpublished). These preliminary results show two peaks of tetra-acetylated H4: one in S phase and one in mid to late G2 phase. The S phase peak correlates both with DNA synthesis and with RNA synthesis; the G2 phase peak correlates with RNA synthesis. DNA synthesis as measured by pulse labelling in *Physarum* shows an initial rise just after metaphase that is much more rapid than the initial rise in tetra-acetylated H4. However, the pulse labelling data are hard to interpret in detail because of rapid changes in pool sizes at that time in the cycle and data

Fig. 8. Turbidity of H1·DNA mixtures as functions of NaCl concentration. (*a*) Calf thymus H1 was phosphorylated *in vitro* at the mitosis-associated sites to give a phosphorylated H1 (PO_4^-H1). This was treated with alkaline phosphatase to give a non-phosphorylated H1 (H1). Each histone was mixed with DNA and the turbidity measured. (*b*) H1 was extracted from *Physarum* plasmodia in prophase ($H1_p$) or S phase ($H1_s$) and purified. Turbidity with DNA was measured as above. H1 or DNA alone gives no turbidity at the concentrations used (10–50 μg ml^{-1}).

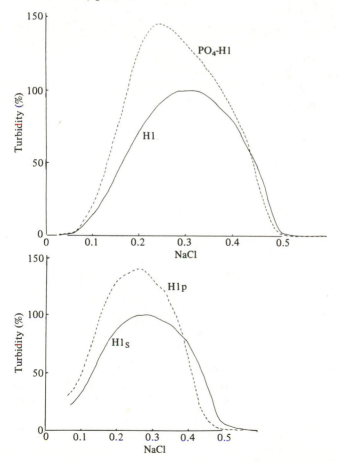

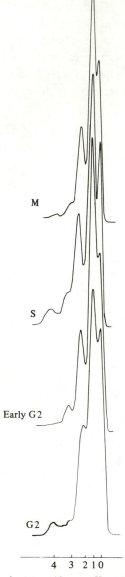

M

S

Early G2

G2

4   3   2 1 0

Acetate residues per H_4 molecule

Fig. 9. Polyacrylamide gel electrophoresis of histone H4 from *Physarum* plas-modia at four stages of the cell cycle. The uneven spacing of the five compo-nents is due to the high concentration of H4 used to give a wide dynamic range for measurement of peak areas.

from isotope dilution experiments are consistent with a slower rise in rate of DNA synthesis, much closer to the observed rise in tetra-acetylated H4 (Cunningham, 1979). RNA synthesis as measured by pulse labelling in whole plasmodia or isolated nuclei shows two peaks in the cell cycle, similar to those observed for tetra-acetylated H4 (Braun *et al.*, 1966). Isotope dilution measurements of rRNA and tRNA synthesis show a low rate of synthesis in S phase with a higher rate of synthesis in G2 phase, especially for rRNA (Hall & Turnock, 1976; Fink & Turnock, 1977). A variety of other experiments suggest that RNA synthesis in S phase is predominantly mRNA and associated sequences while RNA synthesis in G2 phase is predominantly rRNA and associated sequences (Mittermayer *et al.*, 1964; Braun *et al.*, 1966; Cummins *et al.*, 1966, 1967; Mittermayer *et al.*, 1966; Zellweger & Braun, 1971; Grant, 1972; Fouquet & Braun, 1974; Fouquet *et al.*, 1974; Fouquet & Sauer, 1975; Fouquet *et al.*, 1975). The genes for rRNA in *Physarum* are located in the nucleolus which can be isolated substantially free from nucleoplasmic chromatin. The above suggestions predict that S phase nucleolar chromatin is high in tetra-acetylated H4. This prediction is being checked.

The genes for rRNA in *Physarum* have a different or less stable nucleosome structure from most *Physarum* DNA and from adjacent non-transcribed spacer

Fig. 10. Cell cycle variation of the percentage of histone H4 in the tetra-acetylated form, in *Physarum* plasmodia.

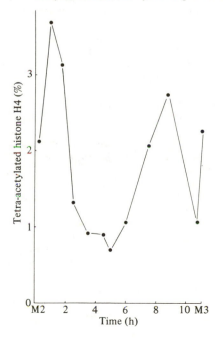

sequences (E. M. Johnson, unpublished, see Johnson, Allfrey, Bradbury & Matthews, 1978). It will be interesting to determine if this is due, at least in part, to acetylation of H4.

I am grateful to my colleagues for permission to describe their unpublished work particularly Mr A. W. Thorne (Fig. 3), Dr S. Corbett (Figs. 7 and 8) and Mr S. Chahal (Fig. 9 and 10). Part of the work described was supported by the Cancer Research Campaign, the Science Research Council and NATO.

References

Adamietz, P., Bredehorst, R. & Hilz, H. (1978). ADP-Ribosylated Histone H1 from HeLa cultures: Fundamental differences to (ADP-Ribose)$_n$-Histone H1 conjugates formed *in vitro*. *European Journal of Biochemistry*, **91**, 317–26.

Adolph, K. W., Cheng, S. M. & Laemmli, U. K. (1977). Role of nonhistone proteins in metaphase chromosome structure. *Cell*, **12**, 805–16.

Allfrey, V. G., Faulkener, R. M. & Mirsky, A. E. (1964). Acetylation and methylation of histones and their possible role in the regulation of RNA synthesis. *Proceedings of the National Academy of Sciences, USA*, **51**, 786–94.

Balhorn, R., Balhorn, M., Morris, H. P. & Chalkley, R. (1972). Comparative high-resolution electrophoresis of tumor histones. Variation in phosphorylation as a function of cell replication rate. *Cancer Research*, **32**, 1775–81.

Berger, N. A., Knaichi, A. S., Steward, P. G., Klerecz, R. R., Forrest, G. L. & Gross, S. D. (1978). Synthesis of poly(adenosine diphosphate ribose) in synchronized chinese hamster cells. *Experimental Cell Research*, **117**, 127–35.

Bradbury, E. M., Carpenter, B. G. & Rattle, H. W. E. (1973). Magnetic resonance studies of deoxyribonucleoprotein. *Nature, London*, **241**, 123–7.

Bradbury, E. M., Inglis, R. J. & Matthews, H. R. (1974). Control of cell division by very lysine rich histone (F1) phosphorylation. *Nature, London*, **247**, 257–61.

Bradbury, E. M., Inglis, R. J., Matthews, H. R. & Langan, T. A. (1974). Molecular basis of control of mitotic cell division in eukaryotes. *Nature, London*, **249**, 553–6.

Bradbury, E. M., Inglis, R. J., Matthews, H. R. & Sarner, N. (1973). Phosphorylation of very lysine rich histone in *Physarum polycephalum*. *European Journal of Biochemistry*, **33**, 131–9.

Braun, R., Mittermayer, C. & Rusch, H. P. (1966). Ribonucleic acid synthesis *in vivo* in the synchronously dividing *Physarum polycephalum* studied by cell fractionation. *Biochimica et Biophysica Acta*, **114**, 527–35.

Candido, E. P. M. & Dixon, G. H. (1971). Sites of in-vivo acetylation in trout testis histone IV. *Journal of Biological Chemistry*, **246**, 3182–90.

Candido, E. P. M. & Dixon, G. H. (1972*a*). Acetylation of trout testis histones in vivo: site of the modification in histone IIb. *Journal of Biological Chemistry*, **247**, 3863–70.

Candido, E. P. M. & Dixon, G. H. (1972*b*). Amino-terminal sequences and sites of in-vivo acetylation of trout testis histones III and IIb$_2$. *Proceedings of the National Academy of Sciences, USA*, **69**, 2015–20.

Candido, E. P. M., Reeves, R. & Davie, R. (1978). Sodium butyrate inhibits histone deacetylation in cultured cells. *Cell,* **14,** 105–14.

Carpenter, B. G., Baldwin, J. P., Bradbury, E. M. & Ibel, K. (1976). Organization of subunits in chromatin. *Nucleic Acids Research,* **3,** 1739–46.

Chambon, P. (1975). Eukaryotic nuclear RNA polymerases. *Annual Review of Biochemistry,* **44,** 613–38.

Chapman, G. E., Hartman, P. G., Cary, P. D., Bradbury, E. M. & Lee, D. R. (1978). A nuclear-magnetic-resonance study of the globular structure of the H1 histone. *European Journal of Biochemistry,* **86,** 35–44.

Compton, J. C., Bellard, M. & Chambon, P. (1976). Biochemical evidence of variability in the DNA repeat length in the chromatin of higher eukaryotes. *Proceedings of the National Academy of Sciences, USA,* **73,** 4382–6.

Corbett, S. (1979). Histone modification in *Physarum polycephalum.* Ph.D. Thesis. Portsmouth Polytechnic.

Corbett, S., Miller, S., Robinson, V. J., Matthews, H. R. & Bradbury, E. M. (1977). *Physarum polycephalum* histones. *Transactions of the Biochemical Society,* **5,** 943–6.

Cremise, C., Chestier, A. & Yaniv, M. (1977). Preferential association of newly synthesized histones with replicating S V40 DNA. *Cell,* **12,** 947–51.

Cummins, J. E., Rusch, H. P. & Evans, T. E. (1967). Nearest neighbour frequencies and the phylogenetic origin of mitochondrial DNA in *Physarum polycephalum. Journal of Molecular Biology,* **23,** 281–4.

Cummins, J. E., Weisfeld, G. E. & Rusch, H. P. (1966). Fluctuations of ^{32}P distribution in rapidly labelled RNA during the cell cycle of Physarum polycephalum. *Biochimica et Biophysica Acta,* **129,** 240–8.

Cunningham, M. (1979). DNA Synthesis in *Physarum polycephalum.* PhD Thesis. University of Leicester.

Dixon, G. H., Candido, E. P. M., Honda, B. M., Louie, A. J., McLeod, A. R. & Sung, M. T. (1975). The biological roles of post-synthetic modification of basic nuclear proteins. In *The Structure and Function of Chromatin,* pp. 220–40. CIBA Foundation Symposium 28. Amsterdam: Elsevier.

Dixon, G. H., Wong, N. & Poirier, G. G. (1976). Adenosine diphospho-ribosylation of basic chromosomal proteins in trout testis nuclei. *Federation Proceedings,* **35,** 1623.

Dolby, T. W., Ajiro, K. & Borun, T. (1979). Physical properties of DNA and chromatin isolated from G_1- and S-phase HeLa S-3 cells. Effects of histone H1 phosphorylation and stage-specific non-histone chromosomal proteins on the molar ellipticity of native and reconstituted nucleo-proteins during thermal denaturation. *Biochemistry,* **18,** 1333–43.

Elgin, S. R. C. & Weintraub, H. (1975). Chromosomal Proteins and Chromatin Structure. *Annual Review of Biochemistry,* **44,** 725–77.

Ernst, G. H. & Sauer, H. W. (1977). A nuclear elongation factor of transcription from *Physarum polycephalum in vitro. European Journal of Biochemistry,* **74,** 253–61.

Finch, J. T. & Klug, A. (1976). Solenoidal model for superstructure in chromatin. *Proceedings of the National Academy of Sciences, USA,* **73,** 1897–901.

Finch, J. T., Lutter, L. C., Rhodes, D., Brown, R. S., Rushton, B., Lewitt, M. & Klug, A. (1977). Structure of nucleosome core particles of chromatin. *Nature, London,* **269,** 29–34.

Fink, K. & Turnock, G. (1977). Synthesis of transfer RNA during the synchronous nuclear division cycle in *Physarum polycephalum. European Journal of Biochemistry,* **80,** 93–6.

Fouquet, H., Böhme, R., Wick, R., Sauer, H. W. & Braun, R. (1974). Isolation of adenylate-rich RNA from *Physarum polycephalum*. *Biochimica et Biophysica Acta*, **353**, 313–22.

Fouquet, H., Böhme, R., Wick, R., Sauer, H. W. & Scheller, K. (1975). Some evidence for replication – transcription coupling in *Physarum polycephalum*. *Journal of Cell Science*, **18**, 27–39.

Fouquet, H. & Braun, R. (1974). Differential RNA synthesis in the mitotic cycle of *Physarum polycephalum*. *FEBS Letters*, **38**, 184–6.

Fouquet, H. & Sauer, H. W. (1975). Variable redundancy in RNA transcripts isolated in S and G_2 phase of the cell cycle in *Physarum*. *Nature, London*, **255**, 253–5.

Freelender, E. F., Taichman, L. & Smithies, O. (1977). Non-random distribution of chromosomal proteins during cell replication. *Biochemistry*, **16**, 1802–8.

Goodwin, G. H., Walker, J. M. & Johns, E. W. (1978). The high mobility group (HMG) chromosomal non-histone proteins. In *The Cell Nucleus*, VI *Chromatin*, ed. H. Busch, Part C, pp. 182–219. London & New York: Academic Press.

Gorovsky, M. A. & Keevert, J. B. (1975). Absence of histone F1 in a mitotically dividing, genetically inactive nucleus. *Proceedings of the National Academy of Sciences, USA*, **72**, 2672–6.

Gorovsky, M. A., Keevert, J. B. & Plager, G. L. (1974). Histone F1 of Tetrahymena macronuclei: unique electrophoretic properties and phosphorylation of F1 in an amitotic nucleus. *Journal of Cell Biology*, **61**, 134–45.

Grant, W. D. (1972). The effect of α-amanitin and $(NH_4)_2 SO_4$ on RNA synthesis in nuclei and nucleoli isolated from *Physarum polycephalum* at different times during the cell cycle. *European Journal of Biochemistry*, **2**, 94–8.

Gurley, L. R., D'Anna, J. A., Barham, S. S., Deaven, L. L. & Tobey, R. A. (1978). Histone phosphorylation and chromatin structure during mitosis in chinese hamster cells. *European Journal of Biochemistry*, **84**, 1–16.

Gurley, L. R., Tobey, R. A., Walters, R. A., Hilderbrand, C. E., Hohman, P. G., D'Anna, J. A., Barham, S. S. & Deaven, L. L. (1978). Histone phosphorylation and chromatin structure in synchronised mammalian cells. In *Cell Cycle Regulation*, ed. J. R. Jeter, I. L. Cameron, G. M. Padilla & A. M. Zimmerman, pp. 37–60. London & New York: Academic Press.

Hall, L. & Turnock, G. (1976). Synthesis of ribosomal RNA during the mitotic cycle in the slime mould *Physarum polycephalum*. *European Journal of Biochemistry*, **62**, 471–7.

Hardie, D. G., Matthews, H. R. & Bradbury, E. M. (1976). Cell cycle dependence of two nuclear histone kinase enzyme activities. *European Journal of Biochemistry*, **66**, 37–42.

Hartman, P. G., Chapman, G. E., Moss, T. & Bradbury, E. M. (1977). Studies on the role and mode of operation of the very-lysine-rich histone H1 in eukaryote chromatin. *European Journal of Biochemistry*, **77**, 43–51.

Hayaishi, O. & Ueda, K. (1977). Poly (ADP-Ribose) and ADP-ribosylation of protein. *Annual Review of Biochemistry*, **46**, 95–116.

Hildebrandt, A. & Sauer, H. W. (1976). Levels of RNA Polymerases during the mitotic cycle of Physarum polycephalum. *Biochimica et Biophysica Acta*, **425**, 316–21.

Hilz, H. & Stone, P. (1976). Poly (ADP-Ribose) and ADP-ribosylation of proteins. *Reviews of Physiology, Biochemistry and Pharmacology*, **76**, 1–59.

Hjelm, R. P., Kneale, G. G., Suau, P., Baldwin, J. P., Bradbury, E. M. &

Ibel, K. (1977). Small angle neutron scattering studies of chromatin subunits in solution. *Cell,* **10,** 139–51.

Hooper, J. A., Smith, E. L., Summer, K. R. & Chalkley, R. (1973). Histone III: amino acid sequence of histone III of the testes of the carp, *Letiobus Bubalus. Journal of Biological Chemistry,* **248,** 3275–80.

Igo-Kemenes, T., Greil, W. & Zachau, H. G. (1977). Preparation of soluble chromatin. *Nucleic Acids Research,* **4,** 3387–94.

Igo-Kemenes, T. & Zachau, H. G. (1977). Domains in chromatin structure. *Cold Spring Harbor Symposia in Quantitative Biology,* **42,** 109–115.

Inglis, R. J., Langan, T. A., Matthews, H. R., Hardie, D. G. & Bradbury, E. M. (1976). Advance of mitosis by histone phosphokinase. *Experimental Cell Research,* **97,** 418–25.

Jackson, V., Granner, D. K. & Chalkley, R. (1975). Deposition of histones onto replicating chromosomes. *Proceedings of the National Academy of Sciences USA,* **72,** 4440–4.

Jackson, V., Granner, D. K. & Chalkley, R. (1976a). Distribution of newly synthesized histones during DNA replication. *Proceedings of the National Academy of Sciences USA,* **73,** 2266–9.

Jackson, V., Shires, A., Tanphaichitr, N. & Chalkley, R. (1976b). Modifications to histones immediately after synthesis. *Journal of Molecular Biology,* **104,** 471–83.

Johns, E. W., Goodwin, G. H., Walker, J. M. & Sanders, C. (1975). In *The Structure and Function of Chromatin,* ed. D. W. Fitzsimons & G. E. W. Wolstanholme, pp. 95–112. CIBA Foundation Symposium 28. Amsterdam: Associated Scientific Publishers.

Johnson, E. M., Allfrey, V. G., Bradbury, E. M. & Matthews, H. R. (1978). Altered nucleosome structure containing DNA sequences complementary to 19S and 26S ribosomal RNA in *Physarum polycephalum. Proceedings of the National Academy of Sciences, USA,* **75,** 1116–20.

Kleinsmith, L. J., Stein, J. & Stein, G. S. (1976). Dephosphorylation of non-histone proteins specifically alters the pattern of gene transcription in reconstituted chromatin. *Proceedings of the National Academy of Sciences, USA,* **73,** 1174–8.

Langan, T. A. (1969). Phosphorylation of liver histone following the administration of glucagon and insulin. *Proceedings of the National Academy of Sciences, USA,* **64,** 1276–81.

Langan, T. A. (1978a). Methods for the assessment of site-specific histone phosphorylation. *Methods in Cell Biology,* ed. G. S. Stein & J. Stein, **19,** 127–42. New York & London: Academic Press.

Langan, T. A. (1978b). Isolation of histone kinases. *Methods in Cell Biology,* ed. G. S. Stein & J. Stein, **19,** 143–52. New York & London: Academic Press.

Leffak, J. M., Grainger, R. & Weintraub, H. (1977). Conservative assembly and segregation of nucleosomal histones. *Cell,* **12,** 837–46.

Levy-W. B., Wong, N. C. W. & Dixon, G. H. (1977). Selective association of the trout-specific H6 protein with chromatin regions susceptible to DNase 1 and DNase II. Possible location of HMG-T in the spacer region between core nucleosomes. *Proceedings of the National Academy of Sciences, USA,* **74,** 2810–14.

Littau, V. C., Burdick, C. J., Allfrey, V. G. & Mirsky, A. E. (1965). The role of histones in the maintenance of chromatin structure. *Proceedings of the National Academy of Sciences, USA,* **54,** 1204–9.

Malette, L. E., Neblett, M., Exton, J. H. & Langan, T. A. (1973). Phosphorylation of lysine-rich histones in the isolated perfused rat liver: Effects of glucagon, cyclic adenosine 3':5'-monophosphate and insulin. *Journal of Biological Chemistry,* **248,** 6289–91.

Marks, D. B., Paik, W. K. & Borun, T. W. (1973). The relationship of histone phosphorylation to deoxyribonucleic acid replication and mitosis during the HeLa S-3 cell cycle. *Journal of Biological Chemistry,* **248,** 5660–7.

Marzluff, W. F. Jr., & McCarty, K. S. (1972). Structural studies of calf thymus F3 histone II. Occurrence of phosphoserine and ε-N-Acetyllysine in thermolysin peptides. *Biochemistry,* **11,** 2677–81.

Matthews, H. R. (1977). Phosphorylation of H1 and Chromosome Condensation. In *The Organisation and Expression of the Eukaryotic Genome,* ed. E. M. Bradbury & K. Javaherian. pp. 67–80. London & New York: Academic Press.

Matthews, H. R. (1979). *Physarum* Chromatin. In *Current Research on Physarum, Proceedings of the 4th European Physarum Workshop,* ed. W. Sachsenmaier, pp. 51–58. Innsbruck University Press.

Matthews, H. R. (1980). Modification of histone H1 by reversible phosphorylation and its relation to chromosome condensation and mitosis. In *Protein Phosphorylation in Regulation,* ed. P. Cohen, pp. 235–54. Amsterdam: Elsevier/North Holland.

Matthews, H. R. & Bradbury, E. M. (1978). The role of H1 histone phosphorylation in the cell cycle: turbidity studies of H1-DNA interactions *Experimental Cell Research,* **111,** 343–51.

Melli, M., Spinelli, G. & Arnold, E. (1977). Synthesis of histone messenger RNA of HeLa cells during the cell cycle. *Cell,* **12,** 167–71.

Mitchelson, K., Chambers, T., Bradbury, E. M. & Matthews, H. R. (1978). Activation of histone kinase in G2 phase of the cell cycle in *Physarum polycephalum. FEBS. Letters,* **92,** 339–42.

Mittermayer, C., Braun, R. & Rusch, H. P. (1964). RNA synthesis in the mitotic cycle of *Physarum polycephalum. Biochimica et Biophysica Acta,* **91,** 399–405.

Mittermayer, C., Braun, R. & Rusch, H. P. (1966). Ribonucleic acid synthesis *in vitro* in nuclei isolated from the synchronously dividing *Physarum polycephalum. Biochimica et Biophysica Acta,* **114,** 536–46.

Nicolini, C. A. (ed.) (1979). *Chromatin structure and function.* New York: Plenum Press.

Nilsson, J. R. (1970). Suggestive structural evidence for macronuclear 'subnuclei' in *Tetrahymena pyriformis* GL. *Journal of Protozoology,* **17,** 548–55.

Noll, M. (1976). Differences and similarities in chromatin structure of *Neurospora crassa* and higher eukaryotes. *Cell,* **8,** 349–55.

Ogawa, Y., Quaglianotti, G., Jordan, J., Taylor, C. W., Starbuck, W. C. & Busch, H. (1969). Structural analysis of the glycine-rich arginine-rich histone 111. Sequence of the amino-terminal half of the molecule containing the modified lysine residues and the total sequence. *Journal of Biological Chemistry,* **244,** 4387–91.

Paulson, J. R. & Laemmli, U. K. (1977) The structure of histone-depleted metaphase chromosomes. *Cell,* **12,** 817–28.

Rattle, H. W. E., Kneale, G. G., Baldwin, J. P., Matthews, H. R., Crane-Robinson, C., Cary, P. D., Carpenter, B. G., Suau, P. & Bradbury, E. M. (1979). Histone complexes, nucleosomes, chromatin and cell-cycle dependent modification of histones. In *Chromatin structure and function,* ed. C. Nicolini, pp. 451–514. New York: Plenum Press.

Rattle, H. W. E., Langan, T. A., Danby, S. E. & Bradbury, E. M. (1977). Studies on the role and mode of operation of the very-lysine-rich histones in eukaryote chromatin. *European Journal of Biochemistry,* **81,** 499–503.

Renz, M., Nehls, P. & Hozier, J. (1977). Involvement of histone H1 in the organization of the chromosome fibre. *Proceedings of the National Academy of Sciences, USA,* **74,** 1879–83.

Riggs, M. G., Whittaker, R. G., Neumann, J. R. & Ingram, V. M. (1977). *n*-Butyrate causes histone modification in HeLa and Friend erythroleukaemia cells. *Nature, London,* **268,** 462–4.

Ruiz-Carillo, A., Waugh, L. J. & Allfrey, V. G. (1975). Processing of newly synthesized histone molecules: nascent histone H4 chains are reversibly phosphorylated and acetylated. *Science,* **190,** 117–28.

Russev, G. & Tsanev, R. (1979). Non-random segregation of histones during chromatin replication. *European Journal of Biochemistry,* **93,** 123–8.

Sautière, P., Tyrou, D., Laine, B., Mizon, J., Ruffin, P. & Biserte, G. (1974). Covalent structure of calf-thymus ALK-Histone. *European Journal of Biochemistry,* **41,** 563–71.

Seale, R. L. (1976). Temporal relationships of chromatin protein synthesis, DNA synthesis and assembly of deoxyribonucleoprotein. *Proceedings of the National Academy of Sciences, USA,* **73,** 2270–4.

Sealy, L. & Chalkley, R. (1978). The effect of sodium butyrate on histone modification. *Cell,* **14,** 115–22.

Steinmetz, M., Streeck, R. E. & Zachau, H. G. (1978). Closely spaced nucleosome cores in reconstituted histone·DNA complexes and histone-H1-depleted chromatin. *European Journal of Biochemistry,* **83,** 615–28.

Sterner, R., Vidali, G., Henrikson, R. L. & Allfrey, V. G. (1978). Post synthetic modifications of high mobility group proteins: evidence that high mobility group proteins are acetylated. *Journal of Biological Chemistry,* **253,** 7601–2.

Stone, P. R., Lorimer, W. S. & Kidwell, W. R. (1977). Properties of the complex between histone H1 and poly(ADP-ribose) synthesised in HeLa cell nuclei. *European Journal of Biochemistry,* **81,** 9–18.

Sung, M. T. & Dixon, G. H. (1970). Modification of histones during spermiogenesis in trout: a molecular mechanism for altering histone binding to DNA. *Proceedings of the National Academy of Sciences USA* **67,** 1616–21.

Tan, K. B. (1977). Histones: metabolism in simian virus 40-infected cells and incorporation into virions. *Proceedings of the National Academy of Sciences USA,* **74,** 2805–9.

Tanphaichitr, N., Moore, K. A., Granner, D. & Chalkley, R. (1976). Relationship between chromosome condensation and metaphase lysine-rich histone phosphorylation. *Journal of Cell Biology,* **69,** 43–50.

Thomson, J., Stein, J. L., Kleinsmith, L. H. & Stein, G. S. (1977). Activation of histone gene transcription by non-histone chromosomal phosphoproteins. *Science,* **194,** 428–31.

Tsanev, R. & Russev, G. (1974). Distribution of newly synthesised histones during DNA replication. *European Journal of Biochemistry,* **43,** 257–63.

Vidali, G., Boffa, L. C. & Allfrey, V. G. (1977). Selective release of chromosomal proteins during limited DNAase I digestion of avian erythrocyte chromatin. *Cell,* **12,** 409–15.

Vidali, G., Boffa, L. C., Bradbury, E. M. & Allfrey, V. G. (1978). Butyrate suppression of histone deacetylation leads to accumulation of multi-acetylated forms of histones H3 and H4 and increased DNase I sensitivity of the associated DNA sequences. *Proceedings of the National Academy of Sciences USA,* **75,** 2239–44.

Vogt, V. & Braun, R. (1976). Repeated structure of chromatin in metaphase nuclei of *Physarum. FEBS. Letters,* **64,** 190–2.

Watson, G. & Langan, T. A. (1973). Effects of F1 histone and phosphory-lated F1 histone on template activity of chromatin. *Federation Proceedings,* **32,** 588.

Weisbrod, S. & Weintraub, H. (1979). Isolation of a sub class of nuclear proteins responsible for conferring a DNase I-sensitive structure on globin chromatin. *Proceedings of the National Academy of Sciences, USA,* **76,** 630–5.

Wilkes, P. R., Bernie, G. D. & Old, R. W. (1978). Histone gene expression during the cell cycle studies by *in situ* hybridisation. *Experimental Cell Research,* **115,** 441–4.

Zellweger, A. & Braun, R. (1971). RNA of *Physarum:* preparation and properties. *Experimental Cell Research,* **65,** 413–23.

Zeuthen, E. & Rasmussen, L. (1972). Synchronised cell division in protozoa. In *Research in protozoology,* ed. T. T. Chem, vol. **4,** pp. 9–145. Oxford: Pergamon Press.

R.G.BURNS

Microtubules and mitosis

Introduction

Mitosis is a key event in the cell cycle: it ensures that there is an exact segregation of the chromosomal genetic material between the two daughter cells, although there are rare instances in which particular chromosomes are selectively lost. The mechanism by which the genetic material is replicated without error is now fairly well understood, but the duplication of DNA followed by an imprecise segregation to the daughter cells is tantamount to suicide. Consequently, as the mortality rate of cells grown under optimal conditions is low, the efficiency and accuracy of mitosis must be very high, and so we can expect a highly sophisticated mechanism for chromosomal segregation.

Although this argument applies equally to both eukaryotic and prokaryotic cells, it is clear that separate mechanisms have evolved in the eukaryotes. This chapter will only deal with the segregation of chromosomes in eukaryotic cells. Alternative mechanisms of chromosomal movement will be examined, although the central problem of *what* determines why each daughter chromosome migrates to the opposite pole is still totally unknown. Considerable attention has been directed in recent years to the in-vitro properties of the structural elements of the mitotic spindle, and it is intended to apply the restrictions which these studies make to the alternative models for chromosomal movement.

The mechanism of mitosis has attracted an immense amount of attention, with the application of a wide variety of experimental techniques. While historically the experimental approach has closely mirrored the development of new methods, it is inappropriate to review merely the sequential development of ideas. It is also important to note that, as with many other areas of cell biology, a number of different cell types have been extensively examined. These cell types have been selected because of their availability, ease of culture or examination, or because they illustrate a particular feature preferentially. It is however assumed that although the different cell types may show particular facets of mitosis, that there is a single common mechanism. This assumption has proved true for many areas of cell biology – for example the mechanisms of DNA replication and protein synthesis, oxidative phosphorylation and photosynthesis – but it may not necessarily be true when considering the mechanism of mitosis. Mitosis is a key event, and one which permits *no* errors, and so it is possible that the cell has evolved multiple ways that are not mutually exclusive, to ensure the exact segration of the DNA to the daughter cells.

Diversity of spindle structure

There are a number of general principles that apply to mitosis in all eukaryotic cells. During prophase, the genetic material condenses to form identifiable chromosomes which, in general, cluster about the centre of the nuclear region during metaphase. All eukaryotic cells yet examined contain microtubules which extend from the metaphase chromosomes towards the two poles. During anaphase the chromosomes migrate towards these poles, and on entering telophase they lose their observed structure and return to their interphase state. However, despite these general guidelines, there are immense and reproducible differences between different cells. In many protozoa and all higher animals, a centriole is present at each of the two spindle poles and each centriole duplicates at the time of mitosis: centrioles are by contrast absent from higher plant and certain (non-motile) algae. Many animal cells have an extensive aster about each pole, the astral rays being formed from microtubules: asters are absent from plant spindles and vary greatly in extent between animal cells. The nuclear envelope usually breaks down during prophase, yet in many fungi and lower green plants the nuclear envelope remains intact and mitosis is intranuclear (Pickett-Heaps, 1972). These differences may be seen as minor variations on a central theme, but even here there are immense variations between cells in the organisation of the microtubules.

At metaphase, the spindle consists of a large number of approximately parallel microtubules extending from the poles to the chromosomes. A very precise parallel arrangement is illustrated in Fig. 1. The parallel arrangement results in the mitotic spindle being highly birefringent when viewed with polarised light, a property which has been extensively exploited as a non-destructive method for studying mitosis. The number of microtubules in the mitotic spindle varies considerably: 1200–1500 in PtK$_1$ cells (McIntosh, Cande & Snyder, 1975; Brinkley & Cartwright, 1971), 1100 in HeLa cells (McIntosh & Landis, 1969), 400 in Chinese hamster cells (Brinkley & Cartwright, 1975), 35 in the diatom *Fragilaria* (Tippit, Schulz & Pickett-Heaps, 1978; and illustrated in Fig. 1) compared with approximately 200 in the diatom *Surirelle ovalis* (Tippit & Pickett-Heaps, 1977), while the yeast *Saccharomyces cerevisiae* diploid has only 40 (Petersen & Ris, 1976). An alternative way of illustrating the diversity between cells is to consider the number of microtubules attached to the chromosome kinetochore. In PtK$_1$ cells there are approximately 40 per chromosome (McIntosh *et al.*, 1975), 90–145 in *Haemanthus* at metaphase (Jensen & Bajer, 1973), while there is only one in the fungus *Thranstotheca* (Heath, 1974) and the yeast *Saccharomyces cerevisiae* (Petersen & Ris, 1976).

There is substantial evidence for at least two classes of microtubules within the spindle: those which extend from pole-to-pole, and others which link the chromosomes to the poles. This is most evident during anaphase when there is a spatial segregation of the two types. In addition, colcemid can prevent the assembly of interpolar tubules while not affecting pole-to-chromosome tubules (Brinkley, Stubblefield & Hsu, 1967) and in some cells similar effects are seen with chloral hydrate while in the same cells cytoplasmic tubules are much less sensitive (Molè-Bajer, 1967, 1969). Furthermore, direct counts of the numbers of microtubules in serial sections cut transverse to anaphase spindles show that in many cells approximately half of the total number of microtubules extend from pole-to-pole (McIntosh *et al.*, 1975; Brinkley & Cartwright, 1975) and at telophase that there are significantly more microtubules at the centre of the spindle than in the immediately adjacent regions (McIntosh & Landis, 1969; McIntosh *et al.*, 1975). This indicates that the interpolar microtubules are not continuous between the two poles but consist of two interdigitating sets, originating at the poles and overlapping at the centre. This interpretation is amply confirmed by ultrastructural studies on diatom spindles. In *Fragilaria,* the interdigitating interpolar microtubules form a central core with the chromosomal material distributed circumferentially, as illustrated in Fig. 1. The chromosome microtubules radiate out from the highly organised spindle poles to the chromosomes and are therefore spatially segregated from the pole-to-pole microtubules. The organisation of the central core can therefore be studied in detail by cutting serial sections. During prophase, there is an approximately constant number of microtubules along the length of the central core, whilst during metaphase and anaphase, there is a 50% decrease in number compared with prophase in the region of the two poles while the number at the centre of the spindle remains constant (Tippit *et al.*, 1978).

Two events can be characterised during anaphase: the shortening of the chromosome-to-pole distance and an increase in the pole-to-pole distance. In general, the chromosomal movement occurs during the first part of anaphase and the elongation of the spindle during the second, although there is considerable variation between cell types. In certain cells, such as the protozoan *Barbulanympha*, the two stages are almost completely separate and have been termed anaphase A and B, respectively (Inoué & Ritter, 1975). Similarly, in the diatom *Fragilaria,* chromosomal movement commences prior to the elongation of the interpolar microtubules (Tippit *et al.*, 1978).

A number of estimates have been made of the amount of force required to move chromosomes during anaphase. The velocity is independent of the size of the chromosomes (Nicklas, 1965), indicating that the mechanism of chromo-

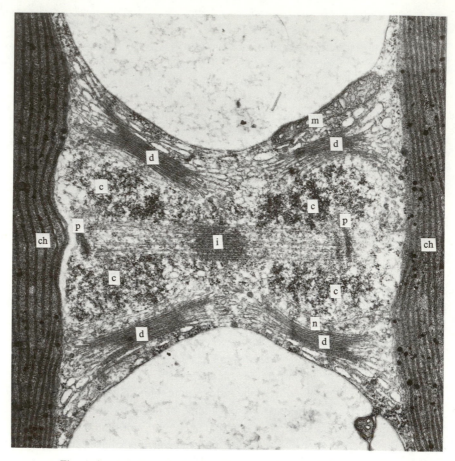

Fig. 1. Section through a late anaphase mitotic spindle of the diatom *Fragilaria* showing spindle microtubules originating in the spindle plaques (p) and inter-digitating at the central core (i). The plaques are surrounded by the electron-dense chromosomes (c). Fragments of the nuclear membrane (n) are also visi-ble. The cytoplasm contains mitochondrial (m), chloroplast (ch) and dictyosome (d) material which is not directly involved in mitosis. Earlier in mitosis the individual pole-to-pole microtubules terminate close to a spindle plaque at both ends, but subsequently at one end they separate from a plaque and move relative to the tubules which have remained attached to the opposite plaque. This move-ment elongates the spindle, which is seen here at anaphase to consist of two half sets of polar microtubules still partly interdigitating at the centre. Move-ment of chromosomes then depends upon the presence of the separate set of kinetochore tubules radiating from the poles to the chromosomes. These tubules are not evident in this section because they are not orientated in its plane. The bowing out of the spindle tubules close to the poles may be caused by compres-sional forces generated during chromosomal movement to the poles. Further details are given by Tippit *et al*. (1978). D. H. Tippit kindly supplied the plate. Magnification: × 25 000.

somal movement is the rate-limiting step rather than factors such as the viscous drag of the chromosomes moving through the cytoplasm. It is important when considering possible models for mitosis to recognise that the rate of chromosomal movement is slow, of the order of 1 μm min^{-1} compared with a rate of 400 μm min^{-1} for the sliding of actomyosin in vertebrate striated muscle (Nicklas, 1971), and that the amount of work is equivalent to only about 0.03 molecules of ATP hydrolysed per second per chromosome (Nicklas, 1975).

Properties of microtubules

Significant advances have been made during the last decade in understanding the molecular structure and mechanism of assembly of microtubules. It has become increasingly evident that microtubules from different sources are very similar although not identical. Although there have been only extremely limited studies of the properties of spindle microtubules *in vitro*, there is now abundant information about microtubules from other sources, particularly on the assembly *in vitro* of brain microtubules. The results of these studies can be used to place certain limits on possible models for mitosis providing it is recognised that there may be subtle, but important, differences between different microtubules. An example of such a difference has already been mentioned, namely that between the interpolar and kinetochore microtubules in terms of their sensitivity to colcemid and chloral hydrate. It is inappropriate to review all of the information known about the assembly of brain microtubules *in vitro*, but it is necessary to understand the general principles before considering the alternative models for mitosis.

Microtubules, whether in mitotic spindles or in other structures, are formed of a protein termed tubulin which has a relative molecular mass (M_r) of 110 000 and is a heterodimer containing two subunits (α and β) each of close to 55 000 M_r (e.g. Bryan & Wilson, 1971; Bibring & Baxandall, 1971). The microtubule is a hollow cylinder of approximately 25 nm diameter. Its walls can be visualised as being built of strands of repeated tubulin units, which are parallel to the long axis, and are referred to as protofilaments. These lie side by side and form a closed circumference usually containing thirteen protofilaments but the number varies somewhat, both *in vivo* and *in vitro*. Spindle microtubules from isolated mitotic apparatuses from the sea urchin have been shown to have thirteen circumferential subunits by staining with tannic acid (Tilney *et al.*, 1973).

Brain microtubules can be purified either directly (Kirkpatrick, Hyams, Thomas & Howley, 1970) or by taking advantage of their cold lability and successively assembling and disassembling them *in vitro* (e.g. Weisenberg, 1972*a*). The purified microtubules contain, in addition to tubulin, a number of other proteins, of which the most prominent, forming 10–20% of the total protein, are a pair of high molecular weight components which have been called Microtubule

Associated Proteins (MAP$_1$ and MAP$_2$) (Sloboda, Rudolph, Rosenbaum & Green-gard, 1975). Mitotic spindles have been shown to contain MAP proteins by staining with anti-MAP antibodies (Sherline & Schiavone, 1977).

When dissociated brain microtubule protein is warmed to 37 °C under optimal conditions, microtubules are formed which are structurally indistinguishable from those observed *in situ*. On warming, there is an initial lag period followed by a rapid increase in polymerisation to a final plateau value. The lag phase can be eliminated by seeding with fragments of non-dissociated microtubules, indicating that the lag is due to a nucleation event (Rosenbaum *et al.*, 1975). Furthermore, the amount of polymerised material is directly proportional to the protein concentration above a critical concentration of approximately 0.2–0.3 mg ml^{-1} (Olmsted *et al.*, 1974; Sloboda *et al.*, 1976), strongly suggesting that the plateau value represents an equilibrium between assembly and disassembly. It has been shown that the rate of polymerisation is directly dependent upon the number of free ends (Borisy, Johnson & Marcum, 1976) and the kinetics of assembly have been interpreted in terms of a linear condensation model. This is an important restriction in understanding the behaviour of microtubules *in vivo*: any model must require that microtubules assemble from their ends rather than by insertion of subunits along their length.

The removal of the MAP from microtubule preparations reduces the ability of the microtubule protein to assemble, while the addition of extra MAPs increases it (Murphy & Borisy, 1975; Kim, Binder & Rosenbaum, 1979). Margolis & Wilson (1978) have studied the disassembly of microtubules under conditions which inhibit assembly, and conclude that assembly is primarily at one end and disassembly at the other. This suggests that the MAPs alter the rate of turnover between free and polymerised tubulin probably by inhibiting the rate of disassembly. Consequently, at equilibrium conditions, a microtubule can be considered to be assembling at one end and disassembling at the other, while the rate of 'treadmilling' of the subunits along the length of the microtubule may depend upon the MAP content. This view of the microtubule as a polarised structure augments the earlier findings that microtubule protein could be polymerised preferentially onto one end of an added seed. In particular, it was found that if flagellar fragments were incubated with low concentrations of microtubule protein that all the assembly occurred at the distal end of the seed, whereas at higher concentrations there was assembly from both ends, although the rate of elongation from the distal end was five to ten times that from the proximal end (Allen & Borisy, 1974; Binder, Dentler & Rosenbaum, 1975; Rosenbaum *et al.*, 1975).

The rate of microtubule assembly onto seeds has been determined by a number of workers and is of the order of 0.5–1.0 μm min^{-1} at a microtubule protein concentration of 1 mg ml^{-1}, and is directly proportional to the concentration of microtubule protein.

These observations have direct relevance to possible mechanisms of mitosis. Assuming that observations *in vitro* can be extrapolated to the events *in vivo*, the assembly occurs primarily at one end with disassembly occurring at the other. The rate of assembly, and therefore the rate of migration under equilibrium conditions, is about 1μm min^{-1}, a value close to the rate of elongation of the mitotic spindle during anaphase B (3 μm min^{-1}; Inoué & Ritter, 1975) and to the rate of pole separation observed in chick tissue culture cells (0.75 μ min^{-1}; Hughes & Swann, 1948). In addition, factors such as the MAPs may influence the rates of association and disassociation of the microtubules. For instance, the observed differences between the interpolar and kinetochore microtubules may be differences in turnover rate rather than representing microtubules formed from different types of tubulin.

It has already been noted that microtubules can be assembled *in vitro* onto heterologous seeds, which include flagellar fragments (Rosenbaum *et al.*, 1975; Sloboda *et al.*, 1976) and basal bodies (Snell *et al.*, 1974). In addition, it has been possible to use isolated chromosomes, centrosomes and the pericentriolar material (Gould & Borisy, 1977; Telzer & Rosenbaum, 1979) and the spindle plaques of yeast (Hyams & Borisy, 1978) as the nucleating structure. These observations have two important implications. First, both the chromosomes and the material at the poles (in many lower eukaryotes, including yeasts and diatoms, the spindle poles are highly organised structures on the nuclear envelope, which are described as spindle plaques or microtubule centres) will nucleate microtubule assembly. All microtubules *in vivo* appear to terminate either on a structure such as a basal body or a spindle plaque or to end in an osmiophilic region such as the pericentriolar material which have been termed Microtubule Organising Centres (MTOCs) (Pickett-Heaps, 1969). This means that the events during the lag phase *in vitro* probably have no significance *in vivo*: nucleation is always onto pre-existing nucleation centres, and by extension, that the location of these nucleating centres determines the cellular location of the microtubules. Microtubule polymerisation *in vivo* would therefore be expected to have kinetic properties reflecting nucleated condensation, such that the equilibrium is independent of the amount of material in the polymerised form.

The second implication of the observation that both chromosomes and the polar material act as nucleation centres is that the interpolar and kinetochore microtubules may have opposite polarity. It has already been noted that microtubules assemble *in vitro* preferentially from one end, and the nucleation of assembly by structures such as chromosomes and spindle plaques suggests that assembly is at the end distal to these structures. Borisy (1978) has compared the rate constants for elongation of porcine brain tubulin onto various nucleating structures and finds a remarkably constant value of approximately 2.5–3.5 M^{-1} s$^{-1} \times 10^{-6}$, equivalent to a rate of 1 μm min^{-1} mg^{-1} tubulin dimer ml^{-1}, and concludes that assembly in all cases is at the favoured end. The implication is therefore that the polar and kinetochore microtubules in each half spindle have opposite polarity, i.e. that they are anti-parallel. Unfortunately, we do not have any methods at present, equivalent to the binding of heavy meromyosin fragments to actin filaments, to determine the in-vivo polarity of microtubules.

The effect of various conditions and factors on microtubule assembly *in vitro* have been examined, and may lead to an understanding of the regulation *in vivo* of microtubule assembly. The optimum pH is in the range 6.4–6.8 (Olmsted & Borisy, 1975) at a concentration of approximately 0.1 M. Polymerisation is inhibited by concentrations of sodium and potassium above about 0.24 M, while the magnesium optimum is dependent upon the nucleotide status. Suitable conditions of pH and solute are commonly found *in vivo* whether or not assembly is occurring and are unlikely to be involved in the regulation of assembly. Similarly, microtubules are highly temperature dependent, depolymerisation occurring on lowering the temperature. Polymerisation is endothermic and entropy-driven with attachment of one tubulin monomer causing disorder of many water molecules ($H = 21$ kcal mol^{-1}, $S = 96$ eu; Gaskin & Gethner, 1976) but temperature is an unlikely regulator *in vivo*.

More significant in terms of the in-vivo regulation is the effect of calcium ions. Under certain conditions, microtubule assembly is only inhibited by high concentrations of calcium (Rosenfeld & Weisenberg, 1974; Olmsted & Borisy, 1975), while under conditions closer to the magnesium optimum, calcium at 10^{-4} M inhibits, yet other workers have observed inhibition at 3.5×10^{-4} M and as low as 7.5×10^{-6} M (Haga, Abe & Kurokawa, 1974). This confusion has been greatly clarified by the demonstration that the addition of calcium-dependent regulatory protein (calmodulin) enhances the calcium sensitivity of microtubule disassembly *in vitro* (Marcum, Dedman, Brinkley & Means, 1978), reducing the minimum concentration for inhibition to well within the physiological range for calcium sensitivity. Of particular interest has been the demonstration that calmodulin is present in the mitotic spindle (Welsh, Dedman, Brinkley & Means, 1978) which was shown by the use of anti-calmodulin fluorescent antibodies (Anderson, Osborn & Weber, 1978).

Another important factor which may regulate microtubule assembly *in vivo* is the GTP concentration. Assembly *in vitro* requires the presence of GTP or a GTP-generating system (e.g. Weisenberg, Deery & Dickinson, 1976; Macneal & Purich, 1978) while certain non-hydrolysable GTP analogues inhibit polymerisation. Tubulin has two guanosine binding sites, one of which binds free GTP (the exchangeable or E-site) while the other is non-exchangeable (the N-site) with the formation of GTP from GDP at that site occurring *in situ* (Jacobs, Smith & Taylor, 1974). Approximately 85% of the N-site is occupied by GTP both before and after polymerisation (Penningroth, Cleveland & Kirschner, 1976), while there is hydrolysis of the E-site GTP. This hydrolysis does not appear to be essential, but microtubules assembled in the presence of GMP-P(NH)P are more stable than those polymerised with GTP (Weisenberg & Deery, 1976), suggesting that hydrolysis is a prerequisite for disassembly. Preparations of microtubule protein contain nucleoside diphosphate kinase (Nickerson & Wells, 1978; Watanabe & Flavin, 1976) which is present in higher concentrations than would be predicted by simple dilution, suggesting a close association between the enzyme and assembled microtubules. The presence and activity of the nucleoside diphosphate kinase may therefore be responsible for 'housekeeping' the microtubules, by modulating the nucleotide status on the E-site and so influencing the turnover rate between free and polymerised tubulin.

A number of post-translational modifications to the microtubule proteins have been described, including the sulphydryl status of the tubulin (Mellon & Rebhun, 1976), the tyrosylation of the C-terminus of the α-chain (Raybin & Flavin, 1975), and the phosphorylation of MAP_2 (Sloboda *et al.,* 1975). However, while these modifications may be extremely important little is known about their possible role *in vivo*, and so they will not be discussed further.

Assembly of the spindle microtubules

Mitosis involves two clear and distinct events: the formation of the mitotic machinery or spindle, and the movement of the chromosomes. The triggering of mitosis has been considered by many authors in this volume and the initial trigger mechanism remains unclear but it is followed by the formation of a large and complex structure: the mitotic spindle. Although the spindle contains many components, we shall restrict our attention to the assembly of the microtubular components.

Mitotic cells contain a high concentration of tubulin, estimated in sea urchin eggs as 5% of the total protein (Raff, Greenhouse, Gross & Gross, 1971) and the estimated minimal tubulin concentration in *Drosophila* eggs is 14 mg ml^{-1} (Green, Raff & Raff, 1979). Although there is evidence for the discontinuous synthesis of tubulin through the cell cycle (e.g. Lawrence & Wheatley, 1975) the tubulin pool is extremely large and the protein is used to form successive spindles (Wilt, Sakai & Mazia, 1967). Consequently there must be changes at the onset of mitosis which permit the assembly of microtubules, and the in-vitro studies identify that these changes may be of two types. First, microtubules assemble onto a seed structure of MTOC, and one possible control could be the presence and activity of such a MTOC. Homogenates of unfertilised *Spisula* eggs do not form microtubules *in vitro,* while within 2.5 min of artificial activation, achieved by suspending eggs in salt solution without fertilisation, homogenates contained dense granular structures capable of acting as MTOCs for microtubule assembly and these subsequently developed a centriole (Weisenberg

& Rosenfeld, 1975). Second, a strong cell cycle dependence is observed in HeLa cells in the ability of heterologous microtubule protein to assemble onto the pericentriolar material (Telzer & Rosenbaum, 1979). Conversely McGill & Brinkley (1975) found, also in HeLa cells, that the interphase pericentriolar material was competent for assembly while in asynchronous cultures of *Saccharomyces cerevisiae* a large number of spindle plaques, were obtained, which were competent to initiate microtubule assembly suggesting strongly that there is no cell cycle dependence (Hyams & Borisy, 1978). These observations indicate that although the activation of the MTOC is an important mechanism of regulating the initiation of mitosis in some cells it is not the sole control.

The mechanism of the activation of the MTOC is unclear, in part because of the difficulty in preparing clean preparations for biochemical analysis. PtK_2 cells that have been sensitized with acridine orange and then irradiated at 488 and 514 nm with a laser exhibit an interesting effect. If the pericentriolar region *at one pole* is irradiated during prophase then no anaphase movement occurs in either half spindle although the nuclear envelope breakdown, chromosome condensation, metaphase plate formation, and cytokinesis all proceed as usual (Berns, Rattner, Brenner & Meridith, 1977). Irradiation was only effective when it included the pericentriolar region, treatment of prophase chromosomes or noncentriolar cytoplasm was ineffective. The spindles of the affected cells contain microtubules, but while those of the non-treated half-spindle extended from the chromosomes to the pole, those on the treated side fell short of the pole. The effect of the irradiation is postulated to show the involvement of a nucleic acid component. One possible way of activating the MTOC would be the synthesis of a specific nucleic acid component which is structurally involved in the switching on of the nucleation site.

This interpretation is reinforced by the finding that purified basal bodies of *Tetrahymena* or *Chlmaydomonas* can, on microinjection, induce aster formation in unfertilised *Xenopus* eggs (Heidemann & Kirschner, 1975), and that this activity is abolished by pretreatment with ribonuclease A, ribonuclease T_1, or nuclease S_1 (Heidemann, Sander & Kirschner, 1977).

The second alternative for regulating the initiation of microtubule assembly is to modulate the size or activity of the microtubule protein pool. While the control of the availability of the nucleating site is an all-or-nothing response, modulation of the microtubule protein pool provides a mechanism for regulating the amount of polymerisation since in-vitro studies indicate that assembly is dependent upon protein concentration. There is strong evidence that this occurs *in vivo*. Stephens (1972) demonstrated that the birefringence and the amount of tubulin in isolated mitotic apparatuses was temperature dependent, and that by altering the temperature at various times following fertilisation he demonstrated that the available pool was determined during early prophase. As the critical time was *before* the onset of microtubule polymerisation, these results indicate that the effect of temperature was on the potentiation of the pool rather than on assembly *per se*.

A number of factors which could modulate the size of the tubulin pool have been identified from *in vitro* studies. Clearly, the modulation could involve both tubulin and MAPs, although there is no evidence at present for the activation of the MAPs in terms of their ability to promote microtubule assembly. Calcium, and the status of the nucleotide and sulphydryl groups on tubulin are all potential mechanisms for the activation of the tubulin pool, while tyrosylation seems unlikely.

In *Spisula* eggs, a significant fraction of the colchicine-binding activity, assumed to be tubulin, is in a particulate form (Weisenberg, 1972*b*). The amount of this particulate tubulin is approximately the same in interphase and metaphase cells, but there is a sharp drop immediately prior to metaphase reaching a minimum at prophase. This suggests that the cell is mobilising tubulin from a structural pool for use during mitosis. The exact nature of this structural pool has not been analysed, but there may be an analogy between this immobilisation of tubulin and that for actin in the acrosome of *Limulus* sperm (Tilney, 1976).

It is important to note that there are likely to be several levels of control on microtubule assembly: a crude control in terms of the size of the available pool (determined by protein synthesis and the activation of the pool as already discussed) would be coupled with finer controls on the degree of polymerisation of the pool (by factors such as the calcium level, the availability of calmodulin, the nucleotide regime, the presence and activity of nucleoside diphosphate kinase, etc). It is also worth noting that while microtubule assembly has been described in terms of an equilibrium between subunits and the assembled microtubules, it is probably more accurate to describe assembly/disassembly as a cycle of (approximately) irreversible events. Consequently, certain of the factors controlling microtubule assembly may have a more marked influence on a particular step of this cycle, so adding to the complexity of understanding the in-vivo regulation.

Models for movement in mitosis

A number of models have been proposed for the mechanism of movement of the chromosomes. These models will be described first, together with a summary of the evidence in their favour, followed by a review of their limitations and of whether the models are mutually exclusive.

The earliest model proposed that chromosome movement results from the controlled assembly and disassembly of the microtubules, although the model was originally formulated in terms of the spindle fibres and their subunits (Ostergren, 1949). This model was refined by Inoué on the basis of extensive studies of the birefringence of living mitotic and meiotic spindles. There is strong evidence that the observed birefringence is a direct measure of the concentration of orientated microtubules (Sato, Inoué & Ellis, 1971). The assembly/disassembly model proposes that there is a dynamic equilibrium between the microtubules and their subunits, such that the kinetochore microtubules shorten while the interpolar microtubules continue to assemble. The effect of this differential assembly and disassembly is that the chromosomes move towards the poles while the poles move further apart. Variations in the precise timing of the assembly and disassembly would permit the observed differences seen in the degree and timing of chromosome movement relative to the poles and of pole segregation between different cell types, such as the clear separation of anaphase A and anaphase B seen in *Barbaranympha* (Inoué & Ritter, 1975). There is considerable evidence in favour of this model. Agents such as low temperature, high pressure and colchicine which inhibit microtubule assembly or promote disassembly block anaphase movement, while agents such as deuterium oxide or glycols, which stabilise microtubules, reverse the effects of low temperature or high hydrostatic pressure. Salmon (1975) has elegantly demonstrated that hydrostatic pressures, which promote microtubule disassembly without causing the immediate total loss of birefringence, cause an *increase* in the rate of chromosomal movement towards the poles, while in a similar study Fuseler (1975) demonstrated that a modest lowering of the temperature also caused an enhanced rate of chromosomal movement. In both cases there was a direct correlation between the initial loss of birefringence and the acceleration of chromosomal movement, strongly indicating that the rate of microtubule dissociation is the rate-limiting step during anaphase movement.

The equilibrium birefringence of *Chaetopterus* metaphase spindles has also been measured under varying conditions of pressure and temperature. Assuming that the birefringence is a direct measure of microtubule polymerisation and that polymerisation occurs by nucleated condensation, there is a close parallel between the thermodynamic parameters measured *in vivo* and those found for polymerisation *in vitro* (Salmon, 1975). The change in enthalpy (ΔH) is estimated as 7.5–22 kcal mol^{-1} *in vivo* compared with 21 kcal mol^{-1} *in vitro,* and the change in the molar volume per subunit on polymerisation (ΔV) as 90 ml mol^{-1} in both *in vivo* and *in vitro*.

A model that proposes chromosomal movement to be dependent upon the assembly and disassembly of microtubules requires that the microtubules should not shear under strain. Inoué & Ritter (1975) have calculated that the longitudinal cohesive force of a microtubule is 1.3×10^{-6} dyne and have compared this to the force necessary to move a chromosome, estimated as 1×10^{-8} dyne (Taylor 1965). Clearly then, a microtubule is unlikely to fragment under the strain, particularly when it is remembered that most chromosomes have a large number of microtubules attached to the kinetochore.

The second model proposes that the mechanisms of movement of the chromosomes involves sliding between microtubules with opposite polarity. In order to understand the principle of this model it is necessary first to outline the mechanism of beating of cilia and flagella. These organelles contain nine peripheral microtubule doublets and a central pair. The outer doublets are linked circumferentially by cross bridges and are linked radially with the central pair through radial spokes. Satir (1968) has elegantly demonstrated, by sectioning serially the tips of cilia from *Elliptio* gills which had been fixed while beating, that the outer doublets slide relative to each other. In parallel with these studies, Gibbons (1965) had shown, using sea urchin flagella, that the cross bridge which links the outer doublets is an ATPase termed dynein. He later demonstrated that, following mild digestion of demembranated axonemes, an outer doublet could be induced to slide relative to its neighbours by the addition of ATP (Summers & Gibbons, 1971).

This observation is central to the model for mitosis proposed by McIntosh, Hepler & van Wie (1969). They proposed that in each half spindle the kinetochore and the interpolar microtubules have opposite polarity, that there are dynein-like cross bridges, and that the cross bridges effect the sliding apart of microtubules with opposite polarity, while having no effect when crosslinking microtubules with the same polarity. The model encompasses the one previously described in that the length of the microtubules is governed by the equilibrium between assembly and disassembly, but it does not attribute the generation of force to this mechanism. Instead, it predicts that microtubules are assembled from the kinetochores and from the centrosomes (pericentriolar material) such that the interpolar and the kinetochore microtubules of each half spindle bear opposite polarity. Dynein-like cross bridges link the microtubules: crosslinks between microtubules with opposite polarity produce movement, while crosslinks between microtubules with the same polarity are ineffective.

There is a substantial amount of evidence in favour of this model. Unlike the model for chromosomal *movement* as a consequence of assembly and disassembly, the kinetochore microtubules would be under tension relative to the interpolar microtubules, resulting in the curved arcs observed in spindles. Furthermore, there is substantial evidence for the presence of cross bridges between spindle microtubules (e.g. Roth, Wilson & Chakraborty, 1966; McIntosh & Landis, 1969; Wilson, 1969; Hepler, McIntosh & Cleland, 1970; Brinkley & Cartwright, 1971; Inoué & Ritter, 1975). The mean distance between adjacent cross bridges in meiotic metaphase cells is estimated as 12.9 and 22.8 nm, while the spacing between the dynein sidearms of sperm tail outer doublets is 24 nm (McIntosh, 1974). The cross bridges measure approximately 14×5 nm (Wilson, 1969) while dynein cross bridges in *Tetrahymena* are $20-22 \times 8-10$ nm (Allen, 1968).

Finally, this model would predict the distribution of microtubules that is observed following the reconstruction of serially sectioned spindles; namely the two sets of interpolar microtubules interdigitating at the centre of the spindle with the parallel alignment of the chromosomal and interpolar microtubules in each half spindle.

The final microtubule-based model to be considered, although there have been numerous others proposed, has been formulated by Margolis, Wilson & Kiefer (1978). They propose that the polarity of the chromosomal and interpolar microtubules in each half of the spindle is the same, unlike the model of McIntosh, Hepler & van Wie, and that microtubule assembly occurs at the centre of the spindle (i.e. the plane of the metaphase plate) whilst disassembly occurs at the poles. Essential to their model is the 'treadmilling' observed *in vitro,* such that a microtubule would 'migrate' from its site of assembly towards the site of disassembly, from the centre towards the pole. Blockage of the assembly of the chromosomal microtubules at the time of metaphase, with the continued assembly of the interpolar microtubules, would effect the movement of the chromosomes at a rate dependent upon the rate of disassembly of the chromosomal microtubules. They further propose that the interdigitating interpolar microtubules are linked by cross bridges, which effect the spindle elongation, as proposed by McIntosh, Hepler & van Wie (1969). This model is in many respects a composite of the two models already outlined, coupled with the 'treadmilling' properties of microtubules.

There are however significant objections to each of these three models. The simple assembly/disassembly equilibrium proposed by Inoué requires that a microtubule which is attached to a chromosome is capable of disassembly under conditions that retain its structural link between the pole and the chromosome. Although this clearly could be achieved if the microtubule was held by a collar at a point behind the disassembly site, no such collar has been observed. Indeed, while the structure of kinetochores is highly complex, the poles appear to lack discernable structure, consisting simply of a profusion of naked microtubule ends. This difficulty is minimised in the 'treadmilling model' since crosslinks between parallel microtubules are permitted (Margolis *et al.,* 1978), although such crosslinks are not proposed to provide the motive force. However, the 'treadmilling' model has a major difficulty in that it proposes that all microtubules, both chromosomal and interpolar, are assembled at the metaphase plane. It has already been shown that microtubules are polar and that under standard conditions all microtubules assemble at their favoured ends (Borisy, 1978), using seeds as diverse as intact microtubules, kinetochores, and spindle plaques. This strongly suggests that microtubule assembly occurs not at the kinetochores but at the free end of the microtubule distal from the MTOC.

Two models implicate cross bridges between the microtubules as providing the motive force in a manner analogous that of cilia and flagella beating. There is however a fundamental difference: the mitosis models propose that *antiparallel* microtubules slide apart, whereas the cilia and flagella models involve the sliding of *parallel* microtubules. While this clearly does not invalidate the proposed mechanism, it does mean that the precise interaction between the cross bridges and the microtubules cannot be identical in the two systems. Finally, since the formulation of the sliding model by McIntosh *et al.,* (1969), Tippit *et al.,* (1978) have shown that in *Fragilaria* the chromosomal microtubules are spatially separate from the interpolar microtubule central core, indicating that interactions between the chromosomal and interpolar microtubules are not mandatory. Indeed, the spatial organisation in *Fragilaria,* together with the observation that chromosome migration precedes the elongation of the central core and that anaphase in *Barbulanympha* can be divided into two phases (Inoué & Ritter, 1975) strongly suggests that two mechanisms may be operating during mitosis, one involving the interpolar microtubules and spindle elongation, and a second which results in the movement of the chromosomes to the poles. The molecular mechanism of neither mechanism is clear at present although it is tempting to assign sliding to the former and microtubule assembly/disassembly to the latter.

The three models which have been presented so far, all directly involve microtubules in the generation of the necessary motive force. A radically different alternative has been proposed by Forer (1974) who argues that the microtubules form a passive cytoskeleton and that the motive force is generated by an actin/myosin microfilament system which interacts with the cytoskeleton, and that the rate of movement of the chromosomes is governed by the rate of disassembly of the microtubule cytoskeleton. There is strong evidence for the presence of actin in the mitotic spindle from experiments using either the heavy meromyosin technique of Ishikawa, Bischoff & Holtzer (1969) (Gawaldi, 1971; Behnke, Forer & Emmersen, 1971; Hinkley & Telser, 1974; Forer & Jackson, 1976) or immunofluorescent labelling of anti-actin antibodies (Cande, Lazarides & McIntosh, 1977). Actin was found, using the latter technique, to be confined to the region between the poles and the chromosomes, and to be absent from the inter-chromosomal region (Cande *et al.,* 1977). The presence of myosin in the mitotic spindle has also been demonstrated by immunofluorescence (Fujiwara & Pollard, 1976), suggesting that the essential components for an actomyosin system are present.

This model predicts an interaction between the microtubule cytoskeleton and the actin microfilaments, and such an interaction has been demonstrated *in vitro* (Griffith & Pollard, 1978) using purified components and low shear viscometry. A number of essential features of this model have therefore been confirmed: the presence of actin and myosin, an interaction between actin and microtubules, and in-situ distribution of the actin. However, a direct test of the functional relevance of the associated myosin strongly suggests that this model is either incorrect or reflects only a part of a more complex mechanism. Anti-myosin antibody was injected into dividing oocytes at the two-cell stage, and the subsequent development of the injected and the control cells observed (Mabuchi & Okuno, 1977). Mitosis continued in the injected cell although cytokinesis, an event known to involve actin/myosin microfilaments, was blocked whereas mitosis and cytokinesis continued unaffected in the untreated cell. These observations suggest that there is not an absolute requirement for myosin during mitosis or that spindle myosin differs from that in the cleavage furrow and is not recognised by the antibody.

The mechanism of mitosis, despite years of detailed study, remains a problem which still has to be solved. While it is evident that microtubules are intimately involved, it is not clear whether they provide, directly or indirectly, the mechano-chemical basis for the generation of force. Little is known about the control of the assembly of the spindle, the regulation of the location and orientation of the metaphase plate, or the factors which ensure that daughter chromosomes migrate to opposite poles. In-vitro studies on purified components have placed important limitations on the possible models for the activation of the tubulin pool and for the basis of chromosome movement, yet such studies have not yet indicated a unique solution. It was argued earlier that two events, chromosomal movement and spindle elongation, may involve two distinct mechanisms. In view of the central importance of mitosis to the survival of a cell, it is possible that mitosis may involve a number of alternative mechanisms, some of which are more pronounced in particular cell types, and that the inhibition of one mechanism may merely accentuate the efficiency of another. Such a fail-safe system would have obvious advantages to the cell, even if it made it more difficult for cell biologists to unravel!

References

Allen, R. D. (1968). A reinvestigation of cross-sections of cilia. *Journal of Cell Biology*, **37**, 825–31.

Allen, C. & Borisy, G. G. (1974). Structural polarity and directional growth of microtubules of *Chlamydomonas* flagella. *Journal of Molecular Biology*, **90**, 381–402.

Anderson, B., Osborn, M. & Weber, K. (1978). Specific visualisation of distribution of calcium dependent regulatory protein of cyclic nucleotide phosphodiesterase (modulator protein) in tissue culture cells by immunofluorescence microscopy – mitosis and intercellular bridge. *Cytobiologie*, **17**, 354–64.

Behnke, O., Forer, A. & Emmersen (1971). Actin in sperm tails and meiotic spindles. *Nature, London*, **234**, 408–10.

Berns, M. W., Rattner, J. B., Brenner, S. & Meridith, S. (1977). The role of the centriolar region in animal cell mitosis. A laser microbeam study. *Journal of Cell Biology*, **72**, 351–67.

Bibring, T. & Baxandall, J. (1971). Selective extraction of isolated mitotic apparatus. Evidence that typical microtubule protein is extracted by organic mercurial. *Journal of Cell Biology*, **48**, 324–39.

Binder, L. I., Dentler, W. & Rosenbaum, J. L. (1975). Assembly of chick brain tubulin onto of flagellar axonemes of *Chlamydomonas* and sea urchin sperm. *Proceedings of the National Academy of Sciences, USA*, **72**, 1122–6.

Borisy, G. G. (1978). Polarity of microtubules of the mitotic spindle. *Journal of Molecular Biology*, **124**, 565–70.

Borisy, G. G., Johnson, K. A. & Marcum, J. M. (1976). Self-assembly and self-initiated assembly of microtubules. In *Cell Motility*, ed. R. D. Goldman, T. Pollard & J. Rosenbaum, vol. C, pp. 1093–108. New York: Cold Spring Harbor Laboratory.

Brinkley, B. R. & Cartwright, J. (1971). Ultrastructural analysis of mitotic spindle elongation in mammalian cells *in vivo*. Direct microtubule counts. *Journal of Cell Biology*, **50**, 416–31.

Brinkley, B. R. & Cartwright, J. (1975). Cold labile and cold stable microtubules in the mitotic spindle of mammalian cells. *Annals of the New York Academy of Science*, **253**, 428–39.

Brinkley, B. R., Stubblefield, E. & Hsu, T. C. (1967). The effect of Colcemid inhibition and reversal on the fine structure of the mitotic spindle of Chinese hamster cells *in vitro*. *Journal of ultrastructural Research*, **19**, 1–18.

Bryan, J. & Wilson, L. (1971). Are cytoplasmic microtubules heterodimers? *Proceedings of the National Academy of Sciences, USA*, **68**, 1762–6.

Cande, W. Z., Lazarides, E. & McIntosh, J. R. (1977). Comparison of the distribution of actin and tubulin in the mammalian mitotic spindle as seen by indirect immunofluorescence. *Journal of Cell Biology*, **72**, 552–67.

Forer, A. (1974). Possible roles of microtubules and actin-like filaments during cell division. In *Cell Cycle Controls*, ed. G. M. Padilla, I. L. Cameron & A. Zimmerman, pp. 319–36. New York: Academic Press.

Forer, A. & Jackson, W. T. (1976). Actin filaments in the endosperm mitotic spindles in a higher plant, *Haemanthus katherinae* Baker. *Cytobiologie*, **12**, 199–214.

Fujiwara, K. & Pollard, T. D. (1976). Fluorescent antibody localization of myosin in the cytoplasm, cleavage furrow, and mitotic spindle of human cell. *Journal of Cell Biology*, **71**, 848–75.

Fuseler, J. W. (1975). Temperature dependence of anaphase chromosome velocity and microtubule depolymerisation. *Journal of Cell Biology*, **67**, 789–800.

Gaskin, F. & Gethner, J. S. (1976). Characterisation of the in-vitro assembly of microtubules. In *Cell Motility*, ed. R. D. Goldman, T. Pollard & J. Rosenbaum, vol. C, pp. 1109–21. New York: Cold Spring Harbor Laboratory.

Gawaldi, N. (1971). Actin in the mitotic spindle. *Nature, London*, **234**, 410.

Gibbons, I. R. (1965). Chemical dissection of cilia. *Archives de Biologie, Liège*, **76**, 317–52.

Gould, R. & Borisy, G. G. (1977). Pericentriolar material in Chinese hamster ovary cells nucleates microtubule formation. *Journal of Cell Biology*, **73**, 601–15.

Green, L. H., Raff, E. C. & Raff, R. A. (1979). Tubulins from a developing insect embryo undergoing rapid mitosis: factors regulating in-vitro assembly of tubulins from *Drosophila melanogaster*. *Insect Biochemistry*, **9**, 489–96.

Griffith, L. M. & Pollard, T. D. (1978). Evidence for actin filament-microtu-
bule interaction mediated by microtubule associated proteins. *Journal of
Cell Biology,* **78,** 958–65.

Haga, T., Abe, T. & Kurokawa, M. (1974). Polymerization and depolymeri-
zation of microtubules *in vitro* as studied by flow birefringence. *FEBS Letters,* **39,** 291–5.

Heath, I. B. (1974). Mitosis in the fungus *Thranstotheca clavata. Journal of
Cell Biology,* **60,** 204–20.

Heidemann, S. R. & Kirschner, M. W. (1975). Induction of aster formation
in eggs of *Xenopus laevis* by isolated basal bodies. *Journal of Cell Biology,*
67, 164a.

Heidemann, S., Sander, G. & Kirschner, M. W. (1977). Evidence for a functional role of RNA in centrioles. *Cell,* **10,** 337–50.

Hepler, P. K., McIntosh, J. R. & Cleland, S. (1970). Intermicrotubular
bridges in mitotic spindle apparatus. *Journal of Cell Biology,* **45,** 438–44.

Hinkley, R. & Telser, A. (1974). Heavy meromyosin-binding filaments in the
mitotic apparatus of mammalian cells. *Experimental Cell Research,* **86,**
161–4.

Hughes, A. F. & Swann, M. M. (1948). Anaphase movements in the living
cell. A study with phase contrast and polarised light on chick tissue cultures. *Journal of Experimental Biology,* **25,** 45–70.

Hyams, J. & Borisy, G. G. (1978). Nucleation of microtubules *in vitro* by
isolated spindle pole bodies of the yeast *Saccharomyces cerevisiae. Journal
of Cell Biology,* **78,** 401–14.

Ishikawa, H., Bischoff, R. & Holtzer, H. (1969). The formation of arrowhead
complexes with heavy meromysin in a variety of cell types. *Journal of Cell
Biology,* **43,** 312–15.

Inoué, S. & Ritter, H. (1975). Dynamics of spindle organisation and function.
In *Molecules and Cell Movement,* ed. S. Inoué & R. E. Stephens,
pp. 3–30. New York: Raven Press.

Jacobs, M., Smith, H. & Taylor, E. W. (1974). Tubulin : nucleotide binding
and enzymatic activity. *Journal of Molecular Biology,* **89,** 455–68.

Jensen, C. & Bajer, A. (1973). Spindle dynamics and arrangement of chromosomes. *Chromosoma, Berlin,* **44,** 73–89.

Kim, H., Binder, L. I. & Rosenbaum, J. L. (1979). The periodic association
of MAP_2 with brain microtubules *in vitro. Journal of Cell Biology,* **80,**
266–76.

Kirkpatrick, J. B., Hyams, L., Thomas, V. L. & Howley, P. M. (1970). Purification of intact microtubules from brain. *Journal of Cell Biology,* **47,**
384–94.

Lawrence, J. H. & Wheatley, D. N. (1975). Synthesis of microtubule protein
in HeLa cells approaching division. *Cytobios,* **13,** 167–79.

Mabuchi, I. & Okuno, M. (1977). The effect of myosin antibody on the division of star fish blastomeres. *Journal of Cell Biology,* **74,** 251–63.

Macneal, R. K. & Purich, D. L. (1978). Stoichiometry and role of GTP hydrolysis in bovine neurotubule assembly. *Journal of Biological Chemistry,*
253, 4683–7.

Marcum, J. M., Dedman, J. R., Brinkley, B. R. & Means, A. R. (1978).
Control of microtubule assembly–disassembly by calcium dependent regulatory protein. *Proceedings of the National Academy of Sciences, USA,* **75,**
3771–5.

Margolis, R. L. & Wilson, L. (1978). Opposite end assembly and disassembly of microtubules at steady state *in vitro. Cell,* **13,** 1–8.

Margolis, R. L., Wilson, L. & Kiefer, B. I. (1978). Mitotic mechanism based on intrinsic microtubule behaviour. *Nature, London,* **272,** 450–2.

McGill, M. & Brinkley, B. R. (1975). Human chromosomes and centrioles as nucleating sites for the *in vitro* assembly of microtubules from bovine brain tubulin. *Journal of Cell Biology,* **67,** 189–99.

McIntosh, J. R. (1974). Bridges between microtubules. *Journal of Cell Biology,* **61,** 166–87.

McIntosh, J. R., Helper, P. K. & van Wie, D. G. (1969). Model for mitosis. *Nature, London,* **224,** 659–63.

McIntosh, J. R., Cande, W. Z. & Snyder, J. A. (1975). Structure and physiology of the mammalian mitotic spindle. In *Molecules and Cell Movement,* ed. S. Inoué & R. E. Stephens, pp. 31–76. New York: Raven Press.

McIntosh, J. R. & Landis, S. C. (1969). The distribution of spindle microtubules during mitosis in cultured human cells. *Journal of Cell Biology,* **49,** 468–97.

Mellon, M. & Rebhun, L. (1976). Studies on the accessible sulphydryls of polymerizable tubulin. In *Cell Motility,* ed. R. Goldman, T. Pollard, & J. Rosenbaum, vol. C, pp. 1149–63. New York: Cold Spring Harbor Laboratory.

Molè-Bajer, J. (1967). Chromosome movements in chloral hydrate treated endosperm cells *in vitro. Chromosoma, Berlin,* **22,** 465–80.

Molè-Bajer, J. (1969). Fine structure studies of apolar mitosis. *Chromosoma, Berlin,* **26,** 427–48.

Murphy, D. B. & Borisy, G. G. (1975). Association of high-molecular-weight protein with microtubules and their role in microtubule assembly *in vitro. Proceedings of the National Academy of Sciences, USA,* **72,** 2696–700.

Nickerson, J. A. & Wells, W. W. (1978). Association of nucleoside diphosphate kinase with microtubules. *Biochemical and Biophysical Research Communications,* **85,** 820–6.

Nicklas, R. B. (1965). Chromosome velocity during mitosis as a function of chromosome size and position. *Journal of Cell Biology,* **25,** 119–35.

Nicklas, R. B. (1971). Mitosis. In *Advances in Cell Biology,* ed. D. M. Prescott, L. Goldstein & E. McConkey, pp. 225–97. New York: John Wiley.

Nicklas, R. B. (1975). Chromosome movement: current models and experiments on living cells. In *Molecules and Cell Movement,* ed. S. Inoué & R. E. Stephens, pp. 97–117. New York: Raven Press.

Olmsted, J. B. & Borisy, G. G. (1975). Ionic and nucleotide requirements for microtubule polymerisation *in vitro. Biochemistry,* **14,** 2996–3005.

Olmsted, J. B., Marcum, J. M., Johnson, K. A., Allen, C. & Borisy, G. G. (1974). Microtubule assembly: some possible regulatory mechanisms. *Journal of Supramolecular Structure,* **2,** 429–50.

Ostergren, G. (1949). *Luzula* and the mechanism of chromosome movement. *Hereditas, Lund,* **35,** 445–68.

Penningroth, S. M., Cleveland, D. W. & Kirschner, M. W. (1976). In-vitro studies of the regulation of microtubule assembly. In *Cell Motility,* ed. R. Goldman, T. Pollard & J. Rosenbaum, vol. C, pp. 1233–57. New York: Cold Spring Harbor Laboratory.

Petersen, J. B. & Ris, H. (1976). Electron microscope study of the spindle and chromosome movement in the yeast *Saccharomyces cerevisiae. Journal of Cell Science,* **22,** 219–42.

Pickett-Heaps, J. D. (1969). The evolution of the mitotic apparatus: an attempt at comparative ultrastructural cytology in dividing plant cells. *Cytobios*, **1**, 257–80.

Pickett-Heaps, J. D. (1972). Variation in mitosis and cytokinesis in plant cells: its significance in the phylogeny and evolution of ultrastructural systems. *Cytobios*, **5**, 59–77.

Raff, R. A. Greenhouse, G., Gross, K. W. & Gross, P. R. (1971). Synthesis and storage of microtubule proteins by sea urchin embryos. *Journal of Cell Biology*, **50**, 516–27.

Raybin, D. & Flavin, M. (1975). An enzyme tyrosylating α-tubulin and its role in microtubule assembly. *Biochemical and Biophysical Research Communications*, **65**, 1088–95.

Rosenbaum, J. L., Binder, L. I., Granett, S., Dentler, W. L., Snell, W., Sloboda, R. & Haimo, L. (1975). Directionality and rate of assembly of chick brain tubulin onto pieces of neurotubules, flagellar axonemes and basal bodies. *Annals of the New York Academy of Science*, **253**, 147–77.

Rosenfeld, A. & Weisenberg, R. C. (1974). Role of magnesium and calcium in microtubule assembly. *Journal of Cell Biology*, **63**, 289a.

Roth, L. E., Wilson, H. J. & Chakraborty, J. (1966). Anaphase structure in mitotic cells typified by spindle elongation. *Journal of Ultrastructural Research*, **14**, 460–73.

Salmon, E. D. (1975). Pressure-induced depolymerisation of spindle microtubules. II. Thermodynamics of *in-vivo* spindle assembly. *Journal of Cell Biology*, **66**, 114–27.

Satir, P. (1968). Studies on cilia. III. Further studies on the cilium tip and a 'sliding filament' model of ciliary motility. *Journal of Cell Biology*, **39**, 77–94.

Sato, H., Inoué, S. & Ellis, G. W. (1971). The microtubular origin of spindle birefringence : experimental verification of Weiner's equation. *Journal of Cell Biology*, **51**, 261a.

Sherline, P. & Schiavone, K. (1977). Immunofluorescent localisation of proteins of high molecular weight along intracellular microtubules. *Science*, **198**, 1038–40.

Sloboda, R. D., Dentler, W. D., Bloodgood, R. A., Telzer, B. R., Granett, S. & Rosenbaum, J. L. (1976). Microtubule associated proteins (MAPs) and the assembly of microtubules. In *Cell Motility*, ed. R. Goldman, T. Pollard & J. Rosenbaum, vol. C, pp. 1171–212. New York: Cold Spring Harbor Laboratory.

Sloboda, R. D., Rudolph, S. A., Rosenbaum, J. L. & Greengard, P. (1975). Cyclic AMP-dependent endogenous phosphorylation of a microtubule associated protein. *Proceedings of the National Academy of Sciences, USA*, **77**, 177–81.

Snell, W. J., Dentler, W. L., Haimo, L. T., Binder, L. I. & Rosenbaum, J. L. (1974). Assembly of chick brain tubulin onto isolated basal bodies of *Chlamydomonas reinhardi*. *Science*, **185**, 357–9.

Stephens, R. E. (1972). Studies on the development of the sea urchin *Strongylocentrotus droebachiensis*. II. Regulation of mitotic spindle equilibrium by environmental temperature. *Biological Bulletin, Woods Hole*, **142**, 145–59.

Summers, K. E. & Gibbons, I. R. (1971). Adenosine triphosphate-induced sliding of tubules in trypsin-treated flagella of sea-urchin sperm. *Proceedings of the National Academy of Sciences, USA*, **68**, 3092–6.

Taylor, E. W. (1965). Brownian and saltatory movements of cytoplasmic granules and the movement of anaphase chromosomes. In *Proceedings of the Fourth International Congress on Rheology*, ed. A. L. Copley, part 4, pp. 175–91. New York: Interscience.

Telzer, B. R. & Rosenbaum, J. L. (1979). Cell cycle dependent, in-vitro assembly of microtubules onto the pericentriolar material of HeLa cells. *Journal of Cell Biology,* **81,** 484–97.

Tilney, L. G. (1976). The polymerisation of actin. III. Aggregates of non-filamentous actin and its associated proteins: a storage form of actin. *Journal of Cell Biology,* **69,** 73–89.

Tilney, L. G., Bryan, J., Bush, D. J., Fujiwara, K., Moosekjer, M. S., Murphy, D. B. & Snyder, D. H. (1973). Microtubules: evidence for 13 protofilaments. *Journal of Cell Biology,* **59,** 267–75.

Tippit, D. H. & Pickett-Heaps, J. D. (1977). Mitosis in the pennate diatom *Surirella ovalis. Journal of Cell Biology,* **73,** 705–27.

Tippit, D. H., Schulz, D. & Pickett-Heaps, J. D. (1978). Analysis of the distribution of spindle microtubules in the diatom *Fragilaria. Journal of Cell Biology,* **79,** 737–63.

Watanabe, T. & Flavin, M. (1976). Nucleotide-metabolising enzymes in *Chlamydomonas* flagella. *Journal of Biological Chemistry,* **251,** 182–92.

Weisenberg, R. C. (1972a). Microtubule formation *in vitro* in solutions containing low calcium concentrations. *Science,* **177,** 1104–5.

Weisenberg, R. C. (1972b). Changes in the organisation of tubulin during meiosis in the eggs of the surf clam, *Spisula solidissima. Journal of Cell Biology,* **54,** 266–78.

Weisenberg, R. C. & Deery, W. J. (1976). Role of nucleotide hydrolysis in microtubule assembly. *Nature, London,* **263,** 792–3.

Weisenberg, R. C. & Rosenfeld, A. C. (1975). In-vitro polymerisation of microtubules into asters and spindles in homogenates of surf clam eggs. *Journal of Cell Biology,* **64,** 146–58.

Weisenberg, R. C., Deery, W. J. & Dickinson, P. J. (1976). Tubulin–nucleotide interactions during polymerisation and depolymerisation of microtubules. *Biochemistry,* **15,** 4248–54.

Welsh, M. J., Dedman, J. R., Brinkley, B. R. & Means, A. R. (1978). Calcium-dependent regulatory protein: localisation in the mitotic apparatus of eucaryotic cells. *Proceedings of the National Academy of Sciences, USA,* **75,** 1867–71.

Wilson, H. J. (1969). Arms and bridges on microtubules in the mitotic apparatus. *Journal of Cell Biology,* **40,** 854–9.

Wilt, F., Sakai, S. & Mazia, D. (1967). Old and new protein in the formation of the mitotic apparatus in cleaving sea urchin eggs. *Journal of Molecular Biology,* **27,** 1–7.

INDEX

269